Sven Nagel

# Design of Cast Steel Components under Cyclic Loading

BAND 15

Versuchsanstalt für Stahl, Holz und Steine
Berichte zum Stahl- und Leichtbau

# Design of Cast Steel Components under Cyclic Loading

by
Sven Nagel

KIT Scientific Publishing

Karlsruher Institut für Technologie
Versuchsanstalt für Stahl, Holz und Steine

Design of Cast Steel Components under Cyclic Loading

Zur Erlangung des akademischen Grades eines Doktor-Ingenieurs
von der KIT-Fakultät für Bauingenieur-, Geo- und Umweltwissenschaften
des Karlsruher Instituts für Technologie (KIT) genehmigte Dissertation

von Sven Nagel, M.Sc.

Tag der mündlichen Prüfung: 10. Dezember 2020
Referent: Prof. Dr.-Ing. Thomas Ummenhofer
Korreferent: Prof. Dr.-Ing. Alain Nussbaumer

**Impressum**

KIT Scientific Publishing

Karlsruher Institut für Technologie (KIT)
KIT Scientific Publishing
Straße am Forum 2
D-76131 Karlsruhe

KIT Scientific Publishing is a registered trademark
of Karlsruhe Institute of Technology.
Reprint using the book cover is not allowed.

www.ksp.kit.edu



Print on Demand 2022 – Gedruckt auf FSC-zertifiziertem Papier

ISSN 2198-7912
ISBN 978-3-7315-1126-7
DOI 10.5445/KSP/1000137373

# Abstract

Cast steel components are theoretically ideal for use in fatigue-stressed constructions due to their high mechanical strength and an almost arbitrary shape adoption. However, fatigue resistance is determined by production-related imperfections. Due to a lack of knowledge and easy-to-use design approaches, the application of cast steel is rejected in most cases or is countered by highest demands in manufacturing quality. This choice is inefficient in many cases and is not necessary from a technical point of view. Even large casting defects do not necessarily reduce the overall performance of a component. The design approach derived in this context allows manufacturing quality to be determined in relation to local stresses.

This approach is validated by extensive experimental and numerical investigations. The resulting fatigue strengths are in line with the findings from research and applications from the last 60 years. The basis for the evaluation are fatigue tests on 60 tensile specimens made of materials G20Mn5 and G22NiMoCr5-6 with real internal casting defects. The transfer to real components is done by examining 19 large-scale specimens with dimensions close to those in practice and realistic geometrical caused stress concentrations. Finally, the damage behavior is mapped and evaluated by numerically supported crack propagation calculations.

The link between resistance and permissible defect size is made using newly defined resistance categories. Based on technically relevant fatigue classes, the maximum allowable defect sizes are defined based on fracture mechanics. The component design is carried out by limiting maximum stress ranges. By following established fatigue assessment methods, the design concept is directly adapted to practical building applications. Influences including mean stress dependency, geometrical tolerances, brittle fracture, stress gradients and the interaction of several defects are considered using reduction factors or additional requirements.

# Kurzfassung

Stahlguss ist aufgrund seiner hohen mechanischen Beanspruchbarkeit und der nahezu freien Formgebung theoretisch optimal für den Einsatz in ermüdungsbeanspruchten Konstruktionen geeignet. Jedoch wird der Bauteilwiderstand gegen Ermüdung durch herstellungsbedingte Ungänzen bestimmt. Aufgrund fehlender Kenntnisse und nicht vorhandener, einfach anwendbarer Bemessungsansätze wird der Einsatz meist abgelehnt, oder durch höchste Anforderungen an die Herstellqualität begegnet. Diese pauschale Wahl ist in vielen Fällen unwirtschaftlich und unter technischen Gesichtspunkten nicht erforderlich, da auch größere Gussfehler die Leistungsfähigkeit eines Bauteils nicht zwangsläufig reduzieren. Um diese Lücke zu schließen, wird in dieser Arbeit ein allgemeingültiger Bemessungsansatz entwickelt und so eine durch lokale Beanspruchungen begründete Festlegung von Herstellqualität ermöglicht.

Dieser Ansatz wird durch umfangreiche experimentelle als auch numerische Untersuchungen abgesichert und deckt sich mit den Erkenntnissen aus Forschung und Anwendungspraxis der letzten 60 Jahre. Grundlage für die Bewertung sind Ermüdungsversuche an 60 Kleinbauteilen mit realen Gussfehlern aus den Werkstoffen G20Mn5 und G22NiMoCr5-6. Der Übertrag auf reale Bauteile erfolgt durch die Betrachtung von 19 Probekörpern mit praxisnahen Abmessungen und realitätsnahen geometrisch bedingten Spannungskonzentrationen. Mit Hilfe von nummerisch gestützten Rissfortschrittsberechnungen kann das Schädigungsverhalten detailliert abgebildet und bewertet werden.

Die Verknüpfung zwischen Beanspruchbarkeit und zulässigen Fehlergrößen erfolgt durch neu definierte Widerstandskategorien. Ausgehend von technisch relevanten Ermüdungsbeanspruchungsklassen werden auf Basis bruchmechanischer Überlegungen maximal zulässige Fehlergrößen definiert. Die Bauteilauslegung erfolgt durch die Begrenzung von maximalen Spannungsschwingbreiten, ist an etablierte Nachweiskonzepte angelehnt und somit direkt auf baupraktische Anwendungen abgestimmt. Einflüsse wie Mittelspannungsabhängigkeit, Formtoleranzen, Sprödbruch, Spannungsgradienten und Interaktion mehrerer Fehler, werden durch Abminderungsfaktoren oder Zusatzbedingungen berücksichtigt.

# Danksagung

Es sind die Menschen, die Erlebnisse und die Impressionen, die uns zu der Person machen, die wir sind. In meinem bisherigen Leben blicke ich auf viele Abenteuer, Eindrücke und Begegnungen mit zahlreichen Menschen zurück, die mich geprägt, unterstützt, ja zu dem gemacht haben, der ich heute bin.

Ich möchte euch allen dafür danken.

Was treibt mich an, zieht mich morgens aus dem Bett und lässt mich abends erst spät einschlafen? Es ist die **Neugier**, die Freude immer Neues zu entdecken, der Drang zu begreifen. Danke, dass ihr dies in mir geweckt habt und mich ständig mit neuen Ansichten, weitreichenderen Ideen oder fordernden Gedankenspielen überrascht. Ihr habt mich auf hohe Berge geschleppt, mit mir entlegene Orte entdeckt, mir interessante Menschen vorgestellt, Dinge erklärt oder mir einen Ort zum Entdecken gegeben.

Der Grundstein für die Umsetzung der aus Neugier entstandenen Vorhaben ist **Optimismus, Mut und Tatendrang**. Ihr habt mir vorgelebt Dinge anzugehen, mutig zu sein und gezeigt, dass es immer eine Lösung gibt.

Momente des Glücks verbinde ich damit, **gemeinsam** etwas geschafft, gemeinsam etwas erlebt zu haben. Ihr habt mir gezeigt, dass wir zusammen die eigenen Möglichkeiten potenzieren und dass die Arbeit, bei ehrlicher Wertschätzung alles andere als eine Last ist. Wir haben Brücken gebaut, Versuche geplant, Vorträge gehalten, Schiffe gesteuert und Ideen gesponnen.

R. Branson sagt: „Wenn dir einer eine großartige **Möglichkeit** gibt, du dir aber nicht sicher bist, ob du es schaffst, sag Ja und lerne später, es umzusetzen". Auch ihr habt mir immer größere Möglichkeiten gegeben. Zudem habt ihr mir euer **Vertrauen** geschenkt, diese Möglichkeiten auch umzusetzen und mich bei meinen Vorhaben unterstützt.

Ihr habt mir gezeigt, dass auch das Bohren von dicken Brettern Freude bereitet, dass jeder Schritt, jeder Beinschlag und jeder Atemzug ihre Wirkung haben und sich **Fleiß** auszahlt. Dies ist aber nur von Dauer, wenn es nicht in Verbissenheit mündet, sondern auch Platz für Anderes ist und die **Ausgeglichenheit und Gelassenheit** sichergestellt ist.

Besonders schätze ich eure **Ehrlichkeit** und eure **Rückmeldungen**. So manch einer kennt mich besser als ich mich selbst, hält mir den Spiegel vor und zeigt mir Irrwege und Eitelkeiten liebevoll auf, sodass eure Ideen auch immer Teil von meinen Handlungen werden.

Ihr habt mich **Freude und Humor** gelehrt und mir so ein mächtiges Werkzeug für schwierige Situationen und die Grundlage für ein gesundes Leben mitgegeben.

Ich danke euch für jeden Millimeter, die ihr meine Lachfalten vertieft, jeden Schweißtropfen, den wir gemeinsam vergießen, jede Entdeckung, die wir miteinander machen, jedes Vorhaben, das wir zusammen angehen, und ich danke euch für euer Vertrauen und eure Unterstützung.

Mit diesem Rückhalt, diesen Erfahrungen und der Summe dieser Fähigkeiten bin ich voller Vorfreude auf all das, was die Zukunft für mich bereithält.

Ich danke meinem Doktorvater Prof. Dr.-Ing. Thomas Ummenhofer, meinem Korreferenten Prof. Dr.-Ing. Alain Nussbaumer, der Promotionskommission mit Prof. Dr.-Ing. Werner Wagner und Prof. Dr.-Ing. Hans-Joachim Blaß, meinen Kollegen, den Industrie- und Forschungspartnern sowie Prof. Dr.-Ing. Peter Knödel, der mich für den Stahlbau begeistert hat.

Ganz besonders danke ich meinen Eltern Ute und Rüdiger, meiner Familie, meinen Freunden und Konny.

Karlsruhe, im Januar 2021

Sven Nagel

# Contents

## Appendix

# Acronyms

AiF . . . . . . German Federation of Industrial Research Associations

ASTM . . . . . American Society for Testing and Materials

FVG . . . . . . German Foundry Association (Forschungsvereinigung Gießereitechnik e.V.)

BNIF . . . . . . British Nuclear Industry Forum

CAD . . . . . . Computer-Aided Design

CT . . . . . . . Computer Tomography

DCTG . . . . . Dimensional Casting Tolerance Grades

DeQaGuss . . . Research Project on the NDT Requirements for Cast Steel Components

DIN . . . . . . German Institute for Standardization

DNV . . . . . . Det Norske Veritas

EC . . . . . . . Eurocode

EMAT . . . . . Electromagnetic Acoustic Transducer

EN . . . . . . . European Standards

ErStaGu . . . . Research Project on the Influence of Casting Defects on the Fatigue Resistance

FAD . . . . . . Failure Assessment Diagram

FAT . . . . . . Fatigue Class

FBH . . . . . . Flat-Bottom Hole

FEA . . . . . . Finite Element Analysis

FEM . . . . . . Finite Element Method

FKM . . . . . . German Research Board of Mechanical Engineering (Forschungskuratorium Maschinenbau e.V.)

FOSTA . . . . Research Association for Steel Application (Forschungsvereinigung Stahlanwendungen e.V. )

GCTG . . . . . Geometrical Casting Tolerance Grades

HCF . . . . . . High Cycle Fatigue

HSE . . . . . . United Kingdom Health and Safety Executive

IWM . . . . . . Fraunhofer Institute for Mechanics of Materials IWM (Freiburg)

IZFP . . . . . . Fraunhofer Institute for Nondestructive Testing IZFP (Saarbrücken)

KIT . . . . . . Karlsruhe Institute of Technology

LCF . . . . . . Low-Cycle Fatigue

LEFM . . . . . Linear Elastic Fracture Mechanics

MT . . . . . . . Magnetic Particle Testing

NDT . . . . . . Non-Destructive Testing

NT . . . . . . . Normalized and Tempered

PT . . . . . . . Penetrant Testing

QT . . . . . . . Quenched and Tempered

RC . . . . . . . Resistance Category

RT . . . . . . . Radiographic Testing

SCF . . . . . . Stress Concentration Factors

SCRATA . . . . Steel Castings Research and Trade Association

SEW . . . . . . Stahl-Eisen-Werkstoffblätter

SIA . . . . . . Swiss Society of Engineers and Architects

SN . . . . . . . Cyclic Stress (S) against Cycles to Failure (N)

TDC . . . . . . Technical Delivery Conditions

UHCF . . . . . Ultra-High-Cycle Fatigue

ULCF . . . . . Ultra-Low-Cycle Fatigue

ULS . . . . . . Ultimate Limit State

UT . . . . . . . Ultrasonic Testing

# Symbols

$A_{act}$ . . . . . . Actual cross-sectional area in investigation plane

$A_{cavity}$ . . . . . Surface of the cavity

$A_{cg}$ . . . . . . . Defect size determined by crack growth simulation

$A_{crack}$ . . . . . Area of stable crack growth

$A_{fail,CT}$ . . . . . Defect area in the failure plane, determined by CT

$A_{fail,mic}$ . . . . Defect area in the failure plane, determined by microscope

$A_{Kth}$ . . . . . . Defect size determined by $\Delta K_{th}$

$A_{max,CT}$ . . . . Defect area at the maximum cross sectional reduction, determined by CT

$A_{mult}$ . . . . . . Maximum summed defect size of multiple defects

$A_{nom}$ . . . . . . Nominal cross-sectional area in the investigation plane

$A_{pore,mic}$ . . . . Pore area in the failure plane, determined by microscope

$A_{proj,CT}$ . . . . Defect area projected in the longitudinal direction, determined by CT

$A_{proof}$ . . . . . Area of inspection cross section

$A_{res}$ . . . . . . Area of residual fracture surface

$A_{sec}$ . . . . . . Gross cross-sectional area

$A_{sim}$ . . . . . . Crack size in failure state determined by crack growth simulation

$A_{sing}$ . . . . . . Maximum defect size of single defects

A . . . . . . . . Elongation at break

C . . . . . . . . Paris parameter also Paris-Erdogan parameter

$D_{max/min}$ . . . . Main axes of the ellipse attributed to stable crack growth

E . . . . . . . . Young’s Modulus

$F_l$ . . . . . . . Lower value of uniaxial force in cyclic loading

$F_o$ . . . . . . . Upper value of uniaxial force in cyclic loading

$K_{eff}$ . . . . . . Effective linear elastic stress intensity factor at stress ratios $R < 0$

$K_{eq}$ . . . . . . . Effective linear elastic stress intensity factor in mixed mode loading

$K_{Ic}$ . . . . . . . Plane strain fracture toughness

$K_{Id}$ . . . . . . . Dynamic plane strain fracture toughness

$K_{JIc}$ . . . . . . Plane strain fracture toughness estimated from $J_i$

$K_o$ . . . . . . . $K_{I,II,II}$ at $\sigma_o$

$K_r$ . . . . . . . Fracture ratio of the applied elastic value K to $K_{Ic}$

$K_{I,II,II}$ . . . . . Mode I, II and III linear elastic stress intensity factor

$L_r$ . . . . . . . Ratio of the applied load to the yield load

$M_{tol}$ . . . . . . Additional bending moment caused by deviations from straightness

$M_{x,y}$ . . . . . . Substitute bending moment

$N_{brit}$ . . . . . . Number of cycles until the failure criterion for brittle fracture is reached

$N_{endu}$ . . . . . Number of cycles of the endurance limit

$N_{fat}$ . . . . . . Number of cycles until the failure criterion for pure fatigue loading is reached

$N_{int}$ . . . . . . Number of cycles until failure at two interacting cracks

$N_{Kth}$ . . . . . . Number of cycles in a crack growth simulation when $\Delta K_{th}$ is reached

$N_{sing}$ . . . . . . Number of cycles until failure from crack propagation calculation of a single crack

$N_{tot}$ . . . . . . Total number of cycles determined by crack propagation calculations

N . . . . . . . . Number of cycles to failure

$R_R$ . . . . . . . Factor to consider mean stress

$R_{eh}$ . . . . . . . Upper yield strength

$R_{endu}$ . . . . . . Endurance ratio

$R_\epsilon$ . . . . . . . Strain ratio

$R_m$ . . . . . . . Uniaxial tensile strength

$R_{p0.2}$ . . . . . . Stress at 0.2 % strain

$R_{SCF}$ . . . . . . Factor to consider stress gradients

$R_{tol,1}$ . . . . . . Factor to consider geometric deviations

$R_{tol,2}$ . . . . . . Factor to consider deviations from straightness

R . . . . . . . . Stress ratio

$SCF_{max,defect}$ . . Stress concentration factor in the area of the defect

SCF . . . . . . Stress concentration factor

SD . . . . . . . Standard deviation

S . . . . . . . . Coordinates of the centroid of the substitute ellipse

$T_{ed}$ . . . . . . . Design temperature

$T_{\sigma}$ . . . . . . . Stress triaxiality

T . . . . . . . . Temperature

$V_{cavity}$ . . . . . Volume of the cavity

$J_i$ . . . . . . . . Crack initiation capacity

$a_{crit}$ . . . . . . Critical crack depth

$a_{res}$ . . . . . . . Depth of a substitute crack at two interacting cracks

$a_{sing}$ . . . . . . Maximum a value for single defects (see below)

a . . . . . . . . Crack depth at surface defects, half the crack depth at internal defects

$b_{\sigma,nom}$ . . . . . Depth at which $\Delta\sigma_{nom}$ is reached

b . . . . . . . . Thickness in the notch area

$c_{res}$ . . . . . . . Half the width of a substitute crack at two interacting cracks

c . . . . . . . . Half the crack width

$dN(i)$ . . . . . . Number of cycles in increment i

da/dN . . . . . Crack growth rate

$da_{max}(i)$ . . . . Maximum crack growth extension in increment i

$\vec{da}(n)$ . . . . . . Local expansion vector

$da(n)$ . . . . . . Local crack extension for node n

e . . . . . . . . Deviation from straightness

$f_{w,\sigma}$ . . . . . . Tensile stress fatigue strength factor

f . . . . . . . . Frequency

$h_1$ . . . . . . . First half axis of the substitute ellipse

$h_2$ . . . . . . . Second half axis of the substitute ellipse

$i_{tot}$ . . . . . . . Total number of crack growth increments

$i$ . . . . . . . . Index for crack growth increment

k . . . . . . . . Slope of the SN curve

l . . . . . . . . Length

$m$ . . . . . . . Total number of crack front nodes

m . . . . . . . Paris exponent also Paris-Erdogan exponent

$n_{bm}$ . . . . . . Fracture-mechanical support index

$n$ . . . . . . . . Node index for describing a crack contour & random coefficient

s . . . . . . . . Ligament between two cracks

$t_{exp}$ . . . . . . Exponent to determine wall thickness effects

$t_{ref}$ . . . . . . . Reference value of wall thickness to determine wall thickness effects

t . . . . . . . . Wall thickness

$z_{max,CT}$ . . . . Longitudinal position of $A_{max,CT}$

$\alpha$ . . . . . . . . Orientation of the substitute ellipse

$\beta_n$ . . . . . . . Effective notch factor

$\chi$ . . . . . . . . Stress gradient

$\Delta F_{test}$ . . . . . Applied axial load range in experimental investigation

$\Delta F_{x,FEM}$ . . . . Applied axial load range in FEM model

$\Delta F$ . . . . . . . Load range in cyclic loading

$\Delta K_{I,c}$ . . . . . Cyclic range in $K_I$ at position c of an ellipse

$\Delta K_{I,a}$ . . . . . Cyclic range in $K_I$ at position a of an ellipse

$\Delta K(n)$ . . . . . Cyclic range in $K_I$ for crack front node n

$\Delta K_{th}$ . . . . . . Threshold stress intensity factor range below which fatigue crack growth does not occur

$\Delta K$ . . . . . . . Cyclic range in $K_I$

$\Delta\sigma_{50\%}$ . . . . . Mean value of the SN curve

$\Delta\sigma_{95\%}$ . . . . . 95% probability of survival of the SN curve

| | |
|---|---|
| $\Delta\sigma_{98\%}$ | 98% probability of survival of the SN curve |
| $\Delta\sigma_{\mathrm{Afail,CT}}$ | Net stress range in failure plane determined by $A_{\mathrm{fail,CT}}$ |
| $\Delta\sigma_{\mathrm{Amax,CT}}$ | Net stress range in the plane of maximum cross-sectional reduction determined by $A_{\mathrm{max,CT}}$ |
| $\Delta\sigma_{\mathrm{core}}$ | Stress range at the edge of the core zone |
| $\Delta\sigma_{\mathrm{R,t}}$ | Wall thickness-dependent fatigue resistance |
| $\Delta\sigma_{\mathrm{endu}}$ | Endurance strength |
| $\Delta\sigma_{\mathrm{c}}$ | Stress range at $2 \cdot 10^6$ cycles according to FAT class |
| $\Delta\sigma_{\mathrm{max,defect}}$ | Maximum stress range under substitute stress in the area of the artificially induced surface defects |
| $\Delta\sigma_{\mathrm{max,N\text{-}only}}$ | Maximum stress range without substitute loads |
| $\Delta\sigma_{\mathrm{max}}$ | Maximal stress range in cyclic loading |
| $\Delta\sigma_{\mathrm{mic}}$ | Net stress range in the failure plane determined by $A_{\mathrm{fail,mic}}$ |
| $\Delta\sigma_{\mathrm{mod}}$ | Modified resistance stress range |
| $\Delta\sigma_{\mathrm{Moor,notched}}$ | Endurance strength at notched Moor specimen |
| $\Delta\sigma_{\mathrm{Moor,unnotched}}$ | Endurance strength at un-notched Moor specimen |
| $\Delta\sigma_{\mathrm{nom}}$ | Nominal stress range in cyclic loading |
| $\Delta\sigma_{\mathrm{Rd}}$ | Design resistance |
| $\Delta\sigma$ | Stress range in cyclic loading |
| $\gamma_{\mathrm{Ff}}$ | Safety factor according to EN 1990:2010 |
| $\lambda_{\mathrm{I}}$ | Factor to consider the accessibility |
| $\lambda_{\mathrm{R}}$ | Factor to consider the stress ratio |
| $\lambda_{\mathrm{d}}$ | Factor to consider the defect class |
| $\lambda_{\mathrm{t}}$ | Factor to consider the wall thickness |
| $\mu$ | Utilization |
| $\nu$ | Poisson's ratio |
| $\sigma_{\mathrm{H}}$ | Hydrostatic stress |
| $\sigma_{\mathrm{Ed,brit}}$ | Stress resulting from ULS loading for poof against brittle failure |
| $\sigma_{\mathrm{eq}}$ | Equivalent tensile stress according to the von Mises yield criterion |

$\sigma_{x,M}$ . . . . . . Axial stresses caused by substitute bending moment

$\sigma_{x,N}$ . . . . . . Axial stresses caused by axial load

$\sigma_{l}$ . . . . . . . Lower stress in cyclic loading

$\sigma_{max}$ . . . . . . Maximal stress

$\sigma_{nom}$ . . . . . . Nominal stress

$\sigma_{o}$ . . . . . . . Upper stress in cyclic loading

$\sigma_{ref}$ . . . . . . . Reference stress used for plastic collapse considerations

$\sigma_{res}$ . . . . . . . Nominal stress in residual fracture surface

$\sigma_{u}$ . . . . . . . Uniaxial tensile strength

$\sigma_{y}$ . . . . . . . Yield strength

# 1 Introduction

The consequences of material fatigue are some of the main causes of damage in load-bearing structures. This applies to applications in mechanical engineering, construction, the offshore industry and numerous other fields. In order to counter these problems, new manufacturing processes are continuously being tested and new materials and joining methods are being developed in order to generate more powerful and consequently, more resource-efficient structures. With the same objective, an increasingly deeper understanding of the physical processes responsible for fatigue is being achieved and modelling and calculation approaches derived in countless research projects on different observational scales.

Casting is one of the oldest metal manufacturing methods and enables the creation of almost any geometric shape. By adapting the shape to the flow of forces, potentials can be exploited as they relate to fatigue-optimized structures. Nevertheless, the application of cast steel components lags significantly behind expectations. Apart from several technical advantages, no economic concessions must be made [70]. The challenges are casting defects occurring in any economical casted component, whose influence on the fatigue strength can hardly yet be quantified in the design process. Due to the lack of specific design methods, it has not been possible to develop a generally valid normative basis so that the application of castings is limited to representative objects.

In order to exploit the full potential of cast steel and to make the design of efficient structures accessible to a wide range of users, the scope of this work is to derive a direct link between fatigue strength and the quality requirements for cast steel components.

## 1.1 Motivation

Cast steel components have both technical and aesthetic advantages due to their almost arbitrary geometric shapes. In terms of mechanical and technological properties, the materials used are equivalent to structural steel as well as high performance materials [105]. Figure 1.1 summarizes the main advantages of cast components in contrast to welded connections subject to fatigue loading, as discussed in [70]. The main advantage results from an optimized component shape adapted to the flow of force. Thus, cast steel components are theoretically ideally suited for use in cyclic-loaded structures.

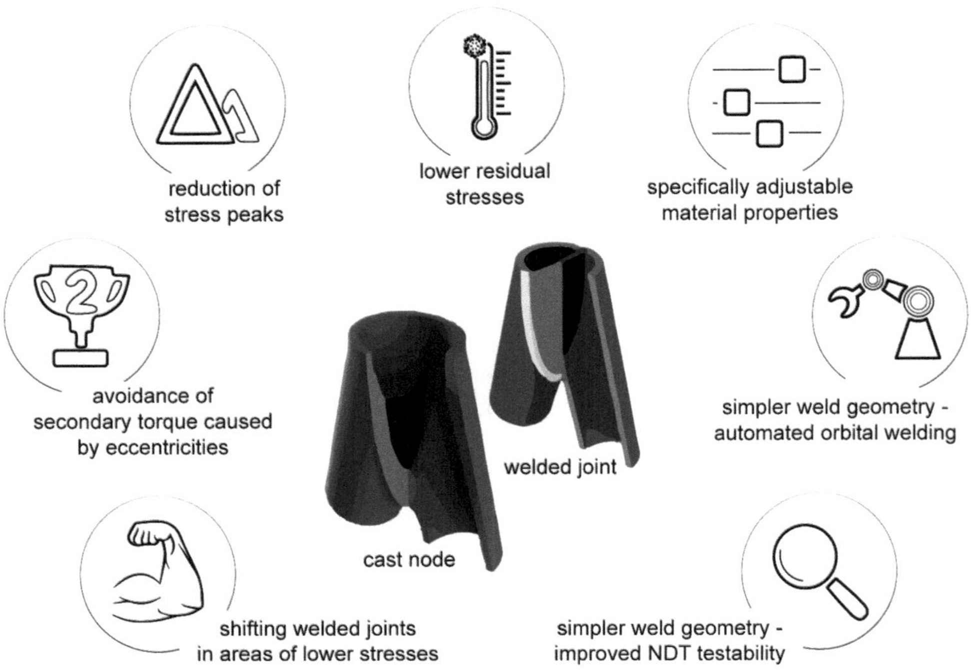


**Figure 1.1:** Advantages of cast over welded joints in cyclic loaded structures analogous to [70]

These advantages are confronted with casting defects occurring in every economic cast steel component [45]. Defects located inside the component as well as on its surface act as additional notches or, in the worst cases, as sharp initial cracks [103] and reduce fatigue resistance compared to sound material [15]. These effects on fatigue strength have not yet been sufficiently quantified. To what extent the advantages discussed in Figure 1.1 outweigh the negative effects of defects is unknown. In Germany casting defects were defined, categorized and included in order contracts in the early 1950s. As their appearance was accepted under certain conditions, they have been termed "discontinuities". These agreed-upon characteristics are assigned to quality classes in Technical Delivery Conditions (TDC) and Non-Destructive Testing (NDT) codes. The particular codes are based on manufacturing and order needs and have not been properly linked to design resistance [70]. With a lack of knowledge on the quantitative performance of each quality class, designers' demands have risen steadily. The foundry industry has responded to these requirements with research into techniques to predict and prevent casting defects [70]. Regardless of the calculative optimization, the economic efficiency of these components decreases with increasing quality requirements as more complex casting systems become necessary, and thus the output, defined as component mass divided by the total mass of steel used, decreases.

Figure 1.2 shows examples of fatigue-loaded structures from different industries. Pont Alexandre in Paris (Figure 1.2a) was built in 1900 and is still in service [70]. All its girders and anchor points are casted and are subject to fatigue loading. Conoco Viking JD was built in 1984 and was equipped with four cast X joints. This project can be regarded as the starting point for casting's application in offshore projects [58]. Figure 1.2b shows a steam turbine, which is subject to high thermal cyclic loads in service. All these mentioned applications indicate that the advantages of cast components outweigh the disadvantages of casting defects. This becomes even clearer through the considerations by fracture mechanics made in [35] for estimating permissible casting defect sizes in real bridge structures. The cast joints used in a truss girder with hollow sections were allowed to have casting defects with an elongation between 30 % and 80 % of the wall thickness in order to not cause fatigue failure under design loads.

(a) (b)

**Figure 1.2:** a) Pont Alexandre, Paris [70]; b) Steam turbine [89]

In addition to these technical uncertainties, further organizational constraints must be overcome in order to apply cast steel in the German construction industry, for example. Up to the present, DIN EN 1993-1-8/NA:2010 [155] is the only European standard for the use of cast steel in construction. Nevertheless, it limits the application of cast steel to static load cases only. The standards of other industrial sectors do not offer any significant additional value for the definition of applicable stress ranges based on of quality classes. Although evaluations by fracture mechanics are possible, their application is complex and generally reserved for a small circle of experts. In addition, the defect locations and sizes must be reliably determined and known. For an estimate in the design process, the designer would have to be able to utilize the experience, calculations and specific casting technologies of the actual producing foundry. According to Figure 1.3 from [70], this is virtually impossible for public building projects in Germany due to German tendering regulations. In order to enable an application under the current legal framework, clear and simple rules must be developed that indirectly serve as a communication medium between the designer and the foundry.

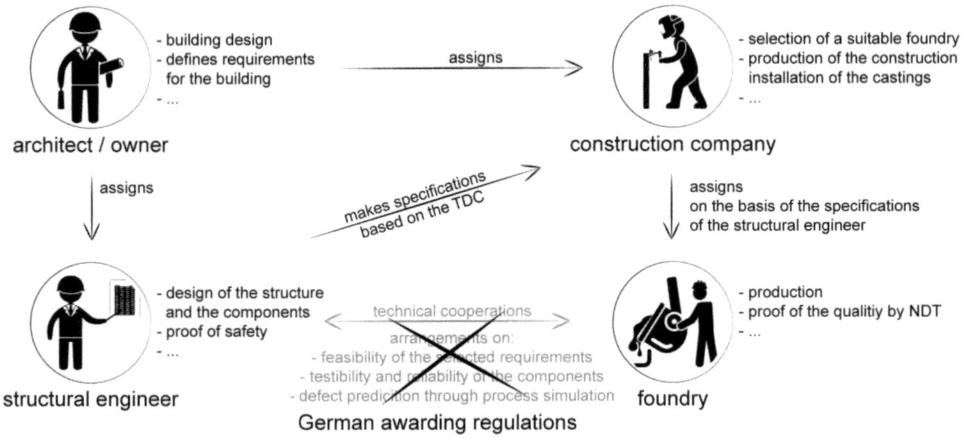


**Figure 1.3:** Cooperation between engineer and foundry is prevented by German building law analogous to [70]

## 1.2 Objectives

The situation described in Section 1.1 is to be improved within the scope of this work and thus facilitate the use of cast steel components in the construction industry. Furthermore, this work contributes to the design of structures that are more efficient and to resource-saving construction. The overall research objective is defined below:

*The aim is to associate the fatigue strength of cast steel components with quality requirements. As a consequence, depending on local stresses, different casting defect sizes are to be accepted. Their influence is to be systematically determined by calculations.*

To achieve this, the previously described unknown technical aspects are to be clarified, and the organizational barriers are to be reduced through concise regulations. The following key aspects are in focus for these investigations:

- understanding the relevant damage mechanisms (shares of crack initialization and growth) caused by casting defects
- quantifying the fatigue resistance depending on defect properties and component influences
- generalizing the findings for the application to common materials and component influences by worst-case considerations
- deriving an easy-to-use design concept

The focus is on the conditions common in civil engineering. In general, the basic mechanisms can be transferred to other fields of application, and the design methods can be adapted to specify their respective requirements. The particular restrictions are defined as:

- Approved materials according to the common relevant technical standards.
- Restriction to the stress range of High Cycle Fatigue (HCF).
- The approaches developed shall follow the design philosophy of the Eurocode (EC). This includes the safety concept, evaluation procedures for fatigue tests, consistency with related design methods (welds) and principles of structural design.
- The design process must be carried out cost efficiently with simple tools in an engineering design office.

This work purely focuses on the cast component itself; structural welds connecting arbitrary steel sections to cast components are not considered. In addition, any kind of repair welding is not within the scope of these investigations. Generally, these are used during the manufacturing process to eliminate casting defects that do not fulfill the quality requirements and to recover damaged components after a certain amount of time in service.

In contrast to the existing quality classes according to EN 1559-2:2014 [152], the design method is to be determined from the point of view of a structural engineer based on load-bearing capacity. If necessary, modified quality requirements that deviate from the current technical delivery conditions in EN 1559-2:2014 [152], including NDT standards, may be defined. Their actual implementation in NDT instructions is not part of these investigations. A comprehensive evaluation of the performance of the relevant NDT procedures in the application of cast components is part of the current research project DeQaGuss [115]. The conservativeness of the assumptions can be reduced and the precision of the procedure can be increased by increasing the accuracy of detecting the size, position and shape of the casting defects in all areas of the component.

This thesis is based on investigations carried out within the German Federation of Industrial Research Associations' (AiF) research project ErStaGu [116] at the Research Center for Steel, Timber and Masonry (Karlsruhe) with its partners, the Fraunhofer Institute for Mechanics of Materials IWM (Freiburg) (IWM) and the Fraunhofer Institute for Nondestructive Testing IZFP (Saarbrücken) (IZFP). The Federal Ministry for Economic Affairs and Energy funded the project based on a decision of the German Bundestag. The promoters of this project were the German Foundry Association (Forschungsvereinigung Gießereitechnik e.V.) (FVG) and the Research Association for Steel Application (Forschungsvereinigung Stahlanwendungen e.V. ) (FOSTA).

The investigations are based on previous results from Spannaus [103] on the influence of casting defects on brittle fracture resistance. The method developed in [103] for the first time allowed designers to select quality classes on the basis of minimal requirements according to technical delivery conditions and the quality classes defined in EN 1559-2:2014 [152] and EN 10340:2008 [142]. With the combination of both methods, it should be possible to define quality requirements under both static and cyclic loading.

## 1.3 Structure

The present work is structured as follows:

**Chapter 2** classifies the research topic in the context of material fatigue and describes the current state of knowledge on the influence of casting defects. For this purpose, the causes, classification and evaluation of different types of casting defects are discussed first. Following this, the main influencing factors are described, and the suitability of current design approaches are assessed. Finally, the state of research on casting defects and their influence on fatigue resistance is presented. The knowledge gained from works in the last 60 years is discussed, and conclusions relevant to the present research question are summarized.

**Chapter 3** describes the experimental investigations and the results that were obtained. Starting with a classification of the flawless base material, the observation scale is successively enlarged. The influence of real casting defects is investigated, and finally the interaction of geometric influences and artificially introduced defects on large components is considered. The resulting findings of the observation scales are summarized using cyclic material behavior, SN curves on real casting defects and component SN curves.

**Chapter 4** presents extended numerical studies of the experimental investigations to better understand the occurring processes. At first, crack propagation in specimens with real casting defects is calculated. In doing so, the shares of crack initiation and propagation on the total service life are estimated, and the hypotheses set up in the evaluation of the experimental investigations are confirmed. Additionally, different calculation procedures are evaluated for the application as a design concept. The second part focuses on the actual stress situation of the large components from Chapter 3.

**Chapter 5** summarizes the major findings of the previous chapters necessary for a simplified design concept. Newly defined resistance categories link applicable stress ranges to actual defect sizes. This connection is based on fracture mechanics and appraises relevant influences such as stress gradients, interactions of defects, brittle fracture, mean stress dependency and geometric deviations. Secondly, the developed design concept is validated by the results of experimental investigations and is discussed in the context of other approaches and research data. Finally, two common cast steel components, a fork head and a K joint, are evaluated using this concept as well as numerical crack growth simulations.

**Chapter 6** summarizes the results obtained in this work and provides an outlook on the potential for further developments in the presented approaches.

# 2 Fundamentals

The origin and technical background of casting defects as well as regulations for their acceptance and detection are initially presented in this chapter. This is followed by basic information on fatigue in metallic materials, design concepts and dominating influencing factors. However, these descriptions are limited to help the reader put the present investigation in the context of fatigue and relevant calculative approaches. Finally, a comprehensive overview of current findings on the influence of casting defects on cast steel components' fatigue behavior is presented, and their implementation in design standards is discussed.

## 2.1 Casting Defects

Cast steel components tend to have casting defects in almost every application, especially for single units and small series [45, 105]. A brief selection of typical casting defects is shown in Figure 2.1. These are caused by mechanical (e.g., stress gradients), technological (e.g., solidification process, chemical composition) or foundry-related (e.g., mold material, mold filling, casting system) influences [86]. In general, these occur inside the component as well as on its surface as geometric deviations, volume defects (e.g., blowholes, gas bubbles, metallic inclusions) or planar, crack-like defects (e.g., hot crack, cold crack). Depending on the type and origin, a classification is made according to the International Atlas of Casting Defects' [86] in seven casting defect classes and subgroups. An exemplary compilation of relevant defect types that reduce load-bearing capacity appears in [103].

(a) (b) (c) (d)

**Figure 2.1:** a) Shrinkage cavities [103]; b) Molding material residuals [5]; c) Non-metallic inclusion [31]; d) Crack [27]

Mechanically, casting defects cause local stress peaks due to the cross-sectional reduction, their notch effects (Stress Concentration Factors (SCF)) and their interaction with geometric notches resulting from the component geometry (amplification of both SCFs). Besides, the strength of the material itself can be negatively affected by casting defects or the casting microstructure. Two processes affect the casting microstructure. First, a dissolved structure surrounding the solidification cavities (see Section 3.2.2.3 or [110]) has a lower density and load-bearing capacity due to reduced bonding between the solidification dendrites. Secondly, the solidification process leads to differences in the concentration of the chemical composition, resulting in fluctuating material properties. This can be observed in all thick-walled metallic components and is referred to as the "technological thickness effect" [70, 103].

Production qualities are agreed upon as a means of communication and form the basis of contracts between foundries and clients. For this purpose, maximum permissible defect sizes and defect accumulations are defined depending on the respective quality class. Manufacturing quality is specified in the technical conditions of delivery (e.g., according to EN 1559-2:2014 [152]) in the form of quality grades and is verified by means of NDT procedures (Penetrant Testing (PT): EN 1371-1:2012 [150]; Magnetic Particle Testing (MT): EN 1369:2013 [148] and EN ISO 9934-1:2017 [158]; Ultrasonic Testing (UT): EN 12680-1:2003 [144] and EN 12680-2:2003 [145]; Radiographic Testing (RT): EN 12681-1:2018 [146] and EN 12681-2:[147]). The classification of casting defect types is primarily relevant for the foundry in terms of defect prevention. As in EN 1559-2:2000 [151], the link between indication size and quality grade is no longer made in the technical delivery conditions but is still made in the respective NDT standards. This means that a consistent evaluation is no longer possible, since different defect sizes are permissible for the same quality grade with different NDT methods [103]. The NDT evaluation is carried out almost independently of the casting defect type and is purely based on the indications detected by the NDT.

Links between quality classes and defects are defined based on good workmanship and have only limited correlation to physical performance under static and cyclical loading. This issue is summarized for welded connections in [49] (p. 1) as: "historically, production technology research and structural design research have only limited dialogue and each group has focused on their own narrow field of interest." This statement can be applied without restriction to cast steel components.

In the past, foundry research has mainly aimed to develop methods for defect prediction and avoidance. In [70], insights into relevant topics such as foundry technology, component design, solidification and casting process simulation are given. Nevertheless, the experimental investigations in [103] showed that, even under almost identical solidification conditions, different solidification cavities with different mechanical effects occur. In these investigations, cylindrical specimens with different thickenings were cast, and Computer Tomography (CT) recordings determined the internal conditions. Figure 2.2 compares the different casting defects that resulted from the same specimen types and nearly identical solidification conditions. Nevertheless, the projected cross-sectional reduction detected by CT ranges from 8 % to 15 %.

With these explanations, it becomes apparent that the coarse occurrence of casting defects can be controlled by the casting technique. However, even defect types that are exclusively assigned to the solidification process cannot be predicted or reproduced unambiguously. In addition, other effects caused by the production process, such as slag inclusions, the chipping of molding material or other contaminations from defects, are not sufficiently considered. The prediction and NDT detection of the dissolved microstructures surrounding cavities is still a major uncertainty. Even complex investigations that do not comply with the industry standard for chilled castings, such as those discussed in Section 3.2.1.2, do not yet allow for an accurate analysis.

Due to the casting process and the decisive impact of casting technology, defect distributions do not follow a statistical distribution (e.g., Weibull) that is evenly spread over the cross-section, as is seen in other materials. Rather, accumulation areas that can be influenced by technological processes occur. The situation described in Section 1.1 of the German Tender Regulation prevents any communication between foundries and structural engineers in the design process. However, it is not possible to estimate and reduce the number of probable defect arrangements without knowledge of the subsequent casting technology. Therefore, in any case, a worst-case analysis must be carried out with the maximum permissible defect sizes and most unfavorable arrangements according to the quality grades. In accordance with the investigations [35, 103], this is done by assuming crack-like initial defects. Especially for internal defects, the industrially applied methods do not provide a direct image of the casting defects but rather displays to be interpreted. The display accuracy is reduced by component geometry (e.g., curved or non-parallel surfaces) and surface roughness to an uncertain extent. Information on the exact shape of the defect is not obtainable [103, 110]. In general, the lower the reliability of the verifiability of defect sizes, positions and shapes, the more conservative the applied design concept needs to be.

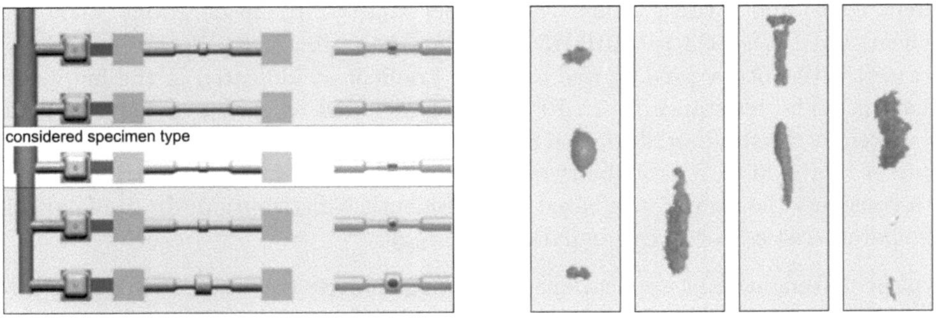


(a) (b)

**Figure 2.2:** a) Specimen geometry and results of a solidification simulation from [103]; b) Different characteristics of the shrinkage cavities from the same specimen type and comparable solidification conditions

## 2.2 Fatigue Behavior of Metallic Materials

The fatigue behavior of metallic materials has been systematically investigated since the 1870s based on the research of August Wöhler [56], and the underlying physical mechanisms, assumptions and influencing factors have been published widely [34, 52, 83, 93, 107, 125]. When considering them, the field of application or research, the method of observation and the associated investigation scale must be taken into account. For a comprehensive description of the processes, experimental investigations and mathematical descriptions, the reader is referred to the technical literature. Only conditions relevant to the classification of this work in the overall context of fatigue research are addressed, and references are provided for a more profound study.

### 2.2.1 Introduction

Primarily depending on the stress level, physical processes of different intensities take place and lead to different relationships between the applied stress range and the number of tolerable load cycles to failure (N). Common definitions of the N ranges for these processes are marked in the SN curve in Figure 2.3a. The mechanisms are subjects of different fields of research and are referred to as Ultra-Low-Cycle Fatigue (ULCF) [20, 55, 99], Low-Cycle Fatigue (LCF) [21, 47, 59], High Cycle Fatigue (HCF) [34, 64, 72, 73, 101, 123] and Ultra-High-Cycle Fatigue (UHCF) [57, 79, 111]. These mechanisms result in different mathematical correlations (e.g., Coffin-Manson [21, 59], Basquin [83]), and for common applications in structural engineering, HCF is decisive. In special cases, such as crane construction, the LCF range gains importance. For the evaluation of the effects of extreme events, such as earthquakes, ULCF is crucial.

The definition of the cycles, which are correlated to the endurance limit, vary between different codes and investigations. All further statistical considerations follow the specifications of EN 1993-1-9:2010 [156] [22, 25]. Accordingly, the linear regression is performed with both a fixed as well as a free gradient as indicated in the individual evaluation. The resistance at $2 \cdot 10^6$ cycles is denoted as $\Delta\sigma_{50\%}$. The scatter is represented by the standard deviation SD with respect to log(N), and the characteristic SN curves represent 95 % probability of survival, are derived from a prediction interval which considers the sample size based on a Student's t-distribution. In the following these characteristic SN curves are marked as $\Delta\sigma_{95\%}$.

In typical civil engineering applications, the damage process is considered in a simplified way of using limit states detectable by conventional NDT methods and phenomenological stages. Their technical background, proportions on the total service life and calculation approaches are illustrated in Figure 2.3b. A schematic failure SN curve (continuous) in the HCF region is shown. The first phase, the crack incubation, reaches until a technical crack forms and is limited by the crack initiation SN curve (dashed line). According to [83], the technical crack is in the range of a crack depth a = 0.5 mm and a crack width 2c = 2 mm. The movements of dislocations, crack nucleation and short crack growths that occur in this phase are discussed in detail in [32, 61, 80]. Depending on the strain amplitude, stress and failure states, this behavior can be estimated using damage models and short crack growth estimation approaches like [24, 30]. In the next phase, stable macroscopic crack growth is observed. This is the main approach used in this thesis and is discussed in detail in Section 2.2.4. Formally, the unstable crack growth and the residual fracture completes the component's service life. Its share of the total service life is negligible due to the over-proportionally increasing crack growth rate (cf. stage 3 in Figure 2.6a). Therefore, a separate visualization in Figure 2.3b is not possible. Considering two different stress ranges, $\Delta\sigma_1$ and $\Delta\sigma_2$, the proportions between crack initiations and growths change. It is also noted that the log-log representation leads to the incorrect assumption that crack growth only accounts for a very small proportion of the service life.

The crack initiation phase is not considered in detail in this work but is understood as the part of the service life that cannot be described by macroscopic crack growth. This is done in order to derive a general design method. As the experimental investigations from Chapter 3 show, crack initiation is decisively determined by the shape and position of the individual discontinuity. An exact prediction is not possible according to the arguments discussed in Section 2.1. Fraunhofer IWM is conducting studies within ErStaGu research project [116] to evaluate the safety level and to utilize further reserves for well-known and reliably detected defects. General aspects such as notch effects, stress ratios, material states, residual stresses or environmental conditions influence one or several of the described fatigue phases and are discussed in [34, 83, 107].

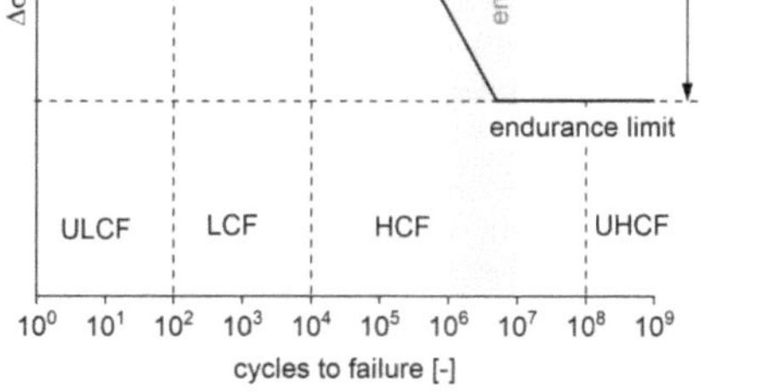


(a)

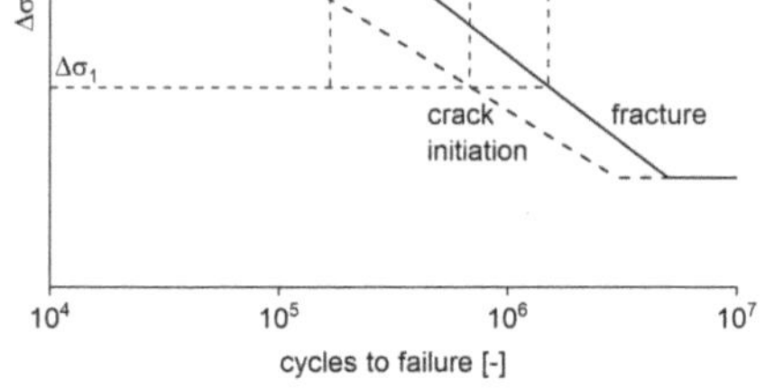


(b)

**Figure 2.3:** a) Classification of fatigue ranges for typical construction steel; b) Stages of fatigue

## 2.2.2 Fatigue Assessment Methods

Table 2.1 provides an overview of state-of-the-art fatigue assessment methods sorted according to their complexity and calculative expense. With the exception of the last two approaches, these methods are established in codes of practice and design guides for the analysis of base materials and welds [46, 156, 160, 162]. Detailed descriptions, instructions and fields of application for each procedure are discussed in [9, 73]. In addition to the basic concepts, Table 2.1 evaluates and classifies these approaches in the context of the present examination.

**Table 2.1:** Overview and evaluation of fatigue assessment methods

| approach | concept | evaluation |
|---|---|---|
| nominal stress | Evaluation of fatigue strength by comparing the acting nominal design stress with tabulated fatigue resistance classes. Stress peaks and notch support effects are taken into account by adjustment factors in cases of un-welded applications [162]. At welds, these classes are representative of common execution details based on experiments [46, 156]. | *goal of this work*: Based on fundamental investigations, the behavior of casting defects should be transferred to this calculation approach. With the current knowledge, this procedure is not applicable due to its a) missing database, b) arbitrary geometric shapes that could be realized by casting and c) lack of a procedure to consider specific defect sizes. |
| structural hot-spot stress | Approximation of the hot-spot stresses present at welds by extrapolating from reference points (surface stresses) in front of the weld. Local stiffness and stresses are directly taken into account. | *not suitable*: The notch radii of cast components are considerably larger than weld transitions, so a direct stress calculation is possible. No approach to consider defects. |
| effective notch stress | Direct comparison of the notch stresses calculated under defined conditions with design SN curves. Consideration of the micro support effects according to [71]. General application for any component geometry according to [162]. Special application for welds according to [46, 160]. | *not considered alternative*: Casting defects could be represented like root defects in welded joints using elliptical keyhole notches. Substitute method with no direct coupling to the physical processes. After a more detailed understanding of the crack initiation phase and a reliable detection method of the casting defect geometry (incl. the resolved microstructure), this approach might be reevaluated. |
| crack propagation | On the basis of an initial crack, its propagation is calculated according to the principles described in Section 2.2.4. | *focus of this research*: According to the discussion in Section 2.1 for general design procedures, casting defects need to be considered conservatively as crack-like defects. |
| strain-based damage model | Depending on the stress state, the plastic strains are accumulated until a material-dependent critical state is reached. The calculation can be coupled with or decoupled from the numerical calculation of the deformation field, considering stiffness changes. | *not in the scope of this work*: The initiation phase of a casting defect can be mapped with this approach. In addition, it can be determined whether one or more imperfections lead to a crack. For their application in a component design procedure, the notch factors of casting defects must be clarified. |
| combination | Combination of damage model and crack growth simulation [9]. | *not in the scope of this work*: Current state of research. Discussed in ErStaGu [116] by Fraunhofer IWM. |

### 2.2.3 Factors Influencing Service Life

The service life of components subject to fatigue loads is influenced by numerous factors. A comprehensive overview is given in [34, 83]. In the following, only the influencing factors essential to the present investigations are discussed. Since the focus of these examinations is on defects, they are analysed in depth in Section 2.2.4.

#### 2.2.3.1 Notches

Notches lead to a local stress increase, which is described in Neuber's notch stress theory [71]. Extensions to a wider range of applications were made in [76, 81, 82, 91]. In addition to geometric notches, material notches and load singularities are generally categorized under the term "notch effect." In this thesis, the presentation is limited to notches caused by discontinuities in component geometries. The stress increase $\sigma_{\mathrm{max}}$ compared to the nominal stress $\sigma_{\mathrm{nom}}$ is described in the linear elastic case by the notch factor SCF:

$$\mathrm{SCF} = \sigma_{\mathrm{max}}/\sigma_{\mathrm{nom}}. \tag{2.1}$$

Due to the existing multiaxial stress state, this stress increase does not have a proportional effect on the service life. This phenomenon is attributed to the micro and macro support effect, depending on the stress gradient, and is described by the effective notch factor $\beta_{\mathrm{n}}$. This relates the fatigue strength of an un-notched component to the nominal stress of the notched component:

$$\beta_{\mathrm{n}} = \frac{\Delta\sigma_{\mathrm{endu}}(\mathrm{SCF} = 1)}{\Delta\sigma_{\mathrm{nom,endu}}(\mathrm{SCF} > 1)}. \tag{2.2}$$

These processes are influenced by the stress level, the type of load, the material and the notch shape, which are described in detail in [53, 75, 83] and are taken into account by additional parameters in regulations such as FKM-Rili-BM:2009 [161]. Within the scope of these investigations, SCFs that are common for a fatigue-optimized cast design are experimentally investigated in Section 3.3 and are taken into account in the design concept in Section 5.2.5 in a simplified way.

#### 2.2.3.2 Component Thickness

Fatigue tests on specimens of different thickness show varying service lives. The differences can be attributed to the mechanisms summarized in Table 2.2. A detailed description is given in [83]. In design standards like HSE:1995 [166] or DNV RP-C203:2001 [159] the influences of thickness are taken into account by Eq. (2.3).

**Table 2.2:** Summary and assessment of the basic mechanisms of size effects

| effect | concept | importance for these investigations |
|---|---|---|
| technological | Thermal gradients in the manufacturing process cause different microstructures. In the case of large component thicknesses, considerable differences can occur between core and rim zones. | Considerable effects of the component thickness on KV values [103]. In the case of high-strength steels, this effect is particularly pronounced. The experimental investigations in Chapter 3 are intended to quantify this influence on fatigue strength. |
| stress gradient | The supporting effect in the context of notches depends on the stress gradient $\chi$. In the case of bending or torsional stress, $\chi = -2/t$. | Material-independent, so the existing concept is directly applicable to any homogeneous, isotropic and linear-elastic material. |
| statistical | A larger volume has a higher probability of microstructural defects occurrence. These defects initiate cracks, and their quantity is usually regarded as evenly distributed (Weibull) throughout the component. | The objective of this work is to determine permissible casting defect sizes. As discussed in Section 2.1, these depend on technological mechanisms, so that worst-case scenarios must be considered. Microstructural weaknesses are less significant and do not play a significant role in these considerations. |

The parameters for Eq. (2.3) were determined by curve fitting on experimental data and thus contain all the mechanisms in Table 2.2. In the application, it is not apparent which portion is attributable to the stress gradient effect. The resulting increase in fatigue resistance at low wall thicknesses is not effective for components subject to normal stress.

$$\Delta\sigma_{\mathrm{R,t}} = \Delta\sigma \left(\frac{\mathrm{t}}{\mathrm{t_{ref}}}\right)^{\mathrm{t_{exp}}} . \tag{2.3}$$

## 2.2.4 Influences of Defects

In this thesis, defects are considered as cracks and are evaluated by fracture mechanics. This follows the descriptions of casting defects in Section 2.1 as well as the findings of [103, 35]. The basic concepts of stress intensity factors, fracture criteria, crack growth and crack interaction are discussed in this section.

### 2.2.4.1 Stress Intensity Factors and Crack Growth

The starting point for any fracture mechanics analysis is a crack-like initial defect in a stressed component. Crack growth depends on the conditions in the process zone surrounding the crack tip and thus on the crack tip stress field. With linear elastic stress intensity factor ($\mathrm{K_{I,II,III}}$), the J-integral or further parameters, the singularity can be counteracted and the states relevant to crack growth in the process zone can be described indirectly. The description of the crack tip field is therefore reduced to a single value in a planar example. Decisive influencing factors are the crack-opening mode, the characteristic of the stress field (depending on the load situation and component geometry), the stress level and the material (linear elastic fracture mechanics, small-scale yielding and elastic-plastic fracture mechanics). Fundamental principles, definitions and the derivation of essential mathematical correlations to fracture mechanics are explained in detail in [13, 16, 17, 28, 120] and are transferred into guidelines for the design and evaluation of components containing flaws in [46, 159], FKM-Rili-BM:2009 [161] or BS 7910:2013 [138].

Within the dominant part of the total lifetime in the HCF regime, the stresses are relatively low, and crack tip stresses follow linear elastic fracture mechanics (LEFM). If not directly stated, all following relations refer to LEFM. In this case, the characteristics of the local stress field can be described by K-factors. According to [28], the crack tip field for crack opening mode I in an infinitely extended disc stressed by uniaxial tensile stress is obtained by:

$$\left\{ \begin{array}{c} \sigma_\mathrm{x} \\ \sigma_\mathrm{y} \\ \tau_\mathrm{xy} \end{array} \right\} = \frac{\mathrm{K_I}}{\sqrt{2\pi r}} \cos(\varphi/2) \left\{ \begin{array}{c} 1 - \sin(\varphi/2)\sin(3\varphi/2) \\ 1 + \sin(\varphi/2)\sin(3\varphi/2) \\ \sin(\varphi/2)\cos(3\varphi/2) \end{array} \right\}. \tag{2.4}$$

The definition of geometric parameters and coordinates is shown in Figure 2.8a. The local stress field, according to Eq. (2.4), is schematically shown. Depending on the global geometric features, $\mathrm{K_I}$ can be calculated as:

$$\mathrm{K_I} = \mathrm{Y(a, geom)}\, \sigma \sqrt{\pi \mathrm{a}} \tag{2.5}$$

with the geometry function $Y$. For a range of representative geometric configurations, the Y-functions were determined analytically (e.g., by complex potentials or derived and tabulated from numerical studies [109, 161]).

As the fracture surfaces in Chapter 3 and the numerical studies in Chapter 4 show, crack-opening Mode I is decisive in the considered cases and generally represents the most unfavorable design case for defects, which are theoretically arbitrarily oriented. Thus, within the scope of this work, an extension to $\mathrm{K_{eq}}$ can be determined according to Eq. (2.6) [161].

$$\mathrm{K_{eq}} = \frac{\mathrm{K_I}}{2} + \frac{1}{2}\sqrt{\mathrm{K_I^2} + 5.34 \cdot \mathrm{K_{II}^2} + 4 \cdot \mathrm{K_{III}^2}} \tag{2.6}$$

Extensions for surface defects as well as internal cracks are presented in [66]. In these cases, $\mathrm{K_I}$ correlates with the defect surface (*area*) and results in:

$$\mathrm{K_I} \cong 0.5 \cdot \sigma_0 \sqrt{\pi \sqrt{\mathrm{area}}} \quad \text{for internal and} \tag{2.7}$$

$$\mathrm{K_I} \cong 0.65 \cdot \sigma_0 \sqrt{\pi \sqrt{\mathrm{area}}} \quad \text{for surface defects.} \tag{2.8}$$

Effects that are not directly depending on K are termed "crack tip stress" in the following descriptions. This states a more general validity of effects caused by cracks, independent of the mathematical concept characterizing the local stress field. Note, this term does not correspond to the stresses at the crack front as given in Eq. (2.4) but to their effects.

Figure 2.6a illustrates a schematic correlation between the range $\Delta K = K(\sigma_o) - K(\sigma_l)$ and the crack growth rate da/dN under cyclic loading. The stress ratio R is defined using the upper $\sigma_o$ and lower stresses $\sigma_l$ as:

$$R = \frac{\sigma_l}{\sigma_o}. \tag{2.9}$$

The process is divided into the three stages: I) no growth, II) stable growth and III) instable growth. The Paris-Erdogan equation (further referred to as Paris' law as common in english literature) describes the mechanisms of stage II as:

$$\frac{da}{dN} = C\,\Delta K_{eff}^{m} \quad \text{if} \quad \Delta K_{eff} > \Delta K_{th} \quad \text{else} \quad \frac{da}{dN} = 0. \tag{2.10}$$

In addition to the coefficients m and C, which need to be determined for a specific stress ratio, the crack growth rate depends on the range of the stress intensity factor and increases with an increasing crack length. According to [161], at $R < 0$, only the part of $\Delta K$ where the crack is loaded by tensile stresses is relevant for propagation. This is considered in:

$$\Delta K_{eff} = (1 - R)K_o \quad \text{if} \quad \Delta K \leq 0 \quad \text{else} \quad \Delta K_{eff} = \Delta K. \tag{2.11}$$

Additionally, Figure 2.6a qualitatively illustrates the changes of the parameters depending on the stress ratio R. According to [34], the Paris exponent m remains constant, whereas C increases and $\Delta K_{th}$ decreases with an increasing R value. Comparing experimental data with the linear approximation in the log-log representation for values $\Delta K \approx \Delta K_{th}$, the crack growth rate could be overestimated. In these cases, a subdivision into two stages, as marked in Figure 2.6a (dashed line) and presented in [163], leads to more realistic and less conservative results. An overview of crack growth parameters from the literature on steel cast material is given in the context of the experimental investigations in Section 3.1.2. Several extensions of the Paris' law, such as [26, 122], have been developed to consider the disproportionate increase of the crack growth rate in stage III. The experimental investigations in Section 3.2.2.8 show that failure is dominated by plastifications and that the LEFM loses its validity anyhow. For a correct description, further procedures become necessary, but due to the negligible number of cycles and low practical relevance, this adaptation and exact consideration of unstable crack growth is omitted in this thesis.

### 2.2.4.2 Iteration Methods in Crack Propagation Calculations

The nonlinear evolution of crack tip stress over an expanding crack length allows no direct calculation of the total number of cycles. Instead, iterative approaches should be applied. Additionally, in the case of 3D problems, the crack shape can change during crack propagation, which directly influences crack tip stresses and must be determined at each increment. Simplified approaches, such as keeping a conservative crack tip shape that is uniformly propagated, could quickly lead to overly conservative results due to the potentiation in Eq. (2.10). For the evaluation, three approaches are compared and related to a sample geometry in Figure 2.4. Additionally, the effects of these methods on the service life are shown schematically with an exemplary number of cycles in Figure 2.5.

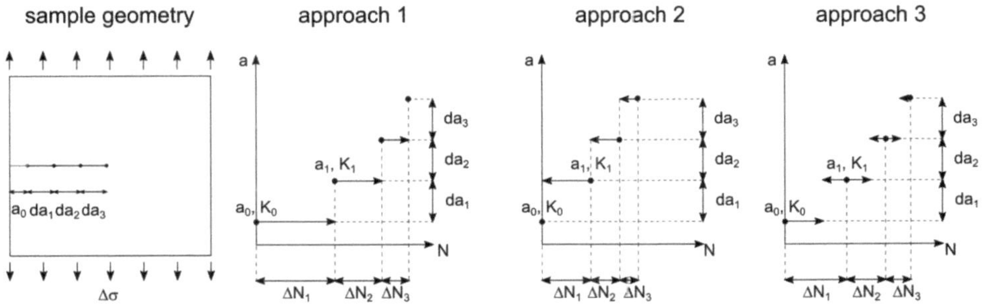


**Figure 2.4:** Iteration approaches: an exemplary geometry for crack propagation

In approach 1, $\Delta K_i$ is determined for an increment $i$ on the crack geometry corresponding to the state $i-1$, and the resulting number of cycles $\Delta N_i$ is calculated for an assumed crack growth increment $da_i$. If large increments of $da_i$ are selected, this procedure considerably overestimates the actual service life.

In approach 2, the other extreme scenario, $\Delta K_i$ is determined on the updated crack geometry for increment $i$ and the number of cycles is determined with the previous crack growth increment $da_i$. This approach is proposed in BS 7910:2013 [138] for calculations without further background investigations and represents the most conservative solution. In this method, the crack geometry must already be known at the end of each increment. This is not a limitation for 2D problems, but for 3D problems, it is not complete and further considerations are necessary.

The third approach is a combination of both previous methods and leads to good approximate solutions even for a small number of increments. Here, $\Delta N_i$ is determined from the averaged stress intensity factor of both states before and after applying the increment $(\Delta K_{i-1} + \Delta K_i)/2$. With a decreasing crack growth increment $da_i$, the results of all three methods converge on a common solution. In the case of all analytical crack propagation calculations in this work, da is set to 0.01 mm, resulting in about $10^4$ to $10^5$ increments.

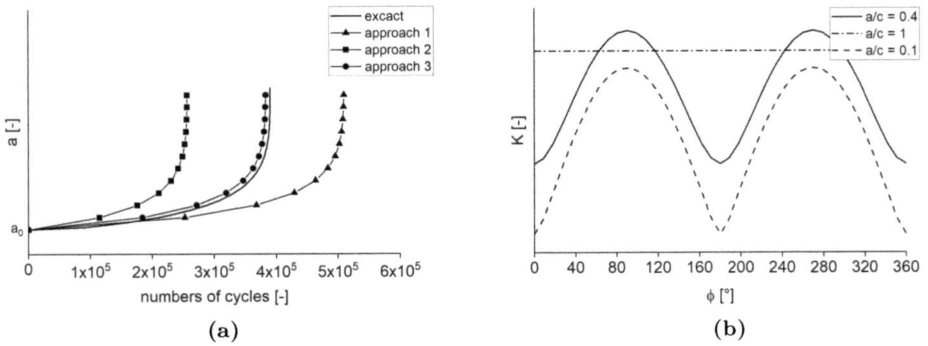


**Figure 2.5:** a) Effects of the iteration approach on the computed number of cycles; b) Crack tip loading along the crack front of an exemplary elliptical internal crack of equal cross-sectional area and different a/c according to Figure B.2

No significant differences between these methods occur, and the straightforward method in approach 1 is applied for reasons of efficiency. In the numerical crack growth simulations of Chapter 4, approach 3 is used if not specifically mentioned.

As shown in Figure 2.5b, even a simple elliptical crack has a variable stress intensity factor along its contour. As can be seen, this is influenced by the ratio of the semi-axes on an identical surface area. For general 3D problems or more complex crack geometries (see Section 4.1.1), the updated crack front geometry must be determined at each increment. In the concrete implementation, this is done by calculating the geometry for increment $i$ according to approach 1 from increment $i-1$ (e.g., using the methods described in detail in Section 4.1.2). To determine the corresponding number of cycles $\Delta N_i$, approach 3 is applied in a subsequent step.

In [19], an approximation method was proposed within the framework of these investigations that estimates values with a deviation of $< \pm 5\,\%$ from the exact solution for a known crack geometry using two evaluation points. The basic idea is the approximation of the relationship dN/da(a) as an exponential function. Finally, the service life results from the integration of this function over the crack length. With an optimized choice of the interpolation points, the error can be reduced to less than 1 %. This procedure offers a considerable reduction in computational effort, especially with regard to numerical investigations of complex geometries. In addition, [19] quantified the effect of differently selected crack shapes at optimally selected points on 3D problems.

#### 2.2.4.3 Failure Assessment Diagram

To assess plastic and brittle fractures based on defects without the need for a complex elastic-plastic calculation, the following fracture mechanics method was developed by [62]. This method only considers static conditions. Nevertheless, it can be used as the definition of the failure state of a cyclic-loaded structure using the maximum stress $\sigma_o$ or for structures that are additionally subjected to static loads. The evaluation is based on a Failure Assessment Diagram (FAD) and the related values $L_r$ and $K_r$. An exemplary diagram is illustrated in Figure 2.6b. A material-dependent limit curve delimits save states against the two failure modes, brittle fracture and plastic collapse as well as the transition between both states. The dominating failure mechanism results from the intersection of the state and limit curve within the assessment. Each state marked in the diagram depends on the applied load, defect size and material properties. The FAD is interpreted as a special form of the J-integral analysis. The procedure used within the framework of this thesis is based on the extension of [62] from BS 7910:2005 [137].

The limit curve is defined according to BS 7910:2005 [137] Level 2B assessment as:

$$\mathrm{K_{r,crit}} = (1 - 0.14 \cdot \mathrm{L_r^2})(0.3 + 0.7 \cdot \exp(-0.65 \cdot \mathrm{L_r^6})) \qquad \text{for } \mathrm{L_r} \leq \mathrm{L_{r,max}} = 1.0. \quad (2.12)$$

In the general procedure according to BS 7910:2005 [138], the no-failure region reaches up to $L_{\max}$, which depends on the ratio between the yield and limit strength. In this context, $L_r$ has to be limited to 1.0, as a pronounced yield plateau cannot be excluded in general applications.

The condition of interest is given by:

$$\mathrm{K_r} = \frac{\mathrm{K_I}}{\mathrm{K_{Ic}}} \quad ; \quad \mathrm{L_r} = \frac{\sigma_{\mathrm{ref}}}{\sigma_{\mathrm{y}}}. \qquad (2.13)$$

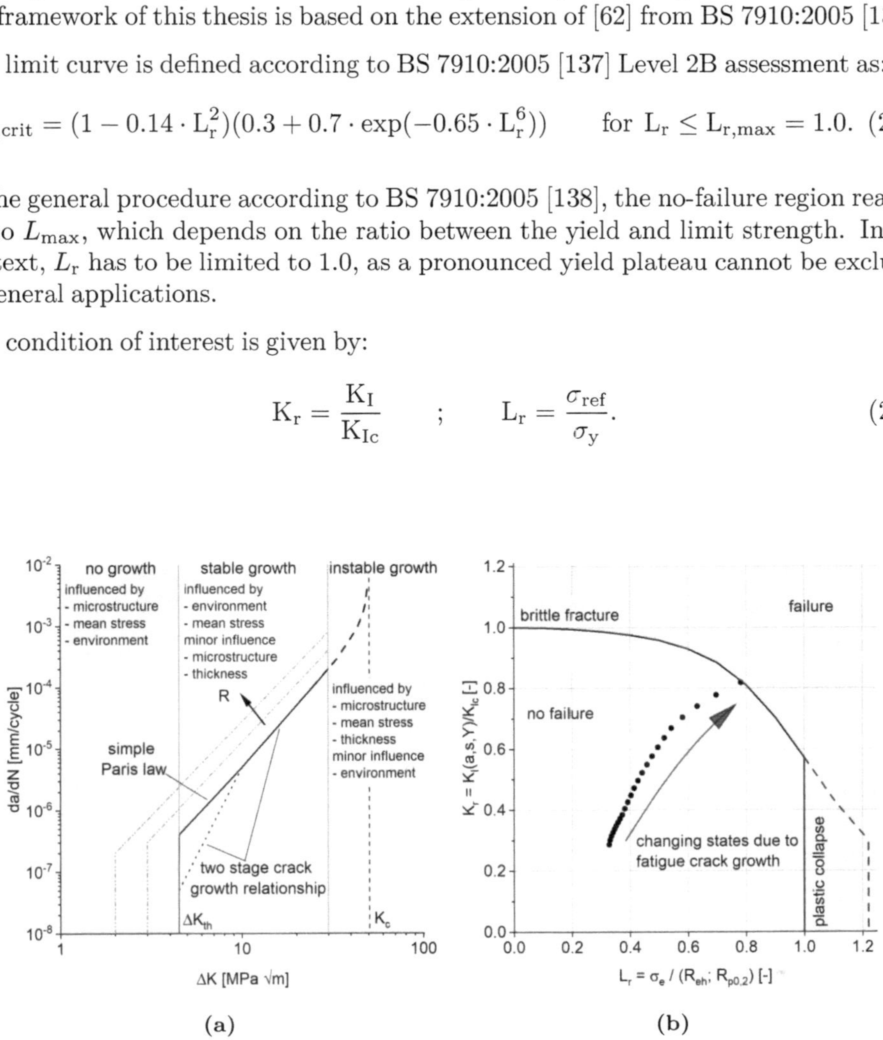


**Figure 2.6:** a) Schematic crack growth diagram including influencing factors according to [85]; b) Exemplary failure assessment diagram

Herein, $\sigma_{\mathrm{ref}}$ results from the consideration of three states:

a) Local plastic collapse: Two conditions relevant for this upcoming context are presented in Eq. (B.1) and Eq. (B.2).

b) Net section failure: Considering the flaw as not load transferring and yielding in the net section.

c) Global plastic failure: Conservatively, it is not considered within this work, as it depends on the individual component. It is only valid for redundant structures where multiple plastic hinges cause failure, and it is mostly captured by a proper design.

In the present context, no safety factors are included. Further, more detailed expansions, which consider the project-specific stress-strain behavior are presented in BS 7910:2005 [137] but are not applied within this investigation due to their lack of universality and limited benefit. In general, residual stress could be considered in the assessment by shifting the origin of the FAD, but as discussed in [103], this is not relevant for steel cast components.

Methods for determining the static ($\mathrm{K_{Ic}}$) and cyclic ($\Delta\mathrm{K_{th}}$, C, m) fracture mechanics parameters are standardized according to ASTM E1820:2018 [127], ASTM E399:2019 [130], ASTM E647:2015 [132] or the master curve concepts [6, 95, 161] based on Charpy tests.

### 2.2.4.4 SN Curves based on Fracture Mechanics

The SN curve shown in Figure 2.7b is derived from a crack propagation calculation. If the stress range $\Delta\sigma_1$ is applied to the exemplary crack situation shown in Figure 2.7a, a crack propagation calculation results in a failure load cycle number $\mathrm{N_1}$.

In the crack propagation calculation, the stress intensity factors are determined analytically or numerically. The local crack growth and number of load cycles are determined

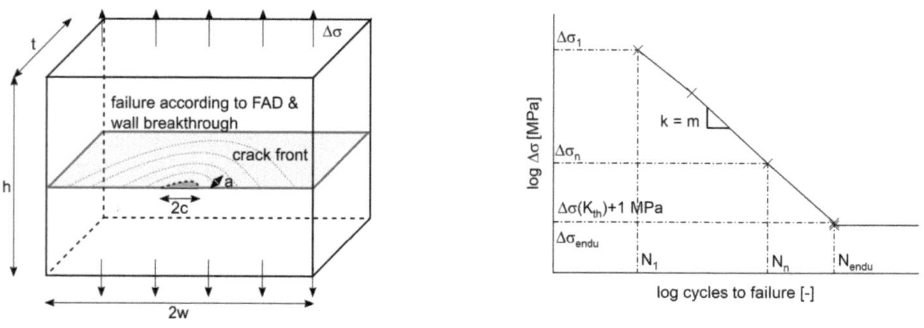


(a) (b)

**Figure 2.7:** a) Exemplary geometric conditions; b) SN curves derived from crack growth incrementally using the Paris law (see Eq. (2.10)) until the limit state defined by the FAD is reached. The same procedure is used for other stress levels $\Delta\sigma_n$. The fatigue strength $\Delta\sigma_{endu}$ is reached when the maximum stress intensity factor in the initial state corresponds to $\Delta K_{th}$. The number of cycles of the endurance limit $N_{endu}$ can be estimated by slightly increasing the stress range (e.g., by $\Delta\sigma_{endu}$ + 1 MPa). In the SN curve based on fracture mechanics, the slope k (in a double logarithmic representation) corresponds to the Paris exponent m.

#### 2.2.4.5 Interaction of Cracks

Multiple cracks can interact in a way that their crack tip stresses and propagation speeds increase or decrease. The underlying mechanical relations have been transferred to simplified classification approaches, such as [138, 161]. For example, the two coplanar cracks in Figure 2.8b need to be treated as a single crack with the dimensions $2c_{res} = 2c_1 + s + 2c_2$ and $a_{res} = \max\{a_1; a_2\}$ when the ligament s is less than the width of the smaller crack $2c_1$ and $2c_2$. Otherwise, the interaction of the cracks could be neglected. As discussed in the following section, this approach is conservative and was intended for monotonic loading. BS 7910:2013 [138] states, without further instructions or references, that these limitations do not need to be considered for cyclic loading. Despite its identical approach, FKM-Rili-BM:2009 [161] does not give any comments on that differentiation. All in all, these limitations need to fulfill different requirements under static and monotonic loading despite the consideration of identical cracks. First, the mechanical behavior of interacting cracks in a 2D plate is discussed. Later, these findings are extended to 3D problems and transferred to fatigue problems. This summary of investigations into coplanar cracks forms the basis for the design concept concerning minimal distances between cracks and a summed up defect size in each evaluation plane in Section 5.2.6. The interaction between two cracks in adjacent planes concludes this section and defines which defects have to be combined into one observation plane in Section 5.4.

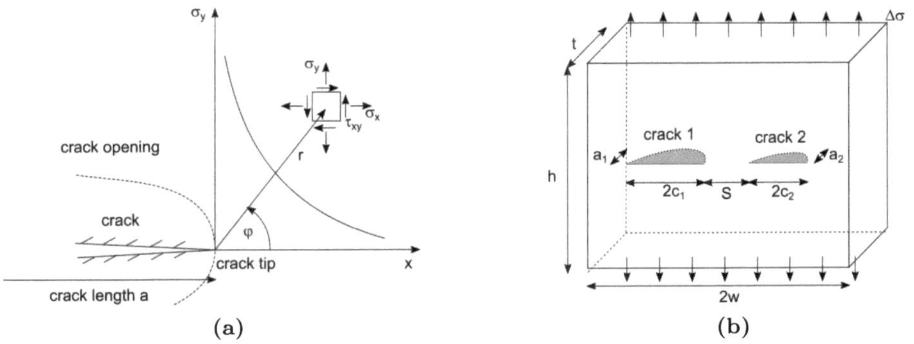


**Figure 2.8:** a) Parameters and properties of the planar mode I crack tip stress field, schematic presentation similar to [28]; b) Interaction of coplanar surface cracks, geometric conditions [67]

*Coplanar Cracks – Static Loading:*
In [28], analytical solutions for a series of cracks in the infinite plate illustrated in Figure 2.9a are presented. The stress intensity factor can be calculated as:

$$K_I = \sigma\sqrt{2b\tan(\frac{\pi c}{2b})}. \tag{2.14}$$

For c →b, it follows from Eq. (2.14) that $K_I \to \infty$. It becomes obvious that the interaction of cracks is only dominant for cracks that are very close; otherwise, the interaction vanishes. The problem of two collinear cracks in a finite plate, which is more practically relevant, is shown in Figure 2.9b, and the solution is derived in [28] according to the procedure from [50]. Textbooks such as [2] refer to catalogs in which solutions are tabulated, like [65, 98].

Figure 2.9c shows the solution for both crack tips according to [65]. When $s \to 0$, the stress intensity factor of the inner tip strives towards infinity, whereas the outer tip results in an amplification factor of $\sqrt{2}$. With Eq. (2.5), this is in line with a crack of twice the original length.

In [102], different crack length ratios were compared with the stress intensity factor of a substitute crack with a length of $2c_{res} = 2c_1 + s + 2c_2$. Further explanations are limited to the most unfavorable case in which both cracks have identical dimensions $c = c_1 = c_2$. The result of this investigation is that for configurations where $s/(2c) < 0.1$, the stress intensity factor of the inner crack tip is higher than that of the substitute crack.

If this stress intensity factor exceeds $K_{Ic}$ under static loading, not only does an abrupt failure of the ligament occur, but the stress intensity factor of the newly formed crack grows unstable as well. This failure mechanism forms the theoretical basis of the classification approaches in BS 7910:2005 [138] or FKM-Rili-BM:2009 [161]. From a purely mechanical point of view, these approaches are quite conservative, as other crack length ratios cause smaller values for a critical $s/(2c)$ value.

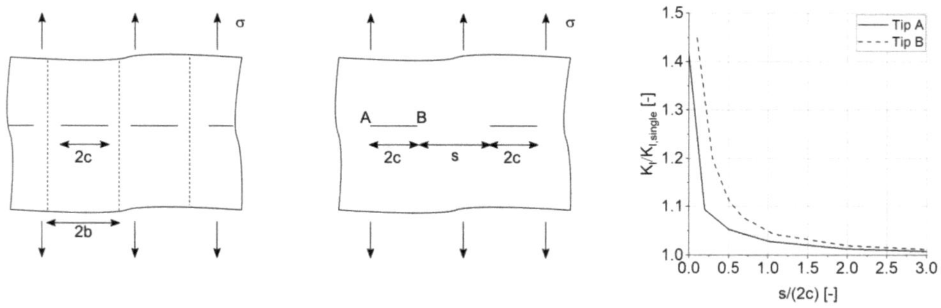


(a) (b) (c)

**Figure 2.9:** a) Line of coplanar cracks in an infinite plane; b) Two coplanar cracks in an infinite plane; c) Sectional linearized stress intensity factors of two coplanar cracks in an infinite plane, compare [65]

According to [28, 66] the interaction effect for 3D cracks is always smaller than that of 2D cracks. One explanation is the cubic decay in the stress field for 3D problems compared to the quadratic decreases for 2D cases. Any 2D consideration is conservative when it is transferred to a 3D problem. This influence was quantified in [67] using the body force method for two coplanar, semi-elliptical surface cracks of different sizes and a/c ratios in a half space. The geometric conditions corresponded to Figure 2.8b where $t = h = w = \infty$. By modifying the a/c ratio for a constant s/(2c) ratio, the greatest increase in stress intensity factor occurs in the extreme case of $a/c \rightarrow \infty$. This result is similar to the previously discussed 2D case. The value of interest was defined as $\gamma$, and it related the stress intensity factor of the interacting crack to a single crack. For the actual geometric relations, $\gamma$ dropped by 10 % from $\gamma = 1.28$ at $a/c = \infty$ to $\gamma = 1.15$ at $a/c = 1$. In a second step, the distances and ratio of the crack sizes were changed to a constant $a/c = 1$.

This led to the following findings. As shown in Figure 2.10a the most unfavorable condition occurs when both cracks $a_1$, and $a_2$, are equally large independent of the s/(2c) ratio. If the cracks are of different sizes, the effects on the smaller one are more significant than those on the larger one. By illustrating amplification factors related to a single crack with similar dimensions, Figure 2.10a falsely suggests that the smaller crack has greater stress intensity factors than the larger one does.

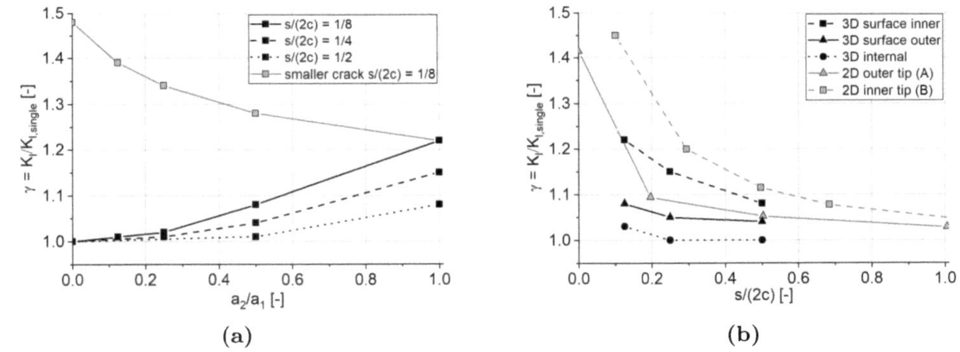


**Figure 2.10:** a) Influence of the crack size ratio on the stress intensity factors of two coplanar cracks [67]; b) Influence of the ligament on the stress intensity factors of two coplanar cracks [67]

Figure 2.10b shows the stress intensity factors at different crack positions for two similar cracks and different s/(2c) ratios. The influence on K at the zenith of the ellipse is negligible. A similar situation exists at the intersection with the component surface opposite to the other crack. For illustration purposes, the data from Figure 2.9c are supplemented for the 2D case. The values in the 3D case decrease rapidly with an increasing s/(2c) ratio and are significantly below the previously discussed 2D values in all cases.

*Coplanar Cracks – Cyclic Loading:*
Under cyclic loading, the crack propagation of two initially non-interacting cracks can be accelerated significantly. As long as $\Delta K < \Delta K_{th}$ applies, this is not the case, and the assessment of whether two cracks interact is analogous to the static case done in [103].

The interaction effects of two growing coplanar semi-elliptical cracks was experimentally investigated in [1, 100]. Four stages were defined in [1] using beach marks as shown in Figure 2.11a. In the first phase, both cracks grew individually and were unaffected by their neighbor(s). The second step is the interaction phase. Both cracks grew faster, as shown in Figure 2.11b. In the coalescence phase, the tips unite and form a new semi-elliptical crack, which grew in the post-coalescence phase as a single larger crack.

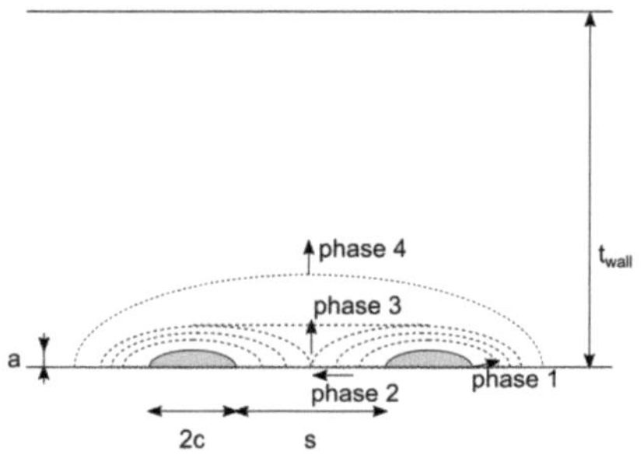


(a)

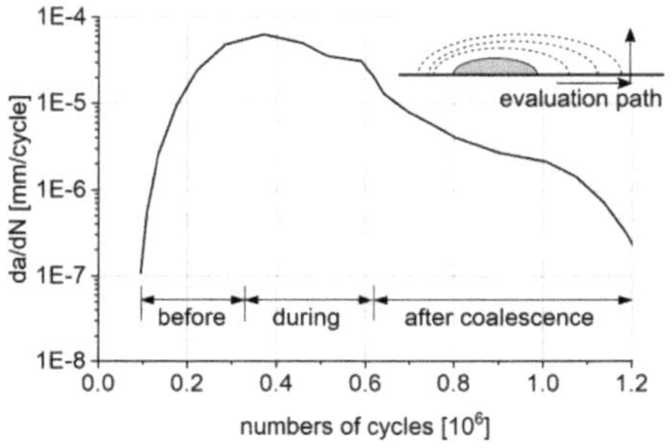


(b)

**Figure 2.11:** a) Phases of crack growth and coalescence for two interacting semi elliptical surface cracks; b) Crack growth rates during coalescence according to [100]

*Non-coplanar Cracks:*
The investigations on planar [98] and three-dimensional [51] problems have shown that the stress intensity factors are reduced for two cracks arranged in parallel planes. Unfavorable cases occur when the distances between the planes and the horizontal distance s (see Figure 2.8b) are small. However, this case is conservatively represented by two coplanar cracks. The approaches in FKM-Rili-BM:2009 [161] and BS 7910:2013 [138] provide an appropriate simplification. According to these approaches, cracks whose crack planes have a distance smaller than the sum of half the crack lengths must be approximated by a substitute crack.

## 2.3 Fatigue of Cast Steel

Figure 2.12 summarizes the most relevant scientific investigations and design approaches to the fatigue design of cast steel components and the influence of casting defects. A series of publications are represented by a single block, and their key aspects are extracted. The overview is subdivided into basic investigations on fundamental correlations and design concepts, respectively. If apparent, the design concepts are linked to the underlying basic investigations. The markers below each block classify whether the investigation focused on

I: the technical flawless base material (FM)

II: the influence of defects (D)

III: the fatigue behavior of full-scale components (C)

IV: special issues like crack initiation or casting process simulations (S).

The design concepts, on the other hand are categorized whether

I: they are based on (a somehow defined flawless) base material (FM) or

II: different defects sizes could be considered specifically (D).

The behavior of welded joints between cast components and steel sections is not relevant within this context. Hence, investigations that purely focus on these welds are excluded.

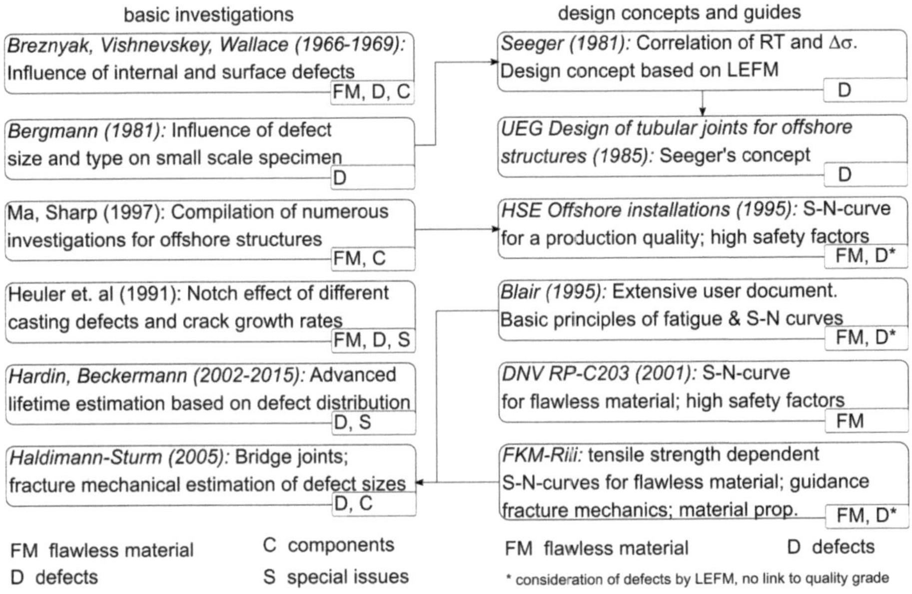


**Figure 2.12:** Relevant investigations and design concepts for fatigue design of cast steel components

Breznak, Vishnevsky and Wallace [15, 117, 118] carried out the first systematic studies on fatigue resistance in cast steel components in the 1960s. These were successively extended and partly transferred into calculation guidelines and codes of practice. In addition, many specific aspects were considered, such as crack initiation, notch factors of discontinuities and the coupling of casting process simulation and fatigue life estimation. Isolated research and assessments of failure cases provided characteristic values for crack propagation in cast steel materials. Investigations that considered casting defects found a partial reduction in fatigue strength compared to components made of defect-free material. Their impact is directly dependent on the stress state, defect location and, in certain cases, the flaw type. However, these findings could not be confirmed experimentally in investigations of actual components, as the examined components (partly taken from series production) had minor or no casting defects in highly loaded areas - despite previously defined poor quality classes. In addition, the experimental investigations were carried out on materials representative of the specific application and era of the research. Nevertheless, this patchwork of individual studies and findings can provide an essential input for deriving a design concept with a focus on discontinuities. Studies that are relevant to the present investigations are presented below. Their key statements are consolidated, their developed knowledge is evaluated and their points relevant to the present work are summarized. In addition, the results of past experimental investigations and existing design concepts provide a basis for validating the statements derived in the framework of Chapter 5.

## 2.3.1 Fundamental Studies

### 2.3.1.1 Breznak, Vishnevsky, Wallace (1966–1969)

The investigations in [15, 117, 118], focused on the properties of **base material**, the influence of casting **defects** as well as real **components**.

In the 1960s, a number of research projects on the resistance of cast steel were carried out at the Case Institute of Technology. These included investigations of structural strength under static loads, including brittle fracture and fatigue resistance. The aim of [117] was to estimate the effects of surface defects on fatigue strength. For this purpose, the influences of the stress state, type of casting defect, production welding and the quenched and tempered condition were examined and compared with the results from flawless material. The geometric characteristics of the defects were larger than permitted by the MT test standard ASTM E125:1963 [126]. In [15], the discoveries were extended to include the effect of internal casting defects. Consequently, internal shrinkage pores that could be assigned by RT to classes 2 and 6 according to ASTM E71:1964 [133] were enforced by inadequate feeding.

Seven types of components (series production) from the railway technology, construction and mechanical engineering fields were investigated in [118]. The parts were subjected to their attributed design loads and were made of five different material grades with carbon contents of 0.21–0.41 M% and tensile strengths in the range of $\sigma_{\mathrm{u}}$ = 480–690 MPa at A = 15−33 %. Component failure was almost exclusively observed in high-stressed areas and was not caused by casting defects. However, the defects were mostly located in low-stress regions.

*Summary:*
The studies on the fundamental influence of casting defects were carried out under bending, torsion and circular bending in the form of Moore testing [63], all at a stress ratio of R = -1. In both studies, the tested material was AISI 8630 with mechanical properties under Normalized and Tempered (NT) conditions of $\sigma_u = 570-620$ MPa, A ≈ 20 % as well as under Quenched and Tempered (QT) conditions of $\sigma_u = 830-1000$ MPa, A = 11−15 %. Surface defects occurred in the form of voids, cold cracks, hot cracks and slag inclusions. Metallurgical treatments (e.g., by reducing the aluminum content to form air voids), modifications in casting technology (e.g., by a local delay of the solidification process to form cavities), mechanical actions (e.g., constraints in the cooling process) or technological steps (e.g., the introduction of slag or the induction of cracks through insufficient quenching) enforced their occurrence. The majority of the internal discontinuities, shrinkage cavities, were located along the specimen's centerlines.

In Figure 2.13a, the test data from [15] and [117] are evaluated and compared. These results are limited to bending loads, as the published information did not allow for more informative evaluations. The effect of surface defects on fatigue strength was much greater than that of internal defects, independent of the heat treatment method. Nevertheless, the effect was stronger for the higher strength material. Different defect sizes of the centerline shrinkage cavities had an insignificant effect on fatigue strength due to their position in the neutral fiber. In this case, the fatigue strength almost corresponded to the resistance of the base material. Besides the lower resistance, the values of surface defects and the gradients of the SN curves were also influenced by the defect location. The gradient for internal defects was higher and corresponded to the base material, whereas the surface defects caused gradients similar to strongly sharped notches or the Paris exponent m = 3.

The endurance ratio $R_{endu}$ according to Eq. (2.15) is the parameter to which attention was directed and is shown in Figure 2.13b. It was concluded that the notched Moore specimen reduced this parameter more strongly than most types of surface defects under bending. The only exceptions were the cavities, which caused significantly lower resistances, especially for the high-strength material. It was also found that the effects of surface defects were greater under bending stress than under torsional stress, whereas the internal defects reduced resistance considerably more under torsional loading.

$$R_{endu} = \frac{\Delta\sigma_{endu}}{\sigma_u} \qquad \text{with } N_{endu} = 1 \cdot 10^7 \text{ cycles} \tag{2.15}$$

The authors recommended that the endurance strength of a cast steel component with typical casting defects could be estimated from the un-notched Moore test [63] of the base material or more simply by the tensile strength according to Eq. (2.16). In their interpretation, surface defects tolerated by ASTM E125:1963 [126] and internal defects according to ASTM E71:1964 [133] were covered by this approach.

$$\Delta\sigma_{\text{endu}} = \Delta\sigma_{\text{Moor,notched}} = 0.7\Delta\sigma_{\text{Moor,unnotched}} = 0.7 \cdot 0.4 \cdot \sigma_{\text{u}} \quad (2.16)$$

*Critical Evaluation:*
The considered loading scenario is justified by the frequent use of cast steel shafts in the 1960s, but it does not allow any generalization. The authors concluded that the surface defects have a negative influence on fatigue strength due to their placement in the region of maximum tensile stress (bending state). It was disregarded that, according to Eq. (2.7), even in a homogeneous tensile stress state, surface defects led to higher stress intensity factor than comparable internal defects. All examined internal defects were placed along the neutral axis; thus, they were not subject to any significant loading, and general statements on their behavior cannot be made. The fracture surface was not documented. In the case of internal defects, the failure could have been initialized from the high-stressed surface areas rather than the internal defects. This might explain the similar gradient of the SN curve compared to that of the base material.

The representation of the results in [15, 117] using the endurance ratio is misleading. The consideration in the actual stress ranges showed that the differences between the quenched and tempered states are much smaller in the presence of surface defects such that the high-strength state is disadvantageous with respect to brittle fracture.

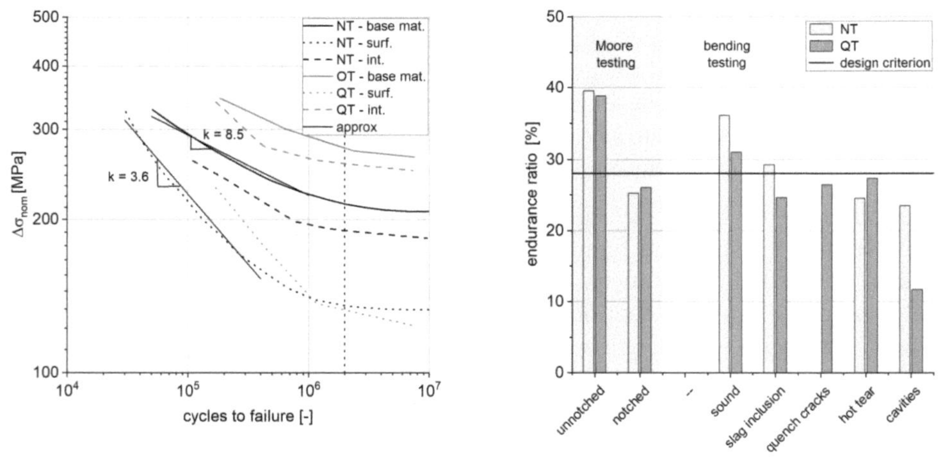


(a) (b)

**Figure 2.13:** a) SN curves for internal and surface defects compared to the base material of [117, 15]; b) Endurance limit of the Moore specimen and different types of surface defects under reversed bending from [117]

A design approach derived from the tensile strength of the base material does not make any statements on behavior within the fatigue strength range. It motivates the use of higher-strength materials despite their more brittle material behavior. Additionally, the results showed that this approach overestimates resistance dramatically for surface cavities.

*Relevance to this work:*

- Surface defects reduce fatigue strength significantly and cause different SN curve gradients compared to the base material.
- Large defects do not affect fatigue strength as long as they only occur in low-stress regions.
- The developed substitute method on strongly notched specimens (e.g., notched Moor tests [63]) might be helpful in material qualification.

#### 2.3.1.2 Bergmann et al. (1981)

The investigations of [10] were done to quantify the influence of casting **defects** on components typical in offshore applications and were carried out within the framework of an industrial project at the Technical University of Darmstadt. For this reason, the primary source is not publicly accessible. Some extracts from these investigations have been published in [35, 112] and form the basis for the only **design concept** linked to production quality for fatigue-stressed cast steel components. The details of this concept developed in [96] are discussed in Section 2.3.2.

*Summary:*
Thirty-two specimens with wall thicknesses of 25 mm were taken from K joint made of the material HOESCH Gs Ark 10, which is comparable to 1.6221 according to DIN 17182:1992 [140]. The specimens contained several types of casting defects (cavities, hot cracks, voids, inclusions) that had been assigned using RT to classes 4 and 5 according to ASTM E446:1972 [131]. In contrast to [15, 117], the tests were carried out under tensile stresses at R = 0 and R = -1. A distinction between the surface and internal defects was not documented. The test results are shown in Figure 2.14, including the related mean value ($\Delta\sigma_{50\%}$) and design ($\Delta\sigma_{98\%}$) SN curve. Both of these were evaluated by [10] with a fixed gradient of k = 5.0. The stress ranges were related to the net section. The defect size had a direct impact on fatigue resistance, whereas the type of defect was determined to have only a subordinate influence.

*Critical Evaluation:*
The experiments for both stress ratios were limited to a different, but small, range. The derived statements on the gradient of the SN curve are not comprehensible. The classification according to RT was done by means of comparison pictures, and the actual defect size and 3D expression are only evaluated indirectly.

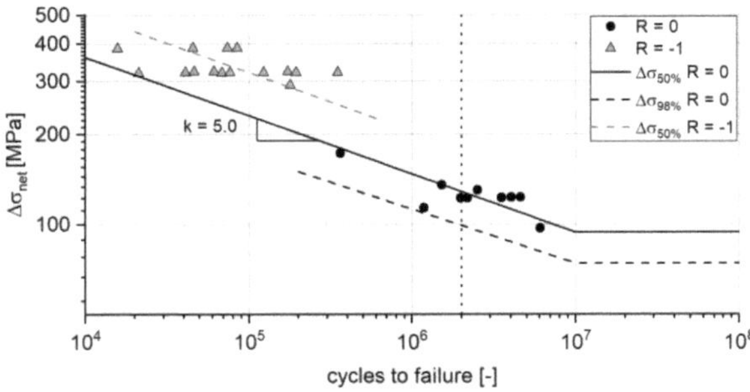


**Figure 2.14:** Results of fatigue tests on specimens with internal casting defects from [10]

*Relevance to this work:*

- The influence on fatigue strength was almost independent of the type of casting defect.
- A not negligible mean stress dependency became apparent.
- The results could be used as reference values for the validation of the design concept in Chapter 5.

### 2.3.1.3 Heuler et al. (1993)

The investigations in [42, 43, 44], focused on the properties of **base material**, the influence of **defects** and considered **special issues** like crack growth and notch effects. The effect of different types of casting defects on the crack initiation characteristics of cast steel components were studied in extensive experimental investigations and supplemented with a comparison of different computational evaluation approaches.

*Summary:*
The starting point for these investigations were flange-like cast components with casting defects (gas voids, tube cavities, slag inclusions, grain boundary cracks, oxide skin, molding material inclusions, hot cracks, micro pores), from which 92 tensile specimens were extracted and tested in strain-controlled fatigue tests at $R_\varepsilon$ = -1. The initial components had a wall thickness of up to 150 mm and were made of GS-18CrMo9-10, which is comparable to 1.7379 according to DIN EN 10293:2015 [141]. The fatigue tests were monitored by an online ultrasonic test setup with up to 16 inclined detectors and were stopped at a crack length of approximately 50 $\mu$m. Afterwards, the specimens were cooled to -170 °C and broken up. The fracture surfaces were examined using fractography. Several casting defects were detected on each of the fracture surfaces and were classified according to the type of defect and its location in relation to the surface and internal defects. The sizes of the 345 analyzed defects were in the range of 0.1 mm$^2$ to 3 mm$^2$, with a specimen cross section of 1050 mm$^2$. The exhibited notch radii are shown in Figure 2.15a. In addition, any fatigue cracks surrounding the defects were

measured and evaluated. From the crack growth ratios of the various individual flaws of a test specimen, information on their respective cracking behavior was derived. Based on the defect properties, an evaluation was carried out based on various extensions of the notch concept according to [71]. These investigations were supplemented by a quantification of fracture mechanics approaches for hot cracks.

In this context, the crack initiation strain life curves were reduced to specimens with casting defects and are shown in Figure 2.15b. The strain range derived from the yield strength of the examined material serves as a reference (here $\Delta\varepsilon = R_{eH}/E$). The large scatter was justified by the varying defect sizes, but it was not considered in the evaluation.

The fracture surface analysis confirmed that surface and near-surface defects led to cracking at an earlier stage compared to internal defects. In addition, it was shown that flaws with a tendency toward smaller notch radii, such as grain boundary cracks and small surface cracks in the welding area, tubular pores, slags and oxide layers, were more notch effective than inclusions, gas cavities and micro pores. Nevertheless, the authors concluded, analogous to the findings of [10], that the influence of the defect type played a minor role. In the authors view, their assessment of crack initiation behavior as a spatial defect using the notch concept correlated satisfactorily with the SN curves of the flawless base material. An mechanical idealization as a sharp crack resulted in a conservative evaluation of the crack initiation time. Finally, the consideration of hot cracks using fracture mechanics was regarded as effective. The crack propagation parameters determined for the base material are summarized in Table 3.2.

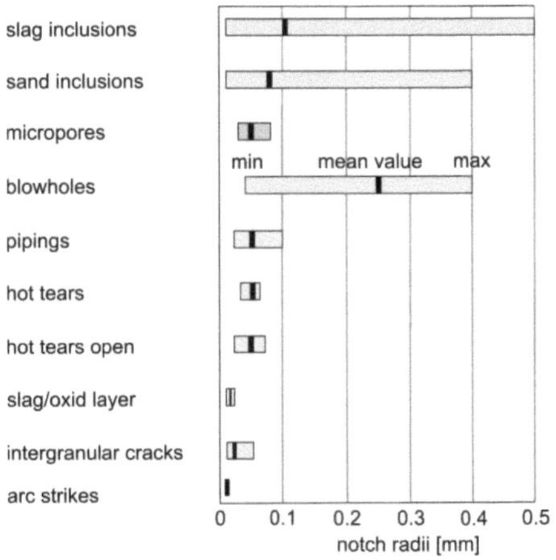


(a)

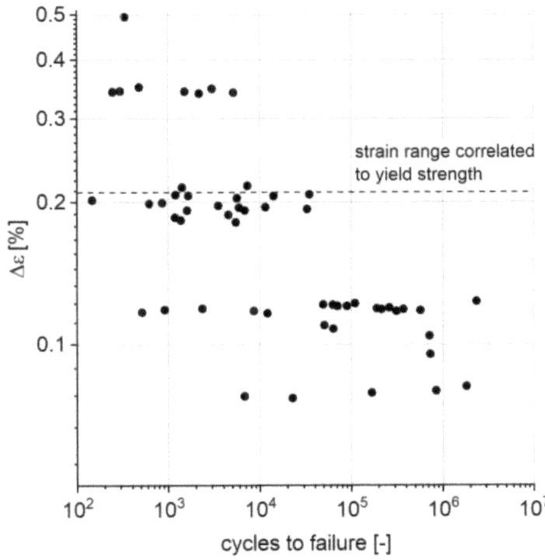


(b)

**Figure 2.15:** a) Radii of different defect types measured on the fracture surface from [42]; b) Strain life curve from [42]

*Critical Evaluation:*
These fundamental investigations showed that the internal casting defects were subject to a crack initiation phase and did not behave as poorly as the assumption of a sharp crack did. Unfortunately, the relationship between these findings and the overall service life of the components were not specified. For a mathematical evaluation of the crack initiation time, the defect size and notch sharpness must be known. In this case, it offers the possibility to determine the total lifetime less conservatively than by using a pure fracture mechanics evaluation. The applied strain amplitudes led to stresses above the yield point and were evaluated together with tests from the HCF range. [21, 59] showed that different damage processes take place in different stress areas and that a joint evaluation is not very useful. Hence, the HCF range is decisive for the questions relevant to this work. The high stresses make a lifetime evaluation based on the crack propagation relation according to Paris [74] doubtful despite approaches that divide an elastic-plastically determined J-integral into elastic and inelastic parts.

*Relevance to this work:*

- Internal casting defects show a distinct crack initiation phase. Surface and near-surface defects cause fatigue cracks at a significantly earlier stage.
- The type of casting defect plays a subordinate role.
- Crack propagation parameters for comparison.

#### 2.3.1.4 Ma, Sharp (1997)

In [58], 80 series of investigations on the fatigue resistance of **base material** as well as **components** from Finland, France, Germany, Japan and the UK were collected and supplemented with additional tests. The results of these investigations form the basis for the **design curves** of the HSE [166] presented in Section 2.3.2.

*Summary:*
All investigations were carried out with a particular focus on the special requirements of offshore applications. This was expressed in studies on the influence of corrosive media, stress concentration factors occurring at representative nodal structures and the materials used. The collected data was divided into three specimen groups:

I: small specimens without geometrically induced stress concentrations.

II: small specimens with geometrically induced stress concentrations (e.g., cross-joints with SCF), which conservatively represent the stress conditions occurring in real environments.

III: large components. In order to force a failure in the casting joints, atypically high SCFs in the range of 3 to 5 were applied for fatigue-optimized casting applications to represent real applications conservatively.

For these three groups, data from different designs and geometries were compared at stress ratios of R = -1, R = 0 and R = 0.1 under tensile and bending loading. The maximum stress range was considered for evaluation. Specific information on the tested materials was not given, only that they harmonize with S355, which is typical for offshore applications.

Selected test results were collected and are shown in Figure 2.16a. The presented selection is limited to experiments carried out in air. Additionally, only specimens of the dominating stress ratio for each group are considered. This dominance is related to the number of tests. Usually, only three to five results were available for the other stress ratios. Therefore, no statistically secured conclusion was possible. On the other hand, a combined presentation would lead to an incorrect interpretation of the scatter. For better comparability, the regression curves were determined according to EN 1993-1-9:2010 [156] and are presented in the diagram. The design curve the authors derived is also given. This resulted from the component tests and a displacement of four times the standard deviation. From the authors' point of view, all samples can be regarded as one population. They conclude that the geometric shape of the notches present in the cruciform joints had no significant influence on the fatigue strength as long as the maximum stress range was considered. Thus, a stress-based design concept can be derived for components without casting defects. The gradients of the regression curves decrease with an increasing SCF, as expected with reference to [90].

In the large-scale component tests carried out in [58], the crack initiation phase was measured by flux leakage. The cycles to failure and to crack initiation and their proportion on the total service life are given in Figure 2.16b. Depending on the stress level, between 50 % and 75 % of the service life was attributed to crack growth. Thus, the crack initiation and fracture lines show different gradients.

The investigations on the effect of discontinuities are limited to two large components. In one case, real defect sizes were determined using mechanical sectioning and were compared to the NDT results from UT and RT. Due to the sharp edges of the discontinuities, these defects were classified as crack-like, despite their volumetric shape. In a second component, a UT-tested component was subjected to fatigue testing until failure. After NDT, slag inclusions and shrinkage pores were found in less-stressed areas of the component. Analogous to the results of [118], a geometrical notch caused the failure. Finally, no significant influence of wall thickness was detected in the considered range of t = 18 mm to 40 mm.

*Critical Evaluation:*
Since the majority of the test specimen types I, II and III were tested at different stress ratios, it is not clear whether an offset was caused by the different stress ratios, materials or influences resulting from the test specimens and SCFs. Different influences on fatigue behavior were mixed in most evaluations. As an example, the authors draw conclusions on the influence of repair welds based on only three specimens. Additionally, one of these three specimens was exposed to a corrosive environment.

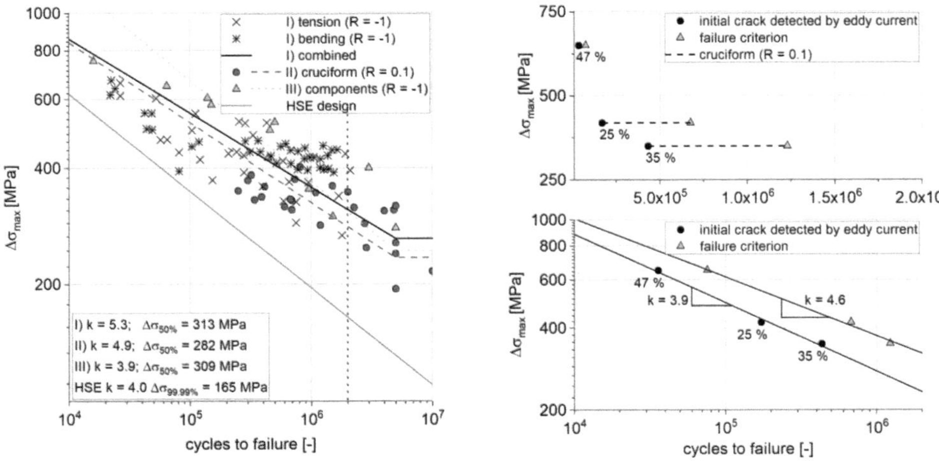


(a) (b)

**Figure 2.16:** a) Summary of selected test data from [58]; b) Evaluation of component tests and crack initiation of [58] displayed both linearly and logarithmically

*Relevance to this work:*

- Material characteristics and design criteria for flawless materials that can be used as comparative values for validation.
- Casting defects can be represented by cracks due to a distinctive crack growth phase in real components without casting defects.

#### 2.3.1.5 Hardin, Beckermann (2002–2015)

Current research on the calculation of steel casting and the effects of casting defects is being carried out, apart from Karlsruhe [78, 87, 103, 114, 119], mainly at the University of Iowa. Hardin and Beckermann focused on linking RT indications with static load-bearing capacity [11, 36, 37, 38, 39, 41, 97] as well as the coupling of casting process simulation and service life evaluation [40]. In a series of publications [11, 36, 37, 97], they investigated the influence of shrinkage pores **(defects)** on the fatigue strength of cast steel components that were supplemented by **special** numerical considerations. These investigations were carried out with the same overall objective as this work: to simplify the use of cast steel by deriving calculation methods and to reduce safety factors with a targeted examination of casting defects. For this purpose, computer tomography was used to determine the pore field on fatigue tensile specimens and to correlate these results with a sequence of simulation procedures. The first step consisted of computing of the pore field using a solidification simulation. This was followed by an elastic Finite Element Method (FEM) stress analysis to derive local stress peaks surrounding the pores and a lifetime evaluation based on a multiaxial damage analysis. In parallel, the methods were transferred to real life components.

*Summary:*
The experimental investigations detailed in [97] were carried out at four stress ranges of $\Delta\sigma = 252, 192, 132$ and 106 MPa on fatigue tensile specimens with a stress ratio of R = -1. These were cast from AISI 8630, similar to the previously discussed results of [117], with $\sigma_u = 1144$ MPa and A = 16 %. The specimens were cylinders cast in a standing position with local thickenings of different characteristics. The diameter of the cast components was 14 mm at a length of 152 mm. The thickness of the local disc-shaped thickening varied in such a way that it resulted in a maximum cross-sectional reduction between 8 % and 21 %. Some of the defects extended to the surface of the specimens machined from these casting blanks. The results of the fatigue tests are shown in Figure 2.17a as nominal stress ranges and as related to the net section. The results are compared to the flawless base material and a supplemented statistical evaluation according to EN 1993-1-9:2010 [156]. The authors set the fatigue strength at $5 \cdot 10^6$ load cycles and assigned the large scatter to the individual defect distribution. A reduction in service life due to the defects was observed and could not only be assigned to the cross-sectional reduction but also to additional notch effects.

The fatigue calculation was performed a) on the basis of the defect distribution determined using CT and b) using casting process simulations. These pore fields were transferred to an FEM mesh using the method described in [41], and the element stiffness properties were represented by modified stiffness and Poisson's ratio. Both parameters depend on an internal variable defined by the porosity volume fraction. Since no plastifications were found in the experimental investigations, the subsequent calculation of the stress field was purely elastic. The lifetime calculation was carried out by evaluating the local stress and strain fields based on the Brown-Miller algorithm [18] or the principal stress hypothesis. The lifetimes calculated in this way overestimated the values determined in the experiments by several orders of magnitude, as shown in Figure 2.17b. An enhanced numerical method, called the "adaptive sub-grid" was developed to achieve a more realistic service life forecast. Still, some specimens were overestimated by orders of magnitude.

*Critical Evaluation:*
The authors referred to the investigations of [66, 117] and assigned a significantly higher reduction in service life to surface defects compared to internal defects. Nevertheless, in the published results, no differentiation was made between these cases. The gradient of the regression curve is significantly lower than that of the base material. This is mainly caused by the large scatter and its effects on the statistical evaluation according to EN 1993-1-9:2010 [156]. A separate evaluation might result in two populations similar to the findings in [15, 117].

The coupling of the cast process simulation and the fatigue calculation is interesting from a scientific point of view but still requires improvements to serve as a standalone design approach. The example discussed in Section 2.1 coupled with the findings of [37] show that the actual pore geometry has significant influence but cannot be sufficiently

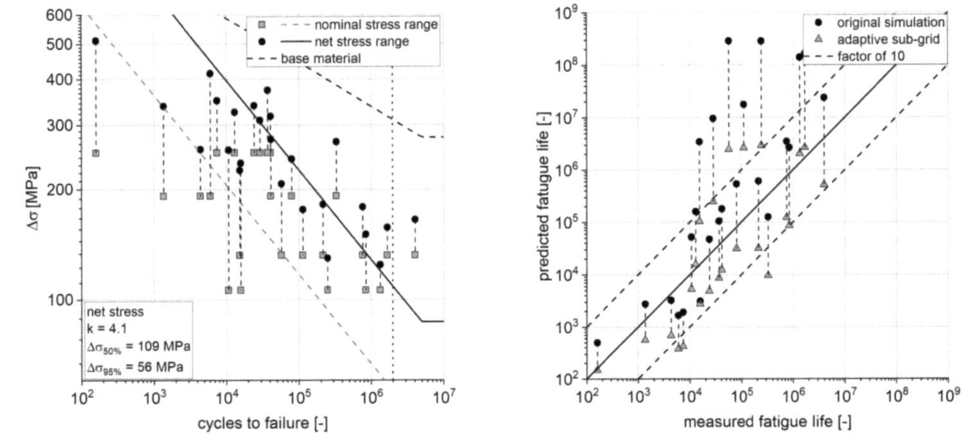


(a) (b)

**Figure 2.17:** a) Tensile fatigue results of [37], including mean values in nominal stress ranges and as related to the net section; b) Relation between measured fatigue life and load cycles calculated by different numerical approaches in [37]

predicted. In addition, this coupling approach is reserved for projects with intense communication between the designer and the foundry.

*Relevance to this work:*

- Well suited specimen geometries can be derived from this investigation.
- The results could be used as reference values for the validation of the design concept in Chapter 5.

#### 2.3.1.6 Haldimann-Sturm (2005)

In [35], cast joints **(components)** for fatigue-loaded truss bridges were evaluated and the influence of **defects** was quantified by fracture mechanics considerations

The starting point was large-scale experimental investigations on truss girders. Since all specimens failed in the connection welds and not in the cast components, their service life was estimated based on fracture mechanics considerations. Crack-like casting defects covering 30 % to 90 % of the wall thickness could be accepted in specific situations in order to achieve the same calculated resistance for the joint and the weld joint. The author's core statement is that the fatigue strength of the connection welds must be significantly improved in order to make use of the existing fatigue strength of the overall structure.

*Summary:*
The experimental investigations were carried out on six truss girders with cast steel joints made of 1.6220 according to EN 10340:2008 [142], which was considered typical for bridge construction. Different joint configurations were considered in order to derive the influence of the local increase in stiffness and the resulting increase in secondary bending moments. In all tests, failure occurred in the weld between the cast component and the circular hollow section.

Possibilities for improving these connections were investigated in a parallel research project [78], and fatigue classes were identified. The experimental investigations thus did not allow any conclusions to be drawn about the effects of discontinuities on the fatigue behavior of cast steel components. In production, defects were permissible due to the relatively low agreed quality requirements according to DIN 1690-2:1985 [139]. However, both a subsequent mechanical separation and fracture surface analysis of segments fractured at low temperatures only revealed casting defects that were smaller than the UT detection limit.

Based on this, the author formulated the objective of an optimum design of the joints from an economic point of view such that the calculated service life of the cast joints should correspond to the calculated service life of the welded connections. In accordance with this principle, permissible initial defect sizes $a_0$ in different component areas were recalculated based on fracture mechanics considerations. The stress intensity factors were determined with the boundary element method. The failure magnitude in the failure state $\mathrm{a}_{\mathrm{crit}}$ resulted from an evaluation using the FAD described in Section 2.2.4.3 and a further limit case that the crack depth was allowed to reach a maximum of 90 % of the wall thickness. The applied loads resulted from the Ultimate Limit State (ULS) and were intended to represent the limit case for brittle failure. In general, a distinction between surface and internal defects was made. The calculation from $\mathrm{a}_{\mathrm{crit}}$ to $a_0$ was carried out using Eq. (2.10) over $2 \cdot 10^6$ cycles, which is assumed to be constant according to the Swiss code SIA 261:2003 [165], and stresses resulting from the fatigue load model. For further simplification, a constant correction factor was derived from numerical investigations to determine the stress intensity factors from the existing stresses. This calculation procedure was transferred into a software code, which was then published and applied to a representative sample bridge. Based on these simplifications, the essential influencing factors were investigated and evaluated in a parameter study. With these findings, the author finally formulated instructions for action and design recommendations.

*Critical Evaluation:*
In Figure 2.18, the principal stresses in the truss joint, weld joints and attached hollow sections of the test specimen from [35] are recalculated and plotted. The stresses are considerably lower in all areas of the joint than they are in the weld. For simplification, continuous transitions between the components were assumed so that the stress hotspots resulting from the weld geometry were not yet included. This and the fact that only discontinuities below the UT detection limit were found results in the specimen and test

configurations used in the experimental investigations being unsuitable to quantify the influence of discontinuities on the fatigue behavior of cast steel components.

The calculation procedure developed in [35] allows for a quick estimation of permissible casting defect sizes for known stress intensity factors. Fatigue and brittle fracture under static loads are evaluated separately for civil engineering applications. This makes the procedure suitable for generalization or transference to other cases. A validation of the fracture mechanics parameters set in [35] and an estimation of how conservative the chosen approaches are were considered useful for a generalization. The selected crack shapes are considered non-conservative as semicircular surface defects or circular internal cracks with a/c = 1 were considered. As explained in [103] and required in [161], a half-axis ratio of a/c = 0.4 leads to significantly higher stress intensity factors.

A closer look at the derived constant correction factors shows that the values numerically determined in [35] correspond very well to a semicircular surface crack or circular internal crack in a flat plate under uniform stress (e.g., according to Annex B.1 of used in this work).

In the derivation of quality classes, the author applied the maximum permissible defect size across the whole of several individual defects with identical total sizes. That a reduction in crack size results in a reduction of stress intensity factors was ignored. Thus, this approach is conservative. As an opposite process, the interaction of several defects is pointed out. This assumption seems sufficient for the underlying application case but not suitable for a general application. From the investigations, it becomes evident that under typical stress conditions and conditions in common practice, casting defects of considerable size can be tolerated without negative effects on the fatigue resistance of the overall structure. In addition, the enormous potential of the use of cast joints in truss structures becomes obvious when fatigue resistance is increased (e.g., by structural adhesive joints [113]) in the connection areas.

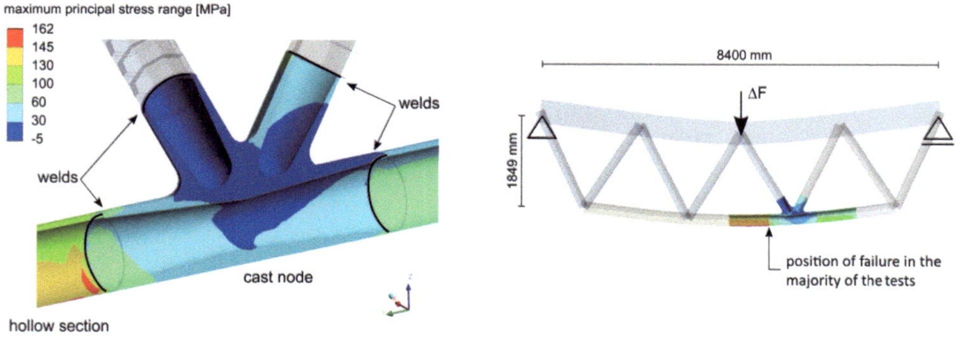


(a) (b)

**Figure 2.18:** a) Reevaluation of maximum principal stresses in a cast joint, welds and connecting sections of the specimen according to [35]; b) Truss girder with joint detail

*Relevance to this work*

- Fracture mechanics evaluation procedure of casting defects can be used as a good example for further studies.
- The author points out the missing link between quality level and resistance.
- A significant influence of dimensional deviations on the stresses was determined.

## 2.3.2 Design Standards

### 2.3.2.1 Procedure for the Direct Consideration of Defect Size

Seeger's [96] method is based on the investigations of [10] presented in Section 2.3.1.2 and adopted in [112]. This procedure is the only known design method in which the defect size is adjustable and directly linked to a design value of fatigue resistance. The starting point is the reference SN curve highlighted in Figure 2.19a. It is defined by Eq. (2.17) and refers to a class 4 defect size according to ASTM E446:1972 [131], ASTM E186:1962 [128] and ASTM E280:1965 [129]; a wall thickness of $\mathrm{t} = 25$ mm; and a stress ratio of $\mathrm{R} \leq 0$. The resistance values were derived using a survival probability of 98 %.

$$\log(\mathrm{N}) = 16.30 - 5\log(\Delta\sigma) \quad \text{for} \quad \Delta\sigma \leq 2 \cdot 10^8 \tag{2.17}$$

The combination of different correction factors, as described in Table 2.3, increases or reduces the applicable stress range according to Eq. (2.18) depending on the situation.

$$\Delta\sigma_{\mathrm{mod}} = \lambda_{\mathrm{R}} \cdot \lambda_{\mathrm{d}} \cdot \lambda_{\mathrm{t}} \cdot \lambda_{\mathrm{I}} \cdot \Delta\sigma \tag{2.18}$$

In addition to the reference SN curve, Figure 2.19a shows the design resistances for the remaining defect classes according to ASTM E446:1972 [131] at $\lambda_{\mathrm{R}} = 1.0$, $\lambda_{\mathrm{t}} = 1.0$ and $\lambda_{\mathrm{I}} = 1.0$. The transfer to defect sizes deviating from ASTM class 4 was based on fracture mechanics considerations.

**Table 2.3:** Correction factors for the design concept according to [96]

| factor | influence of | relation |
|---|---|---|
| $\lambda_{\mathrm{R}}$ | stress ratio | if $\mathrm{R} \leq 0$ then $\lambda_{\mathrm{R}} = 1.0$ else $\lambda_{\mathrm{R}} = 0.85$ |
| $\lambda_{\mathrm{d}}$ | defect class | $\lambda_{\mathrm{d}} = 0.85^{\mathrm{j}-4}$; j = defect class according to [128, 129, 131] |
| $\lambda_{\mathrm{t}}$ | wall thickness | if $\mathrm{t} > 25$ mm then $\lambda_{\mathrm{t}} = (25/\mathrm{t})^{0.15}$ else $\lambda_{\mathrm{t}} = 1.0$ |
| $\lambda_{\mathrm{I}}$ | accessibility | individual weighting (inspection and repair) |

*Evaluation:*
The concept is simple and serves as a guideline. The components' accessibility for inspection and maintenance is particularly relevant for offshore applications the concept was originally aimed at. For complex geometries, a geometry-dependent parameter

considers the reliability of the detection (NDT) of casting defects as important. Since the background documentation is not accessible, the assumptions about defect sizes and fracture mechanics cannot be reproduced. The casting defect classes are based on comparison images, so any quantification and transferability to other NDT methods is difficult. A consideration of the different effects of surface and internal defects in [15, 117] is not apparent. The transferability to current casting materials must be ensured.

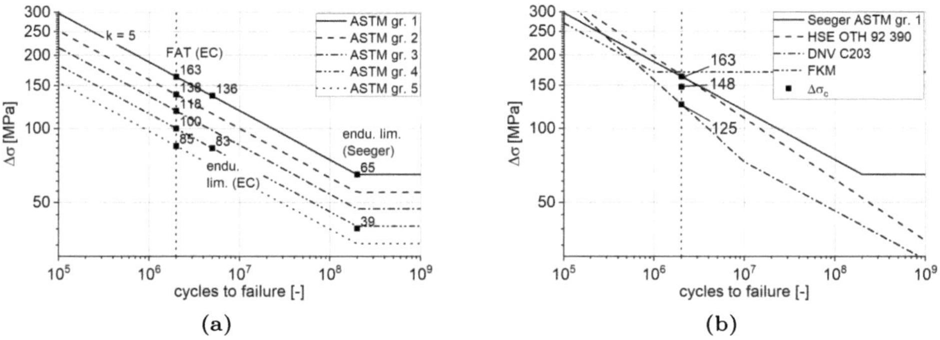


**Figure 2.19:** a) SN curves according to Seeger [96] for different ASTM defect groups, t = 25 mm, R = 0; b) SN curves for flawless material

### 2.3.2.2 Design Curves for Flawless Base Material

"UK Health and Safety Executive Offshore installations: Guidance on design, construction and certification" (HSE) [166], "DNV-RP-C203: DET Norske Veritas AS Fatigue Design of Offshore Steel Structures" (DNV) [159] and German Research Board of Mechanical Engineering (Forschungskuratorium Maschinenbau e.V.) (FKM) FKM-Rili:2012 [162] provide guidance on the fatigue resistance of the base material. The material must fulfill a minimum quality standard with regard to permissible imperfections according to defined NDT standards (e.g., BS 6208:1990 [135] in HSE). However, these are within the detection limit of the NDT methods, so it can be considered a technically flawless material. A limitation to select materials or minimum quality requirements to be met are not specified in any of the guidelines. Figure 2.19b provides an overview of the resistance curves and compares them with Seeger's concept of the highest quality requirements. For a quantitative classification, the respective $\Delta\sigma_c$ is given at $N = 2 \cdot 10^6$ cycles. In the case of resistance from [162], an extrapolation was necessary. The equations for the SN curves are given in Eq. (2.19) to (2.21), and the essential characteristics of the methods are compared in Table 2.4. No curves contain any additional safety factors on the load or on the resistance side. The influence of the wall thickness in components affected by stress gradients is determined using $t_{ref}$ and $t_{exp}$ according to Eq. (2.3).

$$\text{HSE [166]} \quad \log N = 15.169 - 4\log(\Delta\sigma) \tag{2.19}$$

$$\text{DNV [159]} \quad \log N = \begin{cases} 12.592 - 3\log(\Delta\sigma) & \text{for } N \leq 10^7 \\ 16.320 - 5\log(\Delta\sigma) & \text{for } N > 10^7 \end{cases} \tag{2.20}$$

$$\text{FKM [162]} \quad N = 10^6 \left(\frac{f_{w,\sigma}\mathrm{R_m}}{\Delta\sigma}\right)^5 \quad \text{with } f_{w,\sigma}(\text{cast steel}) = 0.34 \tag{2.21}$$

The hidden safety factor in HSE is due to the shift of the design curve by four standard deviations from the mean value of the test results. This is justified in [58] by casting defects that may have been overlooked by NDT and is intended to indirectly take into account that NDT cannot reliably detected or that are below the detection limit. A critical point is that the tests discussed in [58] were specifically carried out with higher SCFs than actually occur in the intended offshore applications. This was done to force a failure in the cast component during testing. As the design values refer to the maximum stress and the higher SCFs lead to larger gradients and higher support effects, the underlying assumptions are not conservative.

In Table 2.4, the resistance for S355 construction steel according to [162], which has similar mechanical properties as 1.6220 according to EN 10340:2008 [142], is added for comparison. Despite its identical mechanical properties, extensive NDT and higher safety factors (not yet fully considered in this comparison), the cast material is assigned a lower resistance. This is justified by the fear of overlooked casting defects.

**Table 2.4:** Comparison of the key parameters of different design concepts for the base material

| | material | prop. of survival | k [-] | $\Delta\sigma_c(2\cdot10^6)$ [MPa] | mean stress | $t_{ref}$ [mm] | $t_{exp}$ [-] |
|---|---|---|---|---|---|---|---|
| HSE | NS | 99.997 % (4SD) | 4 | 165 | R = 0 | 38 | 0.15 |
| DNV | NS | 97.7 % (2SD) | 3 | 125 | NS | 38 | 0.15 |
| FKM | G20Mn5 | 97.7 % (2SD) | 5 | 148 | AP | AP | |
| FKM | S355 | 97.7 % (2SD) | 5 | 165 | AP | AP | |
| Seeger | G13MnNi64 | 97.7 % (2SD) | 5 | 196 | $\lambda_R$ | 25 | 0.15 |

NS = not specified; SD = standard deviation; AP = adaptation possibilities

### 2.3.2.3 Considerations of Casting Defects by Fracture Mechanics

In FKM-Rili-BM:2009 [161], HSE:1995 [166] and [12], general procedures for the fracture mechanics evaluation of cracks are described. These are almost exclusively limited to basic mechanical relationships. In addition, crack propagation parameters and equivalent defect geometries for known defects were given. FKM-Rili-BM:2009 [161] additionally offers conversions from UT indications to equivalent defect sizes. Specific information relevant to the design process, such as defect sizes assignable to the quality classes, were not included in any of the mentioned codes.

### 2.3.3 Summary

The key statements found in the literature on the influence of casting defects on fatigue resistance, as well as existing concepts for the fatigue design of cast steel components were discussed.

From the evaluation of the literature data, essential findings could be drawn for the present work. In all studies, a negative influence of casting defects compared with the defect-free base material was found. The relevance of the defect location (component surface or inside of the component) was discussed but not sufficiently quantified. In many cases, these investigations were combined with other influences, such as bending stresses or a corrosive environment, so that it was almost impossible to derive unambiguous conclusions. The investigations were carried out on various, mostly high-strength materials with $R_m > 800$ MPa. Investigations on real components showed that the presents of larger defects in low-stress areas do not necessarily reduce the overall performance of the component. The casting defect type (solidification cavities, blowholes, gas bubbles, metallic inclusions) was attributed a minor influence. Evaluations of casting defects based on fracture mechanics lead to conservative solutions, but allow the derivation of sufficient fatigue resistances for practical applications.

The review of existing design concepts showed that they were developed almost exclusively for the technically defect-free base material. In order to take account of any by NDT overlooked casting defects, these were subjected to high safety factors. Design resistances and gradients of the SN curves vary between the different guidelines and do not present a clear image. On this basis, it is not possible to define quality requirements in a way that is adapted to the local stresses. Only the concept of Seeger [96] links RT indications for internal defects with fatigue resistance and was developed for offshore applications. Some guidelines provide the basis for fracture mechanics design. For everyday application in an engineering design office, these methods prove to be unsuitable and its application mostly remains restricted to experts; in addition, there is a lack of the tools required for practical application to link quality classes and computational equivalent defect sizes.

# 3 Experimental Investigations

The aim of these experimental investigations was to generate a database for a design concept and to understand the damage mechanisms casting defects cause. For this purpose, the three-level approach shown in Figure 3.1 was chosen.

Section 3.1 (level I) focuses on the properties of the flawless base material. In particular, material models for crack initiation and crack growth were developed, and the necessary parameters were determined experimentally. These investigations were carried out by Fraunhofer IWM as part of the ErStaGu research project [116]. The explanations herein are limited exclusively to the results used in this context.

Section 3.2 (level II) is dedicated to the influence of real casting defects on fatigue strength. Tensile specimens with internal shrinkage cavities were subjected to fatigue loading. As a result, SN curves depending on the defect properties were derived, and the defects as well as the failure mechanisms were investigated.

The investigations in Section 3.3 (level III) are aimed at the influences of other factors common to cast components, such as the interaction of defects with geometric notches or the integral consideration of wall thickness effects. Notched specimens of different wall thickness with artificially induced defects were examined, resulting in component SN curves.

In order to get a more detailed understanding of the damage mechanisms involved, additional numerical simulations, as described in Chapter 4, were carried out. As these studies reproduced the experiments, specific additional data are required, generated and presented within this chapter.

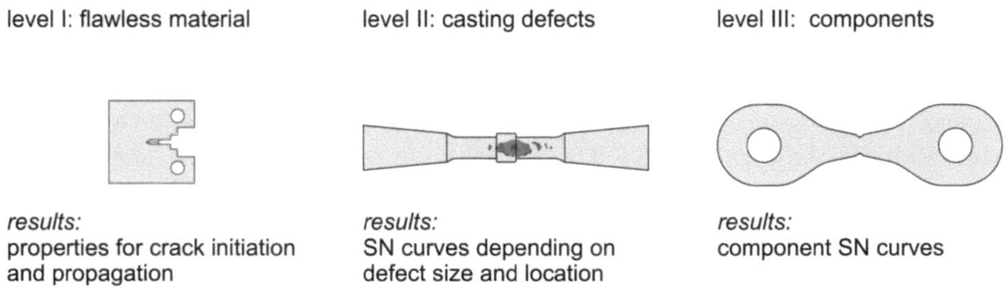


**Figure 3.1:** Three-level approach for experimental investigations

## 3.1 Materials

The experimental investigations were carried out on two materials: G20Mn5+QT (1.6220) according to EN 10340:2008 [142] and a high-strength G22NiMoCr5-6+QT (1.6760) according to SEW 520:2017 [164]. This follows the methodology of [103] and covers a wide range of structural casting applications. The initial hypothesis was analogous to the brittle fracture tests from [103] that 1.6760, being more notch sensitive, follows a different failure mechanism than 1.6220. To gather representative and transferable findings, five different foundries participated in these investigations. Each of the foundries specializes in different lot sizes, industrial sectors, products, melting techniques, alloy concepts and individual heat treatments. Moreover, their molding methods and materials differ significantly. Most of the listed differences are part of the individual companies' knowhow and are kept confidential in the background documentation of ErStaGu [116]. Data from inspection certificates 3.1 and further material investigations are discussed anonymously.

### 3.1.1 Mechanical Properties and Chemical Composition

In structural applications, components of 1.6220 are mostly assembled with profiles of S355 construction steel. In these cases, designers usually define additional requirements to ensure similar yield strengths in both components. To account for this and to stay in line with the investigations of [103], a minimum yield strength of $R_{eH} = 360$ MPa was defined. The chemical composition and mechanical properties according to inspection certificates 3.1 are summarized in Annex A. As the foundries have chosen different test conditions, additional tensile and Charpy tests were carried out and are supplemented in Tables A.1 and A.2. In addition, the modulus of elasticity was determined, resulting in a mean value of 205,000 MPa with a scatter of 195,000 to 210,000 MPa. This is in line with findings on the same materials in [110], where $E = 200{,}000$ MPa was found. In the chemical analysis of the melt in Table A.2, four deviations from the nominal values of individual alloying elements are marked. As these elements have a negative effect on material toughness, the deviations were accepted within these investigations.

Additional tensile tests were carried out by Fraunhofer IWM within the ErStaGu research project [116]. The tensile specimens were taken from cast blocks of 100 mm thickness. As discussed in [103], the influence of the thickness is negligible with respect to the results of the tensile tests. The mean values of the results are summarized in Table 3.1.

**Table 3.1:** Mean values of tensile tests performed by Fraunhofer IWM in [116]

| Mat. [-] | $R_{eH}$; $R_{p0.2}$ [MPa] | $R_m$ [MPa] | A [%] | sample size [-] |
|---|---|---|---|---|
| 1.6220 | 409 | 597 | 26.9 | 8 |
| 1.6760 | 889 | 997 | 13.4 | 8 |

### 3.1.2 Fracture Mechanics Properties

In [116], Fraunhofer IWM determined crack propagation parameters according to ASTM E647:2015 [132] on compact tension specimens from cast blocks of different thicknesses. The cast blocks were designed to be self-supplying and thus free from casting defects. Recordings with the scanning electron microscope confirmed this. The investigations on wall thicknesses of 32 mm, 60 mm and 100 mm showed that the crack propagation properties of both materials are independent of wall thickness. This is in line with the core statements presented in Figure 2.6a. In summary, the fracture toughness $K_{Ic}$ is influenced by wall thickness, but the crack propagation rate da/dN is not. The results are summarized in Table 3.2 and are compared with values from the literature for common cast steel materials.

In Figure 3.2a, the crack propagation parameters in Table 3.2 are classified within the findings of [29]. Herein a linear correlation between the Paris parameters C and m was derived in a semi-logarithmic diagram. This was done based on a regression analysis of an extensive database of structural steel, high-strength steel, weld material and specimens from the heat-affected zone of welds at R = 0. Besides the formal relationship, the scatter of the tests from [29] is plotted. Figure 3.2a indicates that all the parameters collected for cast steel materials follow this relationship.

**Table 3.2:** Crack growth parameters for cast steel compared to the literature

| No. | Ref. | Mat. | Note | R | m | C | $\Delta K_{th}$ | C | $\Delta K_{th}$ |
|---|---|---|---|---|---|---|---|---|---|
| [-] | [-] | [-] | [-] | [-] | [-] | [mm/cycle; $Nmm^{-3/2}$] | | [mm/cycle; $MPa\sqrt{m}$] | |
| (1) | [116] | 1.6220 | | 0.1 | 3.00 | $9.97\ 10^{-14}$ | 228 | $3.15\ 10^{-9}$ | 7.2 |
| (2) | [116] | 1.6220 | | 0.5 | 2.98 | $2.05\ 10^{-13}$ | 149 | $6.06\ 10^{-9}$ | 4.7 |
| (3) | [116] | 1.6760 | | 0.1 | 2.89 | $2.37\ 10^{-13}$ | 221 | $5.13\ 10^{-9}$ | 7.0 |
| (4) | [116] | 1.6760 | | 0.5 | 3.30 | $2.89\ 10^{-14}$ | 139 | $2.58\ 10^{-9}$ | 4.4 |
| (5) | [44] | 1.7379 | | 0.1 | 2.77 | $6.18\ 10^{-13}$ | | $8.83\ 10^{-9}$ | |
| (6) | [35] | - | from [8] | | 3.00 | $2.00\ 10^{-13}$ | 120 | $6.32\ 10^{-9}$ | 3.8 |
| (7) | [161] | 1.6220 | from [48] | 0.1 | 3.80 | $1.20\ 10^{-15}$ | 190 | $0.60\ 10^{-9}$ | 6.0 |
| (8) | [163] | NS | mean | 0.1 | 2.88 | $3.98\ 10^{-13}$ | 149 | $8.32\ 10^{-9}$ | 4.7 |
| (9) | [163] | NS | design | 0.1 | 2.88 | $6.77\ 10^{-13}$ | | $1.41\ 10^{-8}$ | |
| (10) | [163] | NS | mean | 0.5 | 2.88 | $5.86\ 10^{-13}$ | 63 | $1.22\ 10^{-8}$ | 2.0 |
| (11) | [163] | NS | design | 0.5 | 2.88 | $1.29\ 10^{-12}$ | | $2.70\ 10^{-8}$ | |
| (12) | [138] | NS | mean | <0.5 | 2.88 | $3.98\ 10^{-13}$ | 149 | $8.32\ 10^{-9}$ | 4.7 |
| (13) | [138] | NS | design | <0.5 | 2.88 | $6.77\ 10^{-13}$ | | $1.41\ 10^{-8}$ | |
| (14) | [138] | NS | mean | ≥0.5 | 2.88 | $5.86\ 10^{-13}$ | 63 | $1.22\ 10^{-8}$ | 2.0 |
| (15) | [138] | NS | design | ≥0.5 | 2.88 | $1.29\ 10^{-12}$ | | $2.70\ 10^{-8}$ | |
| (16) | [138] | NS | design | - | 3.00 | $5.21\ 10^{-13}$ | | $1.65\ 10^{-8}$ | |

NS = not specified, applicable for a wide range of steels in air, for details consider the individual reference

The dataset marked no. (7) of the FKM-Rili-BM:2009 [161] is identified as an outlier with respect the other listed values but not in relation to the correlation according to [23]. The parameters deviate significantly from the remaining values but still follow the concept of [29]. The values from FKM-Rili-BM:2009 [161] are taken from a single assessment of a damage case [48] on the cone of a ball valve with a diameter of 750 mm and a length of 1870 mm. Qualitatively, this follows the relationship between $K_{Ic}$ and the crack propagation parameters described in [104]. As Table 3.3 entry no. (6) shows, the $K_{Ic}$ of the component from [104] is clearly below the fracture toughness calculated from the minimum requirements of KV values according to EN 10340:2008 [142]. This relation is in line with the findings of [104], where an increasing Paris-exponent m was linked to low $K_{Ic}$ values.

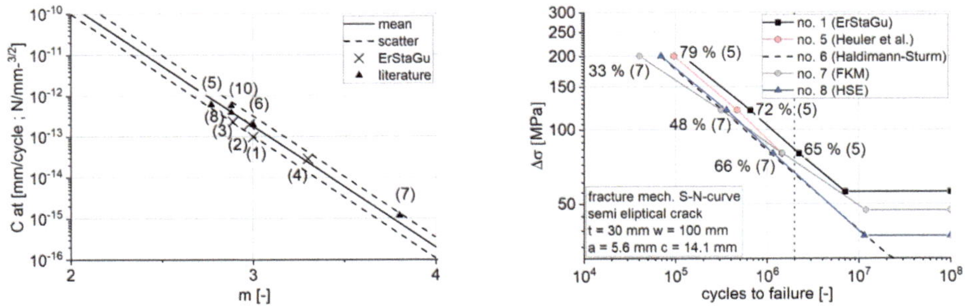


(a) (b)

**Figure 3.2:** a) Classification of crack propagation parameters within the findings of [23]; b) Fracture mechanics-based SN curves for an identical geometric situation, including proportions of the service life calculated according to model no. (1) at different load levels

The effect of the different crack propagation parameters on service life is shown in Figure 3.2b. Fracture mechanics-based SN curves, according to the descriptions in Section 2.2.4.4, are used for this exemplary illustration. The given percentages represent the value's share of the cycles to failure according to model no. (1). The test results lead to slightly longer lifetimes than the literature data do. Table 3.3 provides a collection of various literature data on the fracture toughness of the materials under investigation.

**Table 3.3:** Collection of literature data on the fracture toughness of the materials under investigation

| No. [-] | Ref. [-] | Mat. [-] | T [°C] | $K_{JIc}$ [$Nmm^{-3/2}$] | $K_{JIc}$ [$MPa\sqrt{m}$] | Note [-] |
|---|---|---|---|---|---|---|
| (1) | [103] | 1.6220 | - 30 | 7626 | 241 | $K_{JIc}$ estimation from $J_i$ |
| (2) | | | - 30 | 4223 | 134 | Est. from min KV acc. to [142] |
| (3) | | 1.6760 | - 30 | 4707 | 149 | $K_{JIc}$ Est. from $J_i$ |
| (4) | | | - 30 | 4890 | 155 | Est. from min KV acc. to [164] |
| (5) | [35] | 1.6220 | - 30 | 2400 | 76 | Est. from min KV acc. to [142] |
| (6) | [161] | 1.6220 | - 30 | 1644 | 52 | |
| (7) | [110] | 1.6220 | + 20 | 7748 | 245 | $K_{JIc}$ Est. from $J_i$ |
| (8) | [110] | 1.6760 | + 20 | 4206 | 133 | $K_{JIc}$ Est. from $J_i$ |

## 3.2 Tensile Specimens with Casting Defects

The aim of these investigations was to understand the influence of real casting defects on fatigue resistance as well as the quantification of SN curves depending on defect properties. With regard to a design approach for structural applications, the characteristics of the residual fractures are of special interest. The dissolved metallurgical conditions surrounding the cavities as well as the geometric properties (i.e., shape, size, distribution) were suspected to dominantly influence crack initiation and growth. For this purpose, a series of investigations, as shown in Figure 3.3, were carried out. Conclusions on the objectives of the study could only be drawn from the interplay of the individual approaches. For a simpler understanding, the methodology and objectives of the individual studies are presented in Section 3.2.1. Subsequently, in Section 3.2.2, the evaluation is divided according to individual aspects of the investigation. As shown in Figure 3.3, all results are based on the findings of several individual investigations. An extended analytical and numerical evaluation of the experimental data is given in Section 4.1.

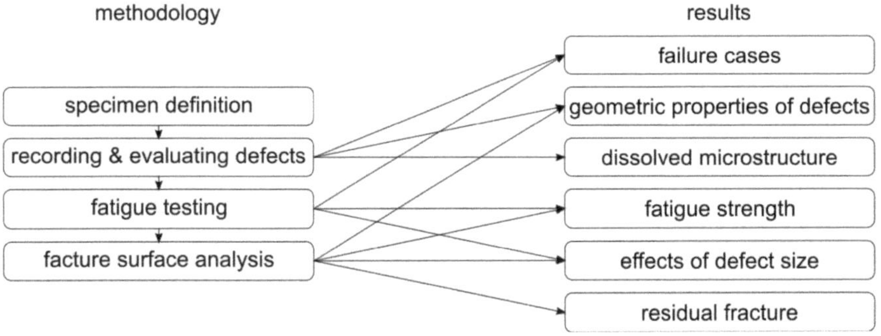


**Figure 3.3:** Overview of investigations on tensile specimens with casting defects subdivided into methodology and results

### 3.2.1 Methodology

#### 3.2.1.1 Specimens

The specimens were designed with the intention to generate reproducible casting defects of different sizes, shapes and distributions. Both 1.6220 and 1.6760 were in the scope of these investigations. The concept of [103] to cast round tensile bars with thickenings in the middle was applied but had to be reevaluated with the result that only two reasonable design variants were identified. The unconsidered casting geometries were excluded due to their large defect sizes, which are not representative of the actual components in relation to the specimen cross-section. In addition, geometries that had led to compact, low-notched defects were also not taken into account. In order to satisfy the boundary conditions of the testing machine as well as to prevent failures in the clamping areas found in preliminary tests, modifications had to be made to the casting geometry.

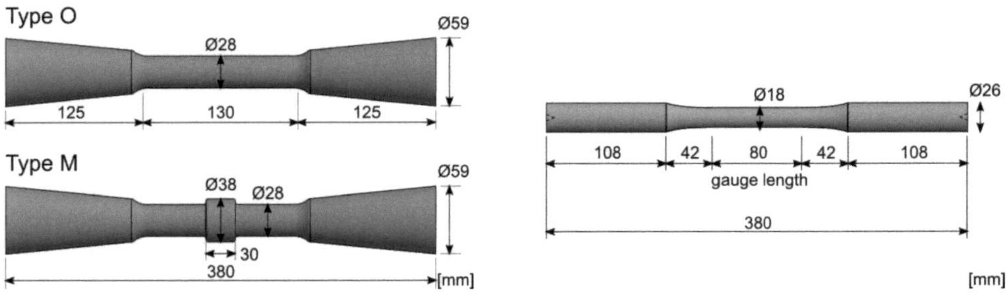


(a) (b)

**Figure 3.4:** a) Casting geometry of the tensile specimens; b) Test geometry machined by turning

The foundries involved in ErStaGu [116] carried out casting process simulations to evaluate the proposed geometries with regard to the formation of pores. Irrespective of the material, the castings were carried out horizontally into the geometries shown in Figure 3.4a. The first specimen shape did not show any thickening and is marked as type O. The thickening of the type M specimen resulted in a locally delayed solidification and thus formed larger shrinkage pores [124]. After casting and heat treatment, the test geometry shown in Figure 3.4b was machined from the cast mold. In order to limit the failure to the gauge length of 80 mm, the transitions were designed as splines. The following notation identifies the specimens:

U-(material)-(type)-(n)

| | |
|---|---|
| U | identifier for tensile specimens with casting defects |
| (material) | 20 = 1.6220; 22 = 1.6760 |
| (type) | O = without thickening; M = with thickening |
| (n) | consecutive number |

#### 3.2.1.2 Recording Internal Casting Defects

The geometric properties of all internal defects were recorded using CT prior to fatigue testing. Subsequently, representative geometric indicators, such as sphericity, were derived, and 3D shape-files were produced on that basis for further numerical simulations. The recording and evaluation were conducted by Fraunhofer IZFP in ErStaGu [116].

A voxel size of 80 $\mu$m x 80 $\mu$m x 80 $\mu$m was chosen to differ between a cavity and the surrounding fungoid microstructure with reasonable effort. In addition, crack-like defects were found with this setting. The requirements of the numerical simulation were lower, and additional simplifications, as described in Section 4.1.1, were necessary. Further details on the recording methodology are described in [69]. Fraunhofer IZFP determined geometric indicators using different pore analysis algorithms. These were defined as:

| | |
|---|---|
| $A_{sec}$ | Gross cross-sectional area of the specimen. |
| $A_{proj,CT}$ | Defect area projected in the longitudinal direction. |
| $A_{max,CT}$ | Defect area at the maximum cross-sectional reduction. |
| $z_{max,CT}$ | Longitudinal position of $A_{max,CT}$. |
| $A_{fail,CT}$ | Defect area in the failure plane. |
| $V_{max}/V_{tot}$ | Proportion of the largest pore to the total pore volume. |
| $A_{cavity}$ | Surface of the cavity. |
| $V_{cavity}$ | Volume of the cavity. |

#### 3.2.1.3 Fatigue Testing

The specimens were tested on a calibrated, 400 kN, high-frequency pulsator at the Research Center for Steel, Timber and Masonry at the Karlsruhe Institute of Technology. All tests were performed under cyclic tension with a stress ratio $R = 0.1$. Fifty-four specimens were tested with nominal stress ranges between 150 MPa and 570 MPa at $f \approx 70$ Hz, resulting in $N = 2 \cdot 10^4$ to $1 \cdot 10^7$ cycles to failure. All specimens were tested until fracture with no predefined limit on cycles. To get beach marks on the fracture surface, eight specimens were tested under load blocks with varying amplitudes. The maximum applied load $F_{u,mark} = F_{u,test}$ was kept constant in the marking and testing cycles, whereas the lower load was adopted and set to $F_{l,mark} = F_{m,test}$. The number of cycles within the marking block was about 15 % to 50 % of the test block. The large variance results from difficulties in imprinting visible beach marks. The high value of 50 % was only applied in one exceptional case. After considering the linear accumulation of damage the cycles in the marking block were not considered in the evaluation. In Annex C, the index R in the stress range marks these specimens. As described in [69], Fraunhofer IZFP used four different NDT techniques to capture crack initialization and propagation in ErStaGu [116]. As these data brought only limited benefit to the overall objective of the present thesis, the results are not considered in this context.

#### 3.2.1.4 Fracture Surface Analysis

Each fracture surface was analyzed using different scales. Visual examinations, as shown in Figure 3.5a were used to characterize the topology of the fracture surface. A distinction was made between one or more crack planes and their characteristics. In addition, the appearance of the residual fracture surface was considered. Secondly, a measurement of the fracture surface was carried out based on light microscope records, as shown in Figure 3.5b. Here, the focus was put on the defects, the regions of stable crack growth and the residual fracture surfaces. The defect outlines were determined by visual changes in the microstructure. This frequently deviated from the area to be attributed to the pure pore volume. In all cases where the area of stable crack growth was clearly recognizable, its shape and size were described using an ellipse. The parameters were defined as:

| | |
|---|---|
| $A_{fail,mic}$ | Defect area in the fracture plane. |
| $A_{pore,mic}$ | Pore area in the fracture plane. |
| $A_{crack}$ | Area of stable crack growth including $A_{fail,mic}$. |
| $D_{max/min}$ | Main axes of the ellipse attributed to stable crack growth. |
| $A_{res}$ | Area of the residual fracture surface $= A_{sec} - A_{crack}$. |

Only selected specimens were investigated with scanning electron microscopy at Fraunhofer IWM. An exemplary recording is shown in Figure 3.5c. These provided high-resolution information on microporosity, solidification dendrites and therefore the fungoid microstructure surrounding the shrinkage cavities.

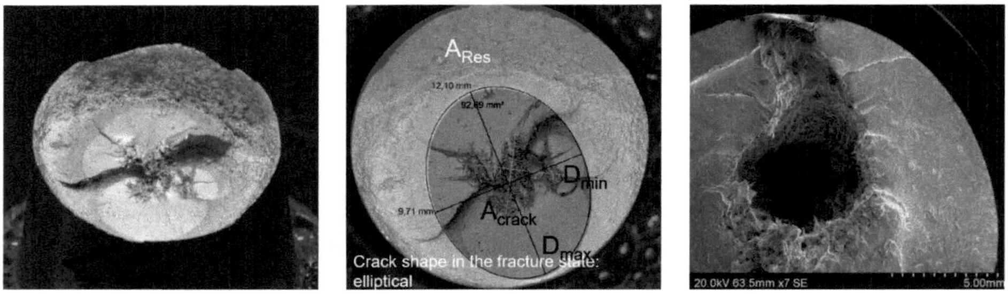


(a) (b) (c)

**Figure 3.5:** Fracture surface analysis: a) Fracture topography using macro pictures; b) Parameter specifications using a light microscope; c) Solidification dendrites and micro pores using scanning electron microscopy

## 3.2.2 Results

The presentation and discussion of the results follow the schemes shown in Figure 3.3 and are limited to essential core statements. Detailed documentation on all individual specimens is given in Annex C.

### 3.2.2.1 Failure Cases

Fractures occurred in different planes with respect to the largest detected defects and the gauge length. To describe these mechanisms, three failure cases were defined as shown in Figure 3.6a. Specimens where failure (the failure plane is indicated by a continuous line) occurred within the gauge length and was caused by the largest detected pore (the plane of maximum cross-sectional reduction is indicated by a dashed line) were defined as case I. In case II, failures, the failure was inside the gauge length but was induced by a smaller defect. This denoted that the maximum cross-sectional reduction was not decisive for failure but that the geometric properties of the defect must have influenced the failure process. In case III, the failure was outside the gauge length. In these cases, no CT data was available within the fracture plane and no further subdivision, as in cases I and II, was possible. The diagram in Figure 3.6b illustrates the distribution of the specimen types and materials between the three cases. The majority of type O specimens were class III failures because there were larger defects in the clamping areas of these specimens, as discussed in the description of Figure 3.9.

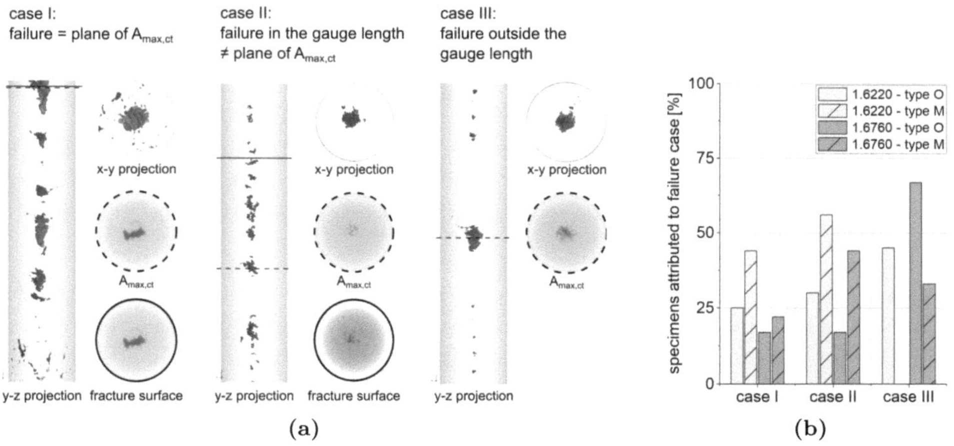


**Figure 3.6:** a) Definition of failure cases; b) Shares of specimen types and materials in the respective failure cases

### 3.2.2.2 Geometric Properties of the Defects

The CT records shown in Figure 3.7a are representative of the full set of specimens and show two defect configurations. These were associated with the specimen type as intended. The results were independent of the material and the foundry. Type O specimens generated several small pores located along the centerline. The pore representative of type M was a single, dominating, large pore with varying longitudinal expansions. The differences were quantified by the maximal cross-sectional reduction and are presented in Figure 3.7b. Defects in type O specimens caused a mean cross-sectional reduction of about 2.5 %, whereas type M specimens showed a reduction of about 10 %. In both cases, the values of the individual specimens (marked by crosses in Figure 3.7b scatter widely. An overlap between both populations caused a continuous distribution of cross-sectional reductions between 1 % and 14 %, as detected by CT.

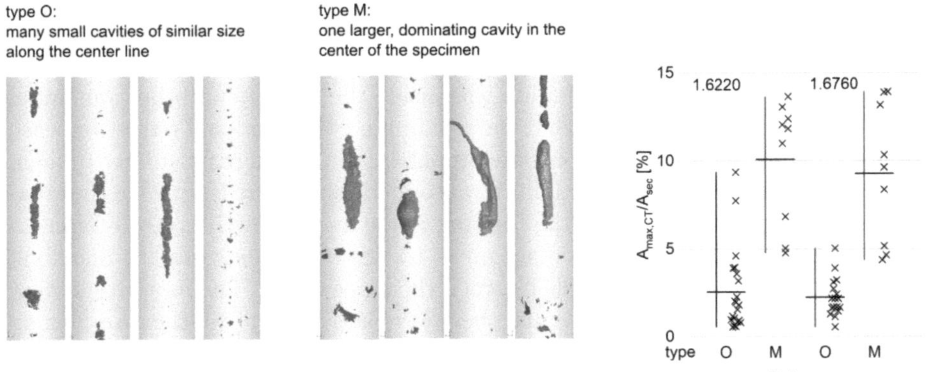


(a) (b)

**Figure 3.7:** a) CT records of representative defect configurations depending on the specimen type; b) Maximum cross-sectional reduction depending on the specimen type and material (mean value, scatter and individual specimen)

In Figure 3.8a, the number of detected pores are compared, and in Figure 3.8b, the volume of the largest pore is related to the total defect volume within the gauge length. Both evaluations were done separately for each specimen type and material and underline the statements above.

In Figure 3.9, the dimensions $A_{proj,CT}$, $A_{max,CT}$, $A_{fail,CT}$ and $A_{fail,mic}$ are compared for all individual specimens. The objective of these diagrams is to check the records and evaluation for plausibility as well as to derive failure relevant patterns. In addition, the differences between CT and microscopy are clarified. A distinction between materials, specimen types and failure cases was made.

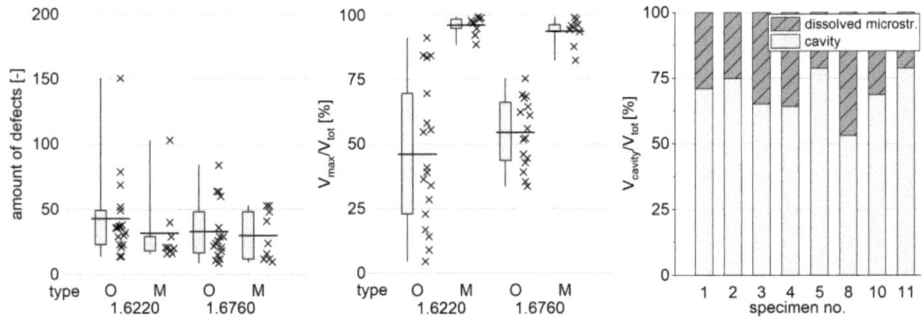


(a) (b) (c)

**Figure 3.8:** Min, max and mean values of the number of detected defects, including values of the individual specimens and 25 % and 75 % percentiles for: a) The amount of defects; b) The cavity and dissolved microstructure volume related to total defect volume for 1.6220, type M specimens; c) The percentage of cavity and dissolved microstructure on the total defect volume

The following aspects became apparent:

- Plausibility check: $A_{proj,CT}$ was larger than $A_{max,CT}$ and $A_{fail,CT}$ in all specimens. $A_{fail,CT}$ was smaller than $A_{max,CT}$ in failure case II. The defect sizes of type M were larger than those of type O.
- The differences between $A_{proj,CT}$ and $A_{max,CT}$ and between $A_{proj,CT}$ and $A_{fail,CT}$ were greater for type O specimens than for type M specimens. This effect was more pronounced for material 1.6760. Especially for type O specimens, there was no proper correlation between $A_{proj,CT}$ and $A_{max,CT}$. This means that defects in parallel planes perpendicular to the specimen's axis were not concentric but rather spread over the specimen's cross section.
- $A_{fail,mic}$ was 10 % to 60 % larger than $A_{fail,CT}$ in failure cases I and II for all specimens. Both measures are relevant for further evaluations. For an understanding of the failure mechanisms, $A_{fail,mic}$ was needed, as the actual reduced cross section was considered. On the other hand, $A_{fail,CT}$ described the defect size detectable by NDT under lab conditions. Thus, it is necessary for the transformation of maximum tolerable defect sizes determined in the design concept of Chapter 5 to NDT displays. This transformation is not within the scope of this work.
- In failure case III, (i.e., failure outside the gauge length), $A_{fail,mic}$ was significantly larger than $A_{max,CT}$. This indicates that large defects occurred in the clamping areas of the specimens.

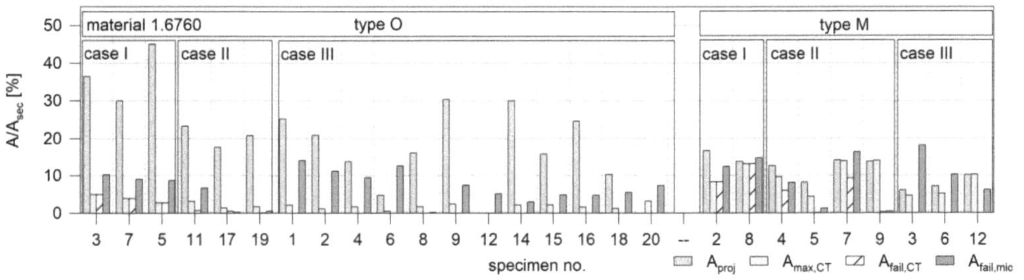


**Figure 3.9:** Description of the defect areas

#### 3.2.2.3 Dissolved Microstructures Surrounding Cavities

Shrinkage cavities are surrounded by a dissolved microstructure. This is caused by dendrites, which grow into the pore void during the solidification process. Due to a lack of liquid material, they are not able to combine into a homogeneous structure [124]. This can be seen in Figure 3.10a, showing an image taken with a scanning electron microscope. In previous studies and design concepts [103], this region was attributed the potential to act as a crack. This assumption was motivated by the sharp-edged areas that can be detected (e.g., in micro sections) (see Figure 3.10b). On the other hand, hardness measurements as shown in Figure 3.10c attributed a much lower hardness to these areas and a reduced load-bearing capacity compared to the base material. This is the basis for a second contradictory hypothesis: stress peaks caused by the pure cavity are smoothed by a locally reduced stiffness, causing a delay in the crack initiation and resulting in higher resistance.

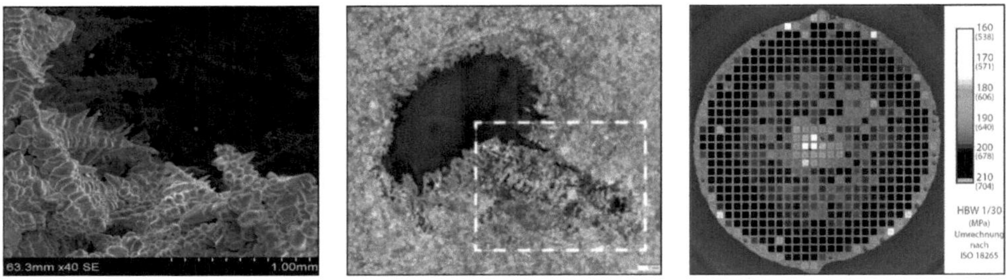


(a) (b) (c)

**Figure 3.10:** a) Solidification dendrites in a scanning electron microscope recorded by Fraunhofer IWM [116]; b) Dissolved microstructure in a micro section from [103]; c) Hardness mapping HBW 1/30 for material G20Mn5, porosity in the middle of the specimen from [103]

In order to estimate the extent of these regions, Fraunhofer IZFP performed a pore analysis of the CT records in ErStaGu [116] with two different configurations. In the first setting, only the pore volume was detected. In the second scenario, the threshold values were modified in such a way that the less dense region of the dissolved microstructure compared to the homogeneous base material was included. Figure 3.11b compares these two volumes. These effects were decisive for type M specimens and are investigated exclusively in this case. Since no significant differences in the defect characteristics could be identified between both materials, the following extended description is limited to material 1.6220.

Figure 3.8c shows the proportions of the pore volume and the dissolved microstructure on the total defect volume. In the considered cases, the dissolved microstructure occupied between 20 % and 40 % of the total defect volume.

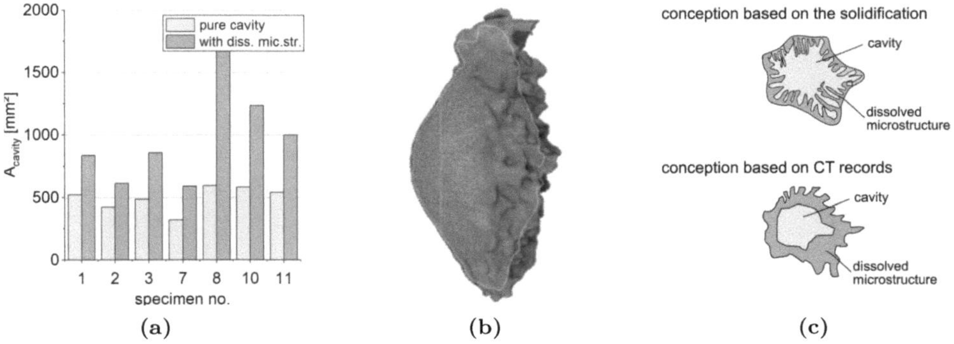


**Figure 3.11:** a) Effects of the dissolved microstructure for material 1.6220 on the defect surface of type M specimens; b) Volume model of pore analysis with and without the dissolved microstructure; c) Models for the description of the dissolved structure

As shown in Figure 3.11a, the defect surface between these approaches increased by 45 % to 110 %. This was partly due to the larger volume but also to a significantly more ramified surface. However, compared to the models shown in Figure 3.11c, its physical correctness is to be doubted. Rather, it is assumed that the boundary areas of the desolved microstructure are blended in the noise of the CT images. The upper illustration in Figure 3.11c represents a limit case, or model presentation. It is assumed that the dendrites grow into the pore and lead to a strongly dendritic pore surface. On the other hand, the transition area between the dissolved structure and the homogeneous material is considered to be smooth more or less, as the solidification front and material supply do not show large local gradients. The other extreme case corresponds to the state derived from the pore analyses. In this case, the surface of the pores is significantly less ramified than the transition area to the homogeneous material.

In the simultaneous observation of fracture surfaces, sections, hardness profiles and CT data, a combination of both model conceptions is assumed to represent reality. For these reasons, further interpretations or quantifications of the dissolved microstructure of parameters derived from the CT data are not considered. Against this background, the representations from Figure 3.8c and Figure 3.11a are not to be understood as absolute values, but rather as qualitative illustrations of the effects. In summary, the desolved microstructure leads to a greater weakening of the cross section, but precise statements on crack initialization are not possible based only on the CT evaluations. Whether the assumption of a sharp crack at the fatigue stress is a sensible engineering model is estimated in Section 4.1. Further investigations are necessary for the precise clarification of the mechanism.

#### 3.2.2.4 Fracture Surfaces

Besides the defect size, the regions of stable crack growth and the residual fracture were investigated on each fracture surface. This section focuses on the crack propagation and the residual fracture. In all specimens, stable crack growth occurred according to crack opening mode I [28] perpendicular to the tensile load. Figure 3.12 compares the fracture surfaces of four specimens at similar load levels. The net stress range was determined by:

$$\Delta\sigma_{\mathrm{mic}} = \frac{\Delta \mathrm{F}}{\mathrm{A}_{\mathrm{sec}} - \mathrm{A}_{\mathrm{fail,mic}}}. \tag{3.1}$$

For material 1.6220, Figure 3.12a shows the fracture surface for an internal defect, and Figure 3.12b shows the fracture surface for a surface defect. In the case of the internal defect, an elliptical or approximately circular area was visible and could be assigned to stable crack growth. In the image, this is marked and can be clearly distinguished from the defect and the residual fracture surface. In the case of a surface defect in Figure 3.12b, this assignment was not obvious. The crack growth was eccentric and originated from the defect. An ellipse could no longer describe the stable crack growth. The transition between the crack growth and the residual fracture was smooth. As shown in Figures 3.12c and d, the situation was comparable for specimens made out of material 1.6760.

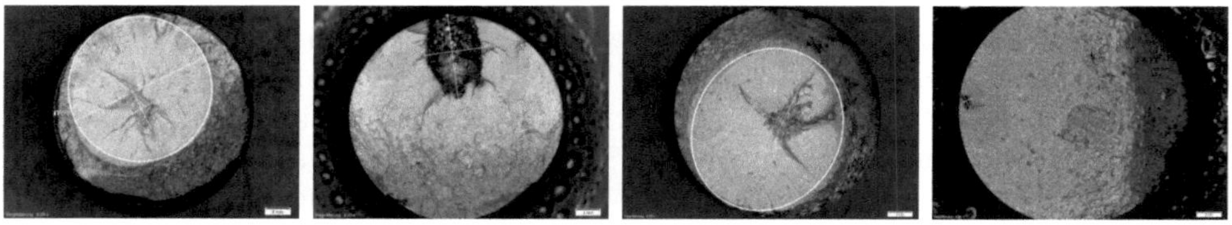

(a) (b) (c) (d)

**Figure 3.12:** Fracture surfaces of: a) U20O2 $\Delta\sigma_{\mathrm{mic}} = 286$ MPa; b) U20O11 $\Delta\sigma_{\mathrm{mic}} = 314$ MPa; c) U22O14 $\Delta\sigma_{\mathrm{mic}} = 268$ MPa; d) U22O8 $\Delta\sigma_{\mathrm{mic}} = 281$ MPa

Figure 3.13 compares the fracture surface topologies of four specimens with internal defects. Besides the failure in one crack plane (cf. Figure 3.13a), several parallel planes of stable crack growth (cf. Figure 3.13b) were observed. The shape of each respective crack growth could be described by a section of an ellipse.

As shown in Figure 3.13c, ragged fracture surfaces were observed in some cases. In these cases, an approximately elliptical area differing from the defect and residual fracture could be identified as well. This failure pattern occurred with branched defects mostly located in the clamping area of the tensile specimen. Generally, the area of stable crack growth could be represented in a good approximation by an ellipse (cf. Figures 3.12a and c and Figure 3.13a). Only in special cases (cf. Figure 3.13d) was stable crack growth limited to one location of the defect. In addition, a comparison of Figures 3.13a and d illustrates the dependence between the expansions of the residual fracture surface and the applied stress level. As expected, the residual fracture area increased with increasing loads.

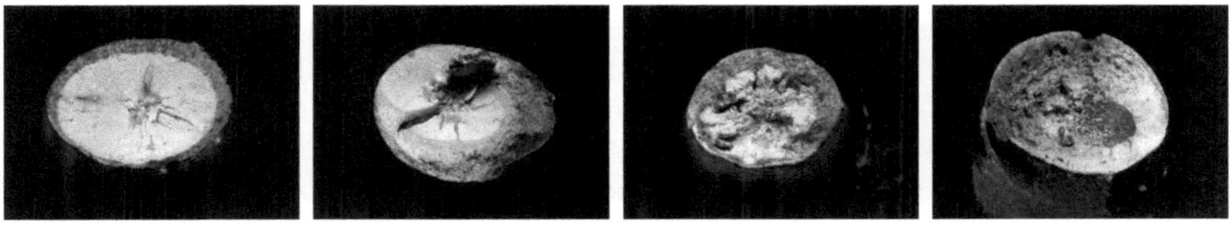

(a) (b) (c) (d)

**Figure 3.13:** Fracture topologies of: a) U20O20 $\Delta\sigma_{\text{mic}} = 241$ MPa; b) U20O17 $\Delta\sigma_{\text{mic}} = 248$ MPa; c) U22O6 $\Delta\sigma_{\text{mic}} = 459$ MPa; d) U22M7 $\Delta\sigma_{\text{mic}} = 384$ MPa

In Figure 3.14, the necking behaviors of both materials are compared. Both specimens had internal defects of similar size and were subjected to similar loading conditions. In the case of material 1.6220, a distinctive necking could be identified, as illustrated by the dashed reference lines. The specific residual fracture surfaces were honeycombed and showed pronounced shear lips. On the other hand, for material 1.6760, no distinctive necking could be observed, and the residual fracture surface was less indicative of ductile failure.

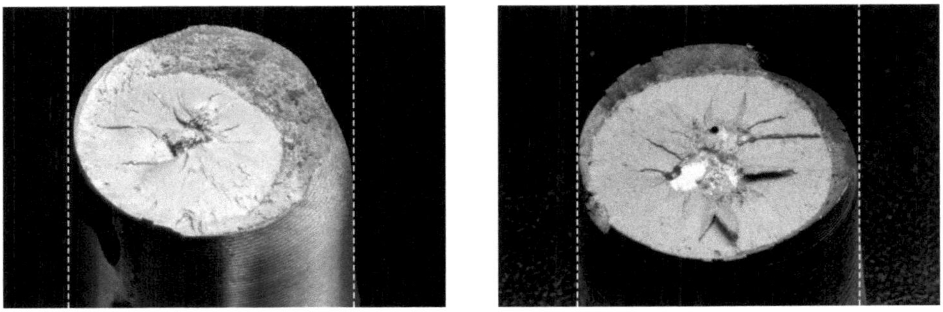

(a) (b)

**Figure 3.14:** a) Necking for 1.6220 specimen U20O2 at $\Delta\sigma = 280$ MPa; b) No pronounced necking for 1.6760 specimen U22M6 at $\Delta\sigma = 280$ MPa

### 3.2.2.5 Fatigue Strength

Emphasis was put on the statements that could be derived directly from statistical analysis. A deeper understanding of the failure mechanisms and an estimation of the proportions of crack initiation and growth were based on numerical simulations in

Section 4.1. For a basic understanding of the decisive correlations, only the regression curves were determined and shown in the following diagrams. These curves were determined according to the evaluation procedure of EN 1993-1-9:2010 [156] with $\Delta\sigma$ as the independent variable. The regression curves were highlighted in the $\Delta\sigma$ range and were extended to the range from $1 \cdot 10^4$ to $2 \cdot 10^7$ cycles (grey dotted lines) for a better comparability. If not specified, the evaluation was performed with a variable gradient k and is indicated within the diagrams. $\Delta\sigma_{50\%}$ corresponds to the mean value of the stress range at $2 \cdot 10^6$ cycles to failure. In addition, the standard deviation SD is given as an indicator of the scatter. The blanket exclusion of individual tests that exceed a limit number of cycles was initially dispensed within these fundamental analyses. Characteristic resistance curves were determined according to EN 1993-1-9:2010 [156].

Figure 3.15a shows the results of all experiments related to the nominal stress range $\Delta\sigma_{nom}$. The data are presented separately for material and specimen type. The regression curve resulted from a common evaluation of all data points. The data showed a large scatter quantified by SD = 0.51. On that basis, no clear dependence on the investigated material could be identified. Type M specimens tended to show lower fatigue resistance. However, numerous outliers disturbed this correlation.

Figure 3.15b shows an evaluation of the experiments separated by material and as a function of the defect size $A_{fail,mic}$. For simplification, the specimens are presented in clusters depending on the reduction of the cross section. The limit values are defined in Table 3.5. The icon size in Figure 3.15b illustrates the defect size associated with the clusters. For each cluster, a regression curve was determined independently from the material and is plotted in Figure 3.15b, including the descriptive parameters.

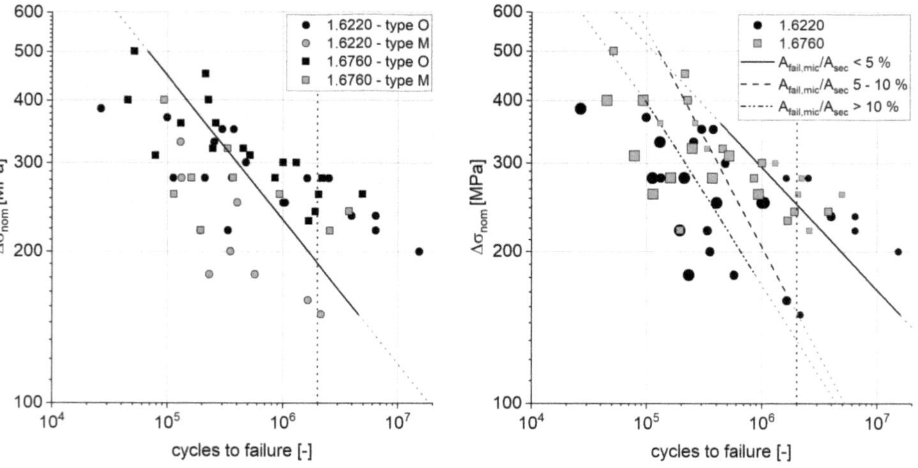


(a) (b)

**Figure 3.15:** Test results of nominal stress SN data related to the material and a) the specimen type; b) the defect size (small icon = $A_{fail,mic}/A_{sec} < 5$ %; medium icon = $A_{fail,mic}/A_{sec}$ 5 % - 10 %; large icon = $A_{fail,mic}/A_{sec} > 10$ %)

**Table 3.4:** Statistical analysis of raw data

| Data [-] | k [-] | SD [-] | $\Delta\sigma_{50\%}$ [MPa] |
|---|---|---|---|
| combined | 3.5 | 0.51 | 190 |
| 1.6220 - type O | 6.5 | 0.46 | 240 |
| 1.6220 - type M | 3.2 | 0.27 | 126 |
| 1.6760 - type O | 5.3 | 0.34 | 244 |
| 1.6760 - type M | 3.9 | 0.52 | 182 |

**Table 3.5:** Clusters of defect sizes

| $A_{fail,mic}$/ $A_{sec}$ [%] | k [-] | SD [-] | $\Delta\sigma_{50\%}$ [MPa] |
|---|---|---|---|
| combined | 3.5 | 0.51 | 190 |
| < 5 % | 4.1 | 0.43 | 249 |
| 5 % - 10 % | 2.3 | 0.42 | 152 |
| > 10 % | 2.9 | 0.37 | 130 |

The loading is indicated by the nominal stress ranges. The hypothesis that larger defects result in lower fatigue resistance could be confirmed. Nevertheless, the data still scattered considerably (SD = 0.37–0.43), and the gradients k of the SN curves deviated significantly from each other (k = 2.3–4.1). A further subdivision into clusters with narrower boundaries did not lead to more distinctive results. Thus, the defect size is not the only influencing factor.

The influence of the reduction in the load-carrying cross section and the resulting increasing stresses were eliminated by the evaluation of the net stress range according to Eq. (3.1). A consideration based on $\Delta\sigma_{mic}$ was useful with regard to real components, since the casting defects that occurred in the test specimens attenuated large parts of the cross sections despite small defect dimensions. This procedure was also used in [10] with a similar justification. The test data were evaluated in this manner, separated by material and presented in Figure 3.16. For both materials, two separate populations could been identified. These populations correlated with the defect location, specifically with internal and surface defects. These differences were not limited to the stress level, but also applied to the gradients k of the SN curves.

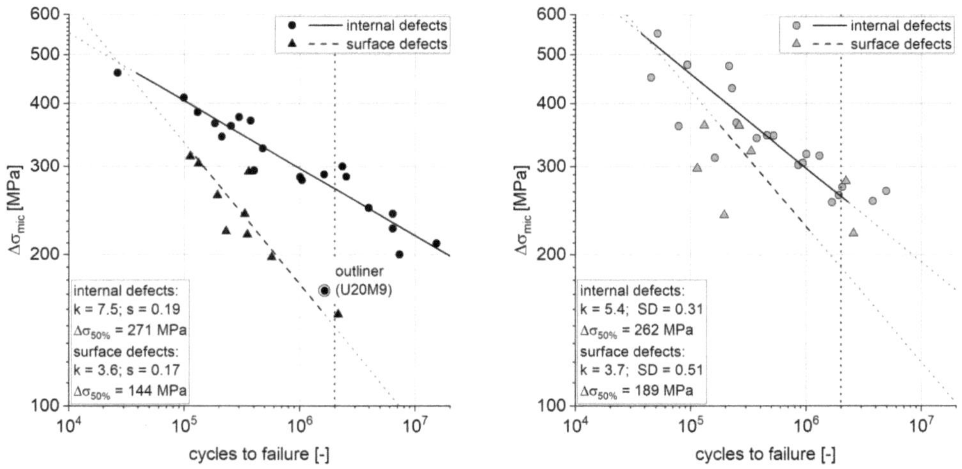


(a) (b)

**Figure 3.16:** Test results evaluated by $\Delta\sigma_{mic}$ for a) 1.6220 and b) 1.6760

This indicates either that different mechanisms dominate the damage process or that notch severities strongly deviate. The gradient k of the surface defects corresponds approximately to the Paris exponent m determined in Section 3.1.2. Given the explanations of [66], it becomes apparent that crack growth is the predominant failure case. A verification is carried out in Section 4.1. The failure of internal defects seems to be coupled with a crack initiation phase, causing the deviation in the gradients k. The scattering was reduced by both the consideration of $\Delta\sigma_{\mathrm{mic}}$ and the evaluation according to the defect location.

It should be pointed out that the "internal" and "surface" defect locations do not refer to the cast components according to Figure 3.4a, but to the machined specimens in Figure 3.4b. Thus, both defect positions have a comparable origin, and the differences in their lifetimes must result from mechanical effects. Nevertheless, the surface defects were often caused by larger defects, which had ventilated vertically due to the horizontal casting position. In some cases, these showed up through gas channels with a narrow cross section. In Figure 3.16a, specimen U20M9 is marked as an outlier. According to the defect location, it was assigned to the internal defect group in the failure plane. However, the pore pierced the surface of the specimen at a plane approximately 20 mm away in the longitudinal axis of the specimen (Annex C). In this specific case, the shape of the fracture surface was similar to those of surface defects. Despite the effects of assigning it to the internal defects being minor (k = 5.6; $\Delta\sigma_{50\%} = 252$ MPa and SD = 0.37), this outlier was not considered in the evaluation. In Figure 3.17a the results of the tests on both materials were evaluated collectively. This showed a material independence while the two populations (internal defects and surface defects) remained almost unaffected.

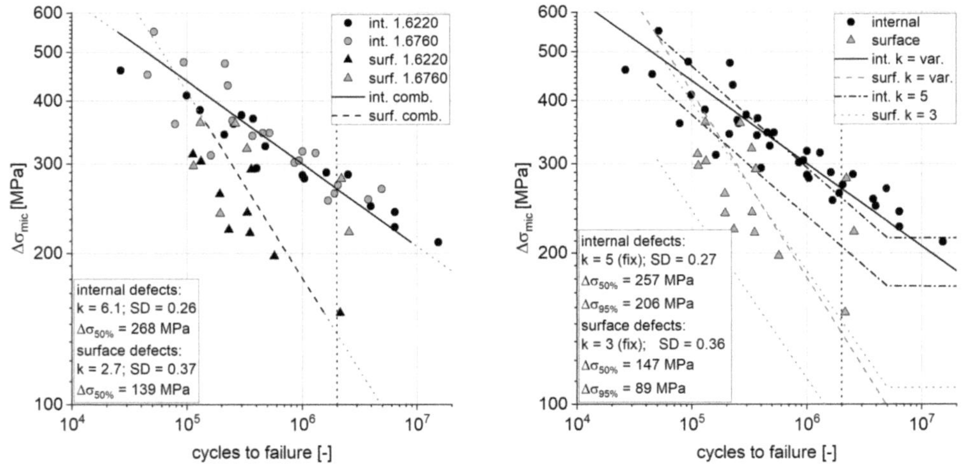


(a) (b)

**Figure 3.17:** a) Combined evaluation of the test results for both materials separated by defect location; b) Statistical evaluation of the test results according to EN 1993-1-9:2010 [156]

The combined evaluation led to a larger scatter compared to the evaluation according to material 1.6220. Compared to the evaluation of material 1.6760, the SD was reduced due to the larger number of samples. Figure 3.17b shows the results of the complete statistical evaluation according to EN 1993-1-9:2010 [156]. Following this concept, endurance strength was defined at $5 \cdot 10^6$ cycles to failure. In addition to the previously determined regression curves with free gradients, the evaluation was carried out with a fixed gradient k. For the evaluations with fixed gradients, the 95 % survival probability is indicated in the diagram as the characteristic resistance. The gradients k were chosen with the following justifications: $k \approx 3$ for surface defects, as crack growth is the failure mechanism that causes failure according to Section 3.1.2, which applies $k = m$. In addition, this corresponds to the gradient of most notch details in EN 1993-1-9:2010 [156] and facilitates its application in a design concept. Internal defects were evaluated using $k = 5.0$, which is in line with EN 1993-1-9:2010 [156] and represents the behavior more realistic than $k = 3.0$ would do.

The significant influence of surface defects is illustrated in Figure 3.18 using the two representative specimens U20M4 and U22M5. In both cases, failure was caused by small surface defects despite larger internal cavities. The CT records are shown in the projection, the section of the maximum cross-sectional weakening and the section in the failure plane. The figure is supplemented by microscopic images of the fracture surfaces. The stress ranges of the test results including the data points marked by "failure" are given as $\Delta\sigma_{\text{mic}}$. Whereas the nominal stress range in the plane with the maximum cross-sectional weakening determined by CT ist marked as "theoretic" and calculated by Eq. (3.2). In this way, a theoretically tolerable number of cycles was determined from the regression curve of the internal defects and is marked in the diagram.

$$\Delta\sigma_{\text{Amax,CT}} = \frac{\Delta \text{F}}{\text{A}_{\text{sec}} - \text{A}_{\text{max,CT}}}. \tag{3.2}$$

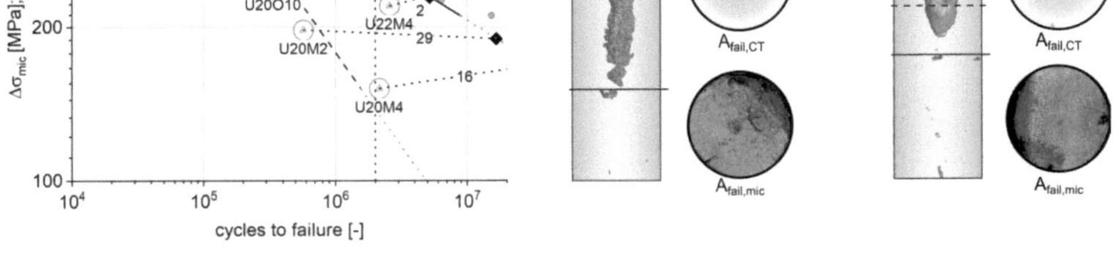


**Figure 3.18:** Dominating effect of surface defects on service life despite larger internal defects

The same procedure was used for further specimens with comparable geometric properties. The dotted lines connected real failure states with states that were theoretically determined for the internal defects. The increasing factor of the number of load cycles was related to the actual failure state and is shown in the diagram. Despite the partly considerably larger cross-sectional weakening of the internal defects, the theoretically determined lifetimes of these internal defects were up to 29 times higher than the cycle numbers that actually occurred. This ratio depended on the load level. The fact that lower stresses occurred in some cases with internal defects resulted from the described differences between $A_{mic}$ and $A_{CT}$. Endurance limit was not considered, when calculating the theoretically tolerable cycles to failure for the internal defects.

In Figure 3.19a, the test results are presented normalized to the tensile strength $R_m$ and seperated according to material and defect location. $R_m$ corresponds to the mean value of the experimental investigations presented in Table 3.1. The gradients k of the regression curves correspond to the evaluations of the individual materials according to Figure 3.16. For a better overview, the comparrison is limited to $P_{50\%}$, limited to the mean values for $2 \cdot 10^6$ cycles to failure. It became apparent that the higher-strength material 1.6760 could only activate a significantly lower proportion of its nominal tensile strength under fatigue loading. The exception was the range of less than $1 \cdot 10^5$ cycles to failure, as the yield strength of 1.6220 was exceeded in this case and specimen of 1.6760 led to higher lifetimes. Additionally, the characteristic resistance according to FKM-Rili:2012 [162] of the flawless base material is added to the chart. In the case, $R_m$ corresponds to the minimum requirements according to the decisive technical delivery conditions. All tests of material 1.6220 led to higher resistance. For material 1.6760, the mean value of the surface defects was below this resistance. "Coupling" the fatigue resistance with the tensile strength is not considered effective if defects are present.

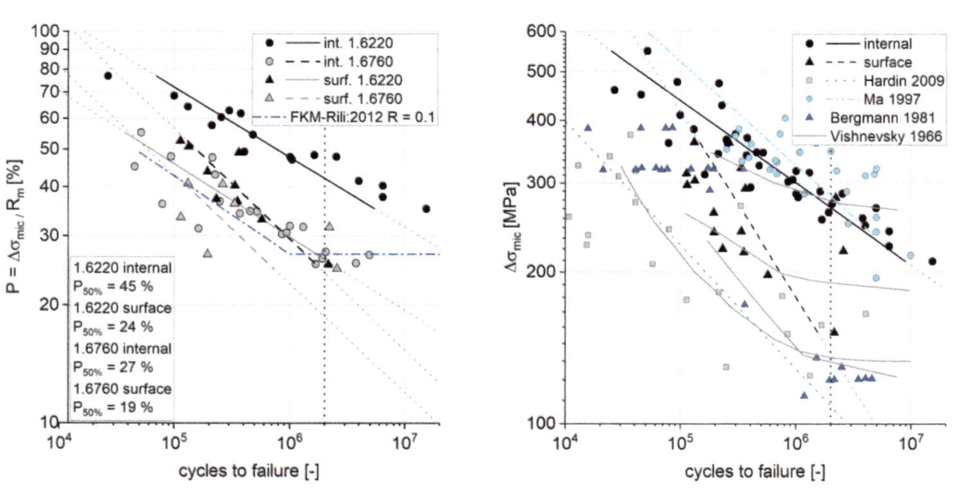


(a) (b)

**Figure 3.19:** a) Test results related to tensile strength; b) Test results compared with literature data

Figure 3.19b classifies the test results in the context of the experimental investigations from literature presented in Section 2.3. The comparison was based on $\Delta\sigma_{\mathrm{mic}}$, corresponds to the representations of published values for faulty components and follows the methodology of the present discussion. If the specimens from the literature showed geometrical notches, the comparison was based on the maximum stress range instead of nominal or net stress ranges to consider stress gradients. This resulted in higher lifetimes due to the supporting effect (see Section 2.2.3). The literature data covered the entire range of results obtained in these experimental investigations. Some of the literature data showed shorter lifetimes. The results of the specimens with internal defects were of the same order of magnitude as the flawless base material in the literature data. The gradients of the regression curves of the internal defects were approximately the same. The difference between the internal and surface defects corresponds to the order of magnitude and ratios of [15], indicated in the diagram as (Vishnevsky 1966). The results of the present investigations are classified as representative to slightly above average. The material independence of fatigue strength determined here is underlined by the comparison, and the necessity of a separate evaluation of internal and surface defects becomes clear.

#### 3.2.2.6 Effects of Defect Size

According to Eq. (2.5), $\Delta\mathrm{K}$ directly depends on the defect size in the order of $\sqrt{\mathrm{a}}$. As an infinite plane is considered in Eq. (2.5), effects caused by reduced cross sections are not decisive. In crack propagations according to Eq. (2.10), a change in crack size a, and thus in $\Delta\mathrm{K}$, has a disproportionate effect on the tolerable number of cycles due to the exponent m. These effects can be quantified according to Eq. (2.7) on three-dimensional problems via the root of the crack surface $\sqrt{\mathrm{area}}$. Using simplified approaches, the effects of these correlations on the experimental results were evaluated. On this basis, the scatter could be explained. In the first step, the investigations were limited to surface defects. Finally, an analogous evaluation of the internal defects was carried out.

Figure 3.20a shows that the detected size and shape of the surface defects varied significantly regardless of the material. If multiple surface defects or both internal and surface defects were present, only the largest surface defect ($\mathrm{A}_{\mathrm{fail,mic,surf}}$) was considered in the evaluation. Minimum, maximum and mean values are marked in the diagram. The situation is illustrated by two representative fracture surfaces of a large and a small defect. Fracture mechanics-based SN curves were determined for an elliptical surface crack with $\mathrm{a/c} = 1$ in a cylinder according to Annex B.1.3. The selected defect sizes were chosen in a way that the minimum and maximum defect sizes $\mathrm{A}_{\mathrm{fail,mic,surf}}$ were mapped separately by material. The lifetime was determined at three load levels with ten iterations each. The integration was carried out according to approach 3 from Figure 2.4. The applied loads were considered by $\Delta\sigma_{\mathrm{mic}}$. This eliminated the effects of a reduced cross section due to a larger initial defect, and the effects of the defect magnitude on the stress intensity factors were considered individually. The crack propagation parameters were set according to Table 3.2 no. (1) and no. (3).

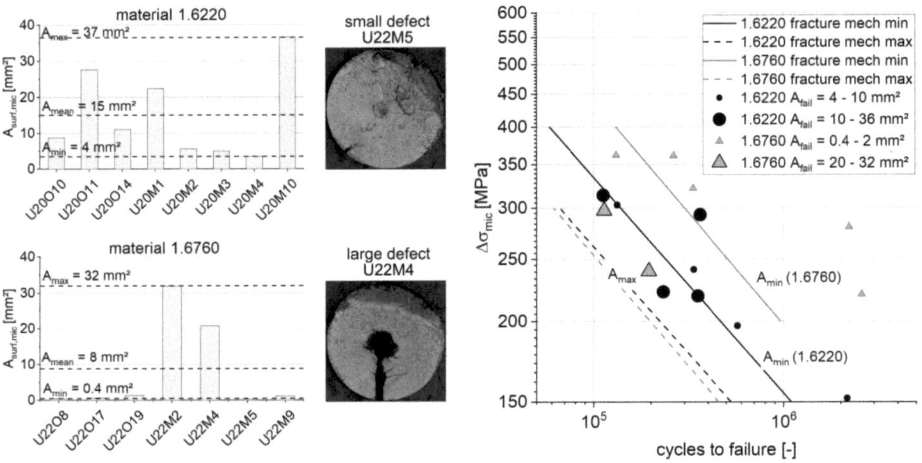


(a) (b)

**Figure 3.20:** a) Sizes of surface defects and representative fracture surfaces for small and large surface defects; b) Comparison of fracture mechanics-based SN curves with test results for surface defects depending on defect size

Figure 3.20b shows the results and compares them with the test data. The specimens are presented as two clusters of smaller and larger defects, respectively, and are considered separately by material. The results of the crack propagation analysis as well as the test data showed a dependence of crack size on service life. The gradients and magnitudes of the calculations showed good agreement with the test data. This indicates that crack growth is the dominant damage mechanism for surface defects under HCF. The deviations between the experiments and calculations are attributed to the simplified model of a semicircular crack. In the real specimens, the major part of the defect was placed in the center of the specimens and not eccentrically on the surface (see Figure 3.20b). For material 1.6220, deviations in the calculated service life showed a factor of two between the small and large defects. These differences increased for material 1.6760 to a factor of five due to the much smaller defect sizes. Again, it is pointed out that the effects of a reduced stress cross section were eliminated and would amplify these differences. The $\sqrt{area}$ law according to [66] cannot be applied, as its validity is limited to surface defects where $\sqrt{area} < 1000\,\mu m$.

For the investigation of the internal defects, a circular concentric crack in a cylinder was considered according to Annex B.1.4. The explanations are limited to material 1.6220, as the investigations on 1.6760 led to similar results due to the comparable defect sizes and crack propagation parameters. Figure 3.21a shows the sizes of the internal defects. Fracture mechanics-based SN curves were calculated for three defect sizes, as given in Figure 3.21b. The experimental data are also presented in three size clusters, and the regression curve is added. The effects of the different defect sizes, which can be seen in the presentation of the test data, were confirmed by this calculation. Nevertheless, the results of the crack propagation and test data differed significantly in both the gradient and order of magnitude. Thus, in the case of internal defects, a dominant share of the total service life must be attributed to crack initiation in addition to the considered crack growth. It is assumed that the extent of the crack initiation, in the HCF range, depends on the load level.

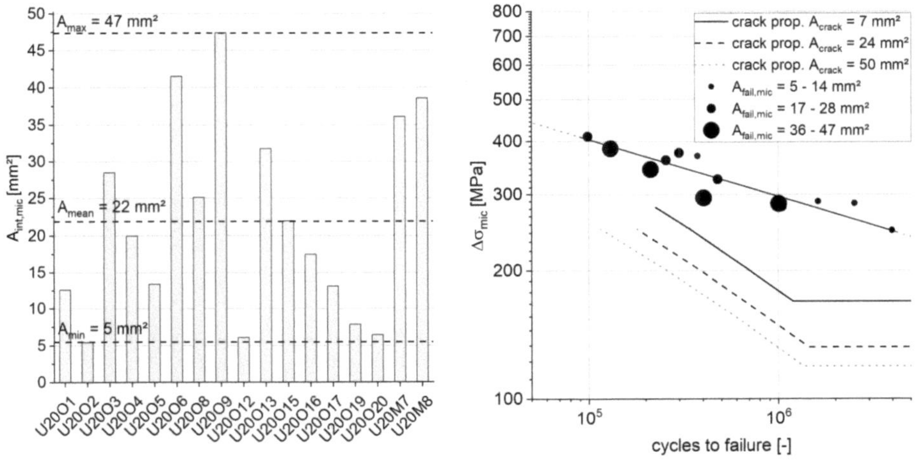


(a) (b)

**Figure 3.21:** a) Sizes of internal defects; b) Comparison of fracture mechanics-based SN curves with the test results

### 3.2.2.7 Effects of Defect Detection Methods

Differences in the defect sizes between CT and microscopy images increase by up to 12 %, as shown in Figure 3.9. The influence of these deviations on stress range and fatigue strength was evaluated. This investigation was limited to failure cases I and II due to the necessity for available CT data of the defect in the failure plane. Only internal defects were considered. As shown in Figure 3.22a, the stress ranges $\Delta\sigma_{\mathrm{mic}}$ resulting from the fracture surface analysis were on average 5 % higher than

$$\Delta\sigma_{\mathrm{Afail,CT}} = \frac{\Delta \mathrm{F}}{\mathrm{A_{sec}} - \mathrm{A_{fail,CT}}}. \tag{3.3}$$

An analysis of the load cycle shown in Figure 3.22b indicated that the variance of 5 % on $\Delta\sigma$ causes a deviation in lifetime of approximately 23 %. This consisted only of the failure cases I and II specimens and was not sufficiently statistically verified. More important was the fact that statements can only be made on a phenomenological level and not a mechanical basis for the examined geometric conditions. Thus, all further evaluations were based on the defect area measured by microscopy.

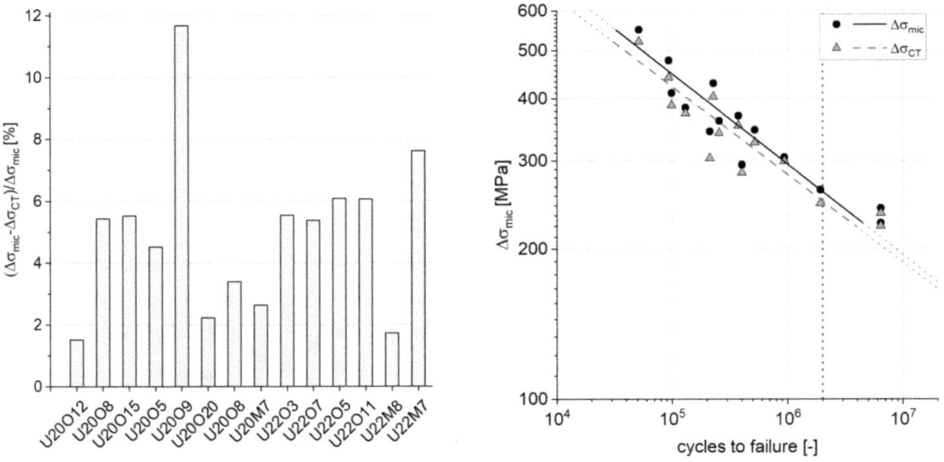


(a) (b)

**Figure 3.22:** a) Deviations between the stress ranges determined by CT data and microscope images; b) Evaluation of the service life according to CT data and microscope images

In later recommendations for the definition of a grading, the projected area of the defect is conservatively considered. When linking defect sizes to NDT displays, this gains special importance but is not within the scope of this work.

### 3.2.2.8 Residual Fracture

The purpose of these investigations was to describe residual fracture behavior. In particular, for applications at low temperatures (e.g., bridges) this has a significant influence on the overall safety level and robustness of a structure. A failure behavior that is indicated by large deformations is preferred to a sudden failure. In Figure 3.23,

$$\text{Net section max stress at failure} = \sigma_{\text{res}} = \frac{\text{F}_{\text{o}}}{\text{A}_{\text{res}}} \tag{3.4}$$

is related to the material's yield and tensile strength mean values according to investigations pressented in Table 3.1. The results are presented separately by steel grade and are restricted to specimens with internal defects. This restriction resulted from the explanations of the fracture surfaces, since it was not possible to make a clear distinction between residual fractures and stable crack growth in surface defects.

Figure 3.23a shows that $\sigma_{\text{res}}$ is 10 % to 90 % higher than the material's mean yield $\text{R}_{\text{eH}}$ strength for all specimens of material 1.6220. Therefore, the failure mechanism was ductile and strongly dominated by inelastic effects. In many specimens, $\sigma_{\text{res}}$ even exceeded the mean tensile strength $\text{R}_{\text{m}}$. In those specimens where this was not the case, the growing crack reached the surface and additional bending occurred that was not captured by this simple approach. Due to the present stress state, $\sigma_{\text{res}}$ could exceed the material uniaxial tensile strength and [33] confirmed that the maximum applicable load of tensile specimens with notches was higher compared to standard tensile tests of the same cross sections. In tensile specimen sharper notches cause higher fracture loads at lower elongations. The influence on the fracture strain is assigned to stress triaxiality ($\text{T}_\sigma = \sigma_\text{H}/\sigma_{\text{eq}}$) and was quantified in [84]. Newer research as [7] identified a discontinuity in this relation at $\text{T}_\sigma \approx 0.4$ and further influencing factors that are neglected in this context. A numerical calculation on a circular internal crack indicated $\text{T}_\sigma \approx 0.95$ at the crack tip. According to [7, 84], this value is associated with low elongation at the fracture and thus higher loads.

For the specimens of material 1.6760, $\sigma_{\text{res}}$ was between 60 % and 80 % of the yield strength, except for three outliers, and did not exceed the tensile strength in any case. It is assumed that the failure mechanism is dominated by crack tip stresses. This finding coincides with the results of the fracture surface analyses. In the specimens of material 1.6220, necking occurred and pronounced shear lips were observed (cf. Figure 3.14). These effects were considerably less pronounced or not recognizable for specimens of material 1.6760. For a more detailed analysis, a circular crack of area $\text{A}_{\text{crack}}$ was assumed and placed centrally in a cylinder with the same diameter as the considered specimen. The appearing stress intensity factors were determined according to Annex B.1.4. The analysis of the failure state was carried out by the FAD as shown in Figure 3.24a. The fracture toughness $\text{K}_{\text{Ic}}$ was selected according to Table 3.3, with $\text{K}_{\text{Ic}} = 7747\,\text{N/mm}^{-3/2}$ for 1.6220 and $\text{K}_{\text{Ic}} = 4206\,\text{N/mm}^{-3/2}$ for 1.6760. All effects resulting from an eccentric defect position, asymmetrical crack growth and deviations from the circular crack shape were not taken into account in these simplified considerations.

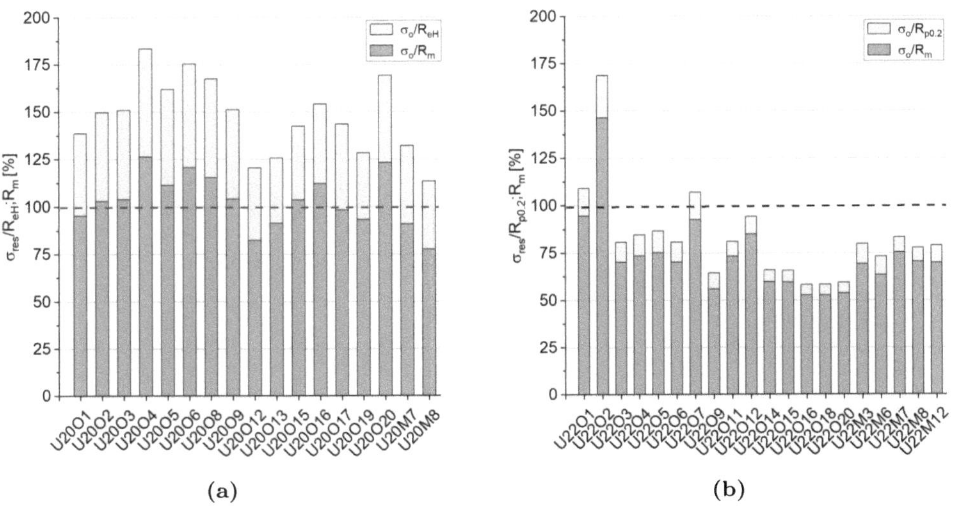


**Figure 3.23:** Stresses in fracture states determined in net sections and related to yield and tensile strength for material a) 1.6220 and b) 1.6760

Figure 3.24a illustrates the assumed differences between the fracture states of both materials. The conditions of the 1.6220 specimens were clearly attributed to the plastic limit state, as they exceeded the right border of the limit curve. States outside the limit curve can occur due to local multiaxial stress states. The determined failure states of the 1.6760 specimen were within the no-failure area of the FAD. Thus, plastic failure was ruled out. One explanation for this is the influencing factors that reduce fracture toughness $K_{Ic}$. As described in [23], a three-dimensional stress state causes a reduction in $K_{Ic}$. This occurred inside thick components or at the crack tips of the internal cracks considered within this context. The qualitative dependency is shown in Figure 3.24b. Although the thickness effect is eliminated by the specimen shape for the determination of $K_{Ic}$, the resulting stress state is fundamentally different from that present at the crack tip under consideration. In addition, a significant influence on the fracture resistance of $K_{Ic}$ is attributed to the temperature and loading speed in [23]. Since the tests were carried out at about 20 °C, temperature effects are negligible at this point. From the tests in [94], differences between static and dynamic $K_{Ic}$ values at a factor in the range of factor 2.0 (S355) to 2.3 (high-strength steel) were derived. According to [23], the dynamic minimum value occurs for steel at about $dK_I/dt > 10^5$ $\mathrm{Nmm}^{-3/2}/\mathrm{s}$. Based on the previous simplifications of a circular crack lying centrically in a shaft, $dK_I$ was calculated in the range of 800 to 1985 $\mathrm{Nmm}^{-3/2}$/cycle. At a testing frequency of $f = 70$ Hz, $dK_I/dt$ was in the range of $0.6 \cdot 10^5$ to $1.4 \cdot 10^5$ $\mathrm{Nmm}^{-3/2}/\mathrm{s}$. The crack resistance present in the specimen must approach its minimum value. The effects of the test frequency on the experimentally determined number of load cycles are considered small, since crack growth in this range increases strongly and over proportionally and very few cycles occur. If stress state and loading speed are taken into account with a reduction factor of 2.8 on $K_{Ic}$, this results in the modified states shown in Figure 3.24b. A clear distinction can be made between the ductile and brittle failure states of the two materials. This corresponds to the results for monotonic loading from [103].

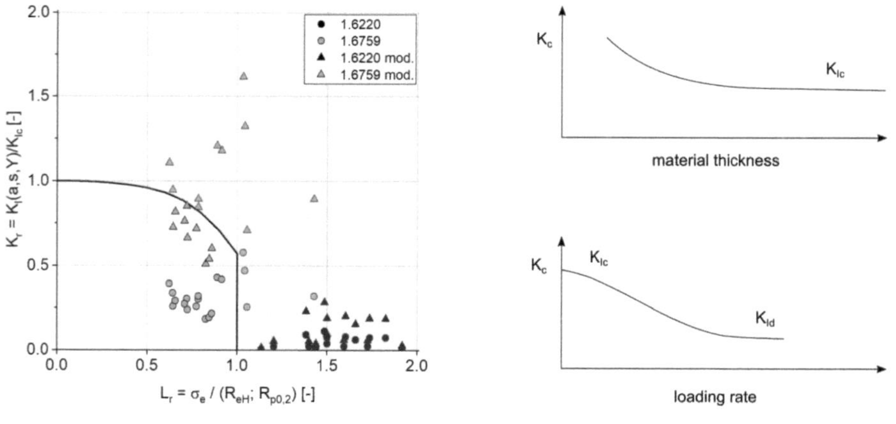


(a) (b)

**Figure 3.24:** a) FAD for tensile specimens in fracture state; b) Influencing factors of $K_{Ic}$ see [23]

### 3.2.3 Summary

These experimental investigations were carried out on tensile specimens from the materials G20Mn5 (1.6220) and G22NiMoCr5-6 (1.6760) with solidification shrinkage cavities. The investigations aimed at a description of casting defect characteristics and the quantification of fatigue resistance dependent on the defect properties. Finally, conclusions were drawn on the relevant damage processes as well as on the residual fracture behavior. The core statements are summarized by:

*Geometric Characteristics of the Casting Defects:*

- The properties of the casting defects were determined based on computed tomography records made by Fraunhofer IZFP as well as by microscopy.
- Depending on the specimen type, the specimens show defects with maximum cross-sectional weakening for "Type O" specimen of 1 % - 10 % and for "Type M" specimen of 6 % - 20 %.
- The shape of the defects varied, depending on the specimen type and larger defects were surrounded by a dissolved microstructure of different intensity.

*Fracture Surfaces:*

- In all cases, fatigue crack growth followed crack-opening mode I.
- In the failure planes, it was possible to detect purely internal defects as well as defects that penetrated the surface of the test section. In the case of internal defects, the areas of stable crack growth could be clearly identified in the form of an elliptical crack front. In the case of surface defects, the region of stable crack growth was not so clearly visible and differed from that of pure internal defects.
- A significant influence of the casting defect shape on the fracture surface was found. There were clear areas of stable crack growth dependent on the stress level. Here, the branched casting defects grew progressively into an elliptical crack front. In the case of strongly branched defects with a pronounced dissolved microstructure, a ragged fracture surface with a 3D topology occurred; in addition, multiple parallel crack planes were found in individual cases.

*Influence of Casting Defects on Fatigue Resistance:*

- The location of the defect (internal or surface defect) had a decisive influence on the fatigue resistance. On average, the resistance $\Delta\sigma_{50\%}$ at the reference value $2 \cdot 10^6$ load cycles of the internal defects $\Delta\sigma_{50\%,\text{internal}} = 268$ MPa (with resp. to the net section) was 1.9 times higher than that of the surface defects $\Delta\sigma_{50\%,\text{surface}} = 139$ MPa. Distinct differences in the gradients of the regression lines were present. For internal defects $k \approx 6$ whereas for surface defects it resulted in $k \approx 3$ and let to the assumption of different dominating damage mechanisms.

- No significant deviation of the fatigue resistance was found for the different materials tested. In these cases, a correlation to tensile strength was found to be inappropriate in the presence of defects.
- Simplified analytical fracture mechanics investigations underlined the assumption that a crack initiation phase occurs in the case of internal defects. The proportion of crack initiation in the total service life correlated with the load level. For surface defects, crack growth appeared to be the dominant damage mechanism.
- The fracture mechanics investigations showed a correlation between the scatter of the test results (net stress range) and the defect size for the surface defects.
- In the case of residual fracture behavior, clear differences were found between the materials. The residual fracture of material G20Mn5 was ductile, whereas the failure states of the G22NiMoCr5-6 specimens were more brittle.

## 3.3 Component Tests with Artificial Defects

The three main objectives of the component tests were to:

- gain knowledge on the influence of the technological thickness effect on fatigue strength using an integral consideration
- quantify the influences on fatigue strength caused by the interaction of defects and stress gradients
- create a data-driven basis for the validation of the design concept in Chapter 5.

With these intentions, components of different wall thicknesses and representative geometric notches were tested under cyclic tension. The influence of the interaction of geometrically caused stress concentrations and casting defects was examined by adding representative artificial defects to the roots of the notches. Within this chapter, only the experimental investigations and directly derived results are discussed. An extended numerical evaluation is presented in Section 4.2. This presentation follows the same structure as the description of the tensile fatigue specimen in Section 3.2, meaning that the methodology and results are discussed separately.

### 3.3.1 Methodology

#### 3.3.1.1 Specimens

Figure 3.25 gives an overview of the investigated specimens. Three wall thicknesses were chosen, and the specimens were cast by three different foundries. These investigations were limited to material 1.6220. The chemical composition and mechanical properties according to inspection certificates 3.1 are summarized in Annex A.

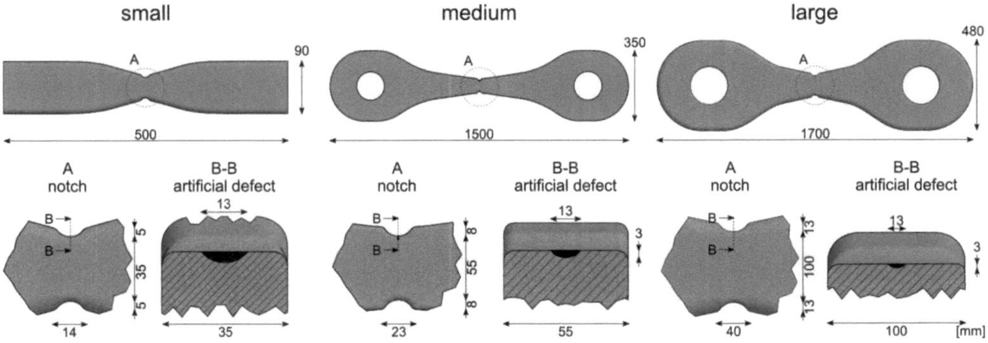


**Figure 3.25:** Geometry of the component test specimens

The specimens were identified using the following concept:

G-(size)-(foundry)-(n)

| | |
|---|---|
| G | identifier for component tests |
| (size) | K = small; M = medium; G = large |
| (foundry) | A, B, C according to Annex A |
| (n) | consecutive number |

*Geometric Notch:*
The notch parameters were chosen based on Finite Element Analysis (FEA) in a way that the resulting SCF was approximately 2.0. This definition was based on different published examples of executed cast components and further engineering requirements for an adequate fatigue design. In [112], SCFs are presented in a range from 1.2 to 4.7, and 2.7 is not exceeded in the majority of cases. These examples resulted from nodes typical in offshore structures in the 1950s. Using state-of-the-art Computer-Aided Design (CAD) and FEM-supported design, SCFs lower than 2.0 shall be aimed at in fatigue-loaded components nowadays. [35] came to a similar conclusion that the SCFs of optimized cast steel nodes are almost exclusively lower than 2.0 and supported this statement with published examples.

The geometric proportion of the notches as well as their relation to the wall thickness were kept constant for all specimen sizes, as shown in Figure 3.25. Doing so, identical stress distributions, gradients and thus supporting effects were achieved independent of component size. This condition is illustrated in Figure 3.26a. It shows identical axial stresses in the center of the specimens that were related to $\Delta\sigma_{\text{nom}}$ along an evaluation path connecting both notches. Additionally, the stress distribution on the component surface is shown. Compared to the center, the stresses as well as the magnification were lower due to the multiaxial stress state.

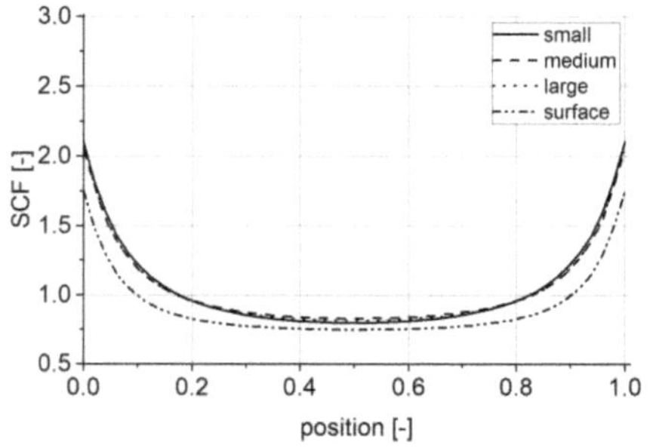


(a)

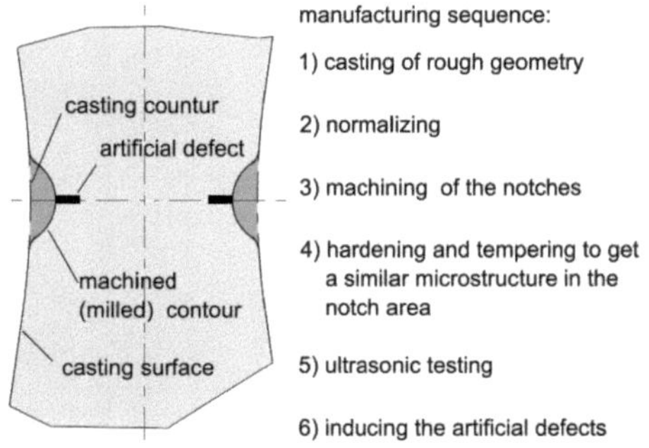


(b)

**Figure 3.26:** a) Comparison of the normal stress distributions depending on the component type; b) Manufacturing sequence of the geometric notch

Since the geometric notch is representative of cast components in terms of its effect but not its actual dimensions, it could not be cast directly. This is due to heating of the molding material in the corner areas and considerable deviations from the nominal geometry in the notch area. In order to avoid this but at the same time achieve the typical surface condition of a cast component in the notch area, the production sequence was adjusted. The raw casting contour (see Figure 3.26b) coarsely reproduced the geometric shape. After normalization, mechanical processing was carried out in which the final notch shape was milled. The subsequent hardening and tempering created the required conditions in the notch area. Subsequently, quality class 1 according to EN 12680-1:2003 [144] was verified in the notch area. The design tolerance class DCTG 12 according to ISO 8062-3:2008 [157] was defined. Finally, the artificial defects were symmetrically induced as semi-elliptical surface defects in both notches of each specimen.

*Artificial Defects:*
The defects were induced by die-sinking erosion perpendicular to the principal stress. The size and shape of the artificial defect was derived to represent the largest and most unfavorable crack that can be overlooked by NDT. Caused by thickness of electrode used in the die-sinking erosion process, the defects had a lateral extension of about 1.1 mm.

According to FKM-Rili-BM:2009 [161], near-surface defects must be reproduced conservatively by surface defects due to an interaction with the free surface and thus increased crack tip stresses. This justified a translation of the UT requirements for the rim zone to artificial surface defects (definition of core and rim zone acc. to [144]). According to EN 12680-1:2003 [144] (UT), the maximum permissible diameter of an equivalent Flat-Bottom Hole (FBH) is limited to 3 mm both in the core and in the rim zone for quality grade 1. FKM-Rili-BM:2009 [161] provides an approximation method for converting the equivalent FBH to a worst-case crack. The dimensions of this semi-elliptical surface crack are in the ratio $a/c = 0.4$, where the crack depth a is determined according to Eq. (3.5). This resulted in a crack size of $a = 2.7$ mm and $c = 6.7$ mm.

$$a = \frac{2FBH}{\sqrt{5}} \tag{3.5}$$

According to EN 1369:2013 [148] (MT), the dimension of the smallest MT-detectable defect depends on the surface condition. In this case, the comparative surface pattern according to BNIF or SCRATA is weighed. According to EN 1370:2012 [149], the sand molding process is assigned to SCRATA class A2/A3 for medium and small castings. Thus, the dimension of the smallest recognizable indication is 2 mm. If the component is ground, a surface condition of 1S2 (BNIF) and thus a detection limit of 0.3 mm could be assumed according to EN 1370:2012 [149]. In summary, the crack defined by UT became decisive.

#### 3.3.1.2 Measuring Concept

In addition to the recording of common parameters, such as nominal stress range respectively load range and cycles to failure, all specimens were equipped with additional measuring techniques. Both conventional strain gauges and Electromagnetic Acoustic Transducers (EMATs) were used to get additional information on the actual stress state and the crack initiation and propagation behavior.

*Strain Gauges:*
Two uniaxial strain gauges and a rosette strain gauge were arranged as shown in Figure 3.27a on both sides of each specimen. This enabled the stress state shown and aimed for in Figure 3.26a to be checked and enabled any deviations to be quantified. Besides, additional bending moments resulting from geometric deviations or eccentric clamping in the testing machine could be mapped. Due to the high stress gradients in the notch area, the evaluation was carried out as an average value over the respective measurement width. Recording over the entire test duration made it possible to map not only the initial stress situation but also the crack propagation, load redistributions and the description of which initial defect dominates. The definition of the relevant geometric parameters and coordinate specifications are listed in the introductory remarks of Annex D. The types and exact positions of the strain gauges as well as the development of the strain ranges over the test period are documented in Annex D for each specimen separately.

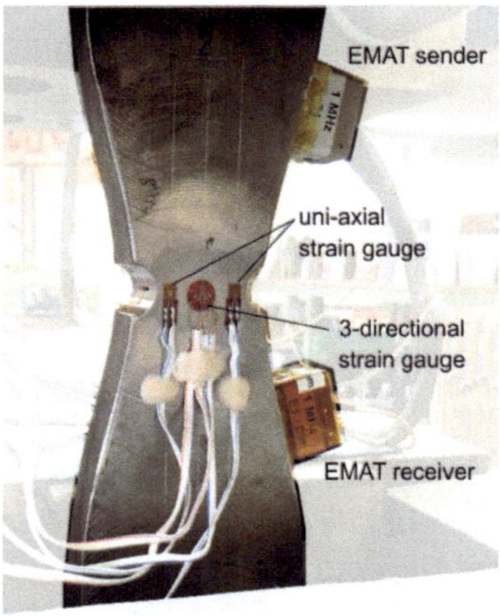


(a)

(b)

**Figure 3.27:** a) Arrangement of strain gauges and EMATs; b) Identification of representative regions on the fracture surface

*Electromagnetic Acoustic Transducers (EMATs):*
The aim was to track the ratio of the crack initiation phase to the total service life and the crack propagation velocities. For this purpose, a measurement method developed by Fraunhofer IZFP based on the propagation time measurement of electromagnetically induced surface waves [68, 108] was applied. Fraunhofer IZFP carried out this conception and evaluation in the context of ErStaGu [116]. The principal idea is based on the time of flight of surface waves. These are excited by the transmitter by an electromagnetic field in the component and have a direction. The time of flight and the intensity of the waves until they reach the receiver are recorded. If a crack propagates in the observed area, the propagation path is extended, and a change in the propagation time and amplitude can be measured. The measuring rate was chosen in accordance with the test frequency so that approximately 5 to 10 measurements were made per load cycle. In this specific case, the transmitter and receiver were arranged around the notch area as shown in Figure 3.27a. In select cases, two units were used, and thus the behavior of both notch areas resp. both artificial defects was observed. As these data brought only limited benefit to the overall objective, the results are not considered within this context.

#### 3.3.1.3 Fatigue Testing

The force-controlled fatigue tests were carried out with a stress ratio of R = 0.1 under cyclic tension. Depending on the specimen size, calibrated servo-hydraulic testing machines at the KIT Research Center for Steel, Timber and Masonry of different load ranges were used. These are summarized in Table 3.6, including the corresponding test frequencies. The considered nominal stresses $\Delta\sigma_{nom}$ relate the applied force range to the individually measured cross sections in the notch areas; the defect area was deduced. In specimen GKC3, beach marks were induced according to the procedure described in Section 3.2.1.3.

**Table 3.6:** Details of the component tests

| specimen [-] | thickness [mm] | quantity [-] | max load of testing machine [kN] | testing frequency [Hz] |
|---|---|---|---|---|
| small | 35 | 8 | 1000 | 3.7 - 5.0 |
| medium | 55 | 5 | 3000 | 0.2 - 1.0 |
| large | 100 | 6 | 6300 | 2.0 - 2.5 |

#### 3.3.1.4 Fracture Surface Analysis

A fracture surface analysis was carried out for all specimens and is documented in Annex D. Figure 3.27b shows the key areas that were identified. Special focus was put on the quantification of the residual fracture as well as the non-uniformity of the crack growth.

## 3.3.2 Results

### 3.3.2.1 Fatigue Strength

Figure 3.28 shows the nominal stress range $\Delta\sigma_{nom}$ separately for each component size. As major differences were observed between the specimens of foundry A and those of foundries B and C, these were evaluated separately. The mean value $\Delta\sigma_{50\%}$ was calculated with no restrictions to the gradient k according to EN 1993-1-9:2010 [156] at $2 \cdot 10^6$ cycles and is presented along with k in the diagrams.

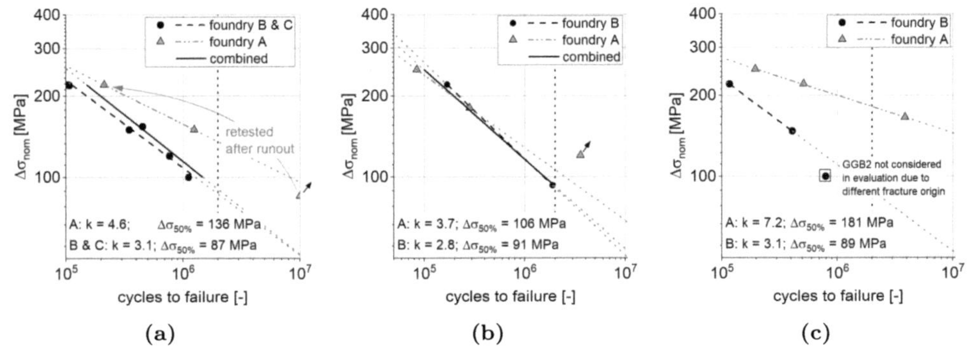


**Figure 3.28:** Tests results for a) small, b) medium and c) large specimens

Both populations showed considerable differences in service life independent of component size. This was particularly pronounced at low load levels and resulted in large deviations in the gradients of the regression curves. The comparison of a population over the different component sizes showed an almost identical load level. $\Delta\sigma_{50\%}$ of foundries B and C spread between 87 MPa and 91 MPa, while for foundry A, this dispersion was greater. The reason for this scatter in the population itself is not considered a primarily physical phenomenon but is rather due to the small sample size and missing data points due to outliers in the low load level. The gradients of the regression curves behaved analogously. In the specimens from foundries B and C, this was in the range $k = 2.8$ to 3.1 and thus corresponds very well with the Paris exponent $m = 3.0$ determined in Section 3.1.2. The connection established in Section 3.2.2 with the effect of surface defects as crack-like defects is underlined.

Residual stresses resulting from a missing or ineffective stress relief heat treatment in foundry A were assumed to be the reason for the differences between the populations. Their determination and evaluation according to ASTM E837:2013 [134] is documented in Annex F. The measurements carried out on large components showed major differences. Significant compressive residual stresses in the relevant areas were identified for the specimens from foundry A. In addition, differences in the manufacturing processes and thus different metallurgical compositions in non-regulated elements could increase the effects of residual stresses.

Due to the limited number of specimens, component GKA1 was tested after reaching $1 \cdot 10^7$ load cycles without any detectable crack ($\Delta\sigma_{nom} = 85$ MPa) at an increased load level ($\Delta\sigma_{nom} = 220$ MPa) until failure. In the case of the large specimens, GGB2 had to be excluded from the evaluation based on $\Delta\sigma_{nom}$, since the origin of failure was a grinding groove on an edge, which as shown in Section 4.2.2 showed additional stresses due to large deviations from straightness.

Figure 3.29a shows all component tests collectively. Based on EN 1993-1-9:2010 [156] and the correlation with the Paris exponent, the statistical evaluation of the regression curves for the specimens from foundries B and C was carried out with a predefined gradient k = 3.0. Despite the same defect size, all curves are almost identical for different component cross sections. Even the collective evaluation shown in Figure 3.29b did not increase the scatter so significantly that the physical processes taking place must be defined. Figure 3.29b also shows a comparison with the results of the tensile specimens with real casting defects from Section 3.2. The influence of the notch effect became apparent in this context. Compared to the tensile specimens with surface defects, the results of the component tests showed an offset to lower strength with a comparable overall behavior. The effects of the SCF on fatigue behavior were determined after the evaluation of the strain gauges and the numerical determination of the actual stress states in Section 4.2.

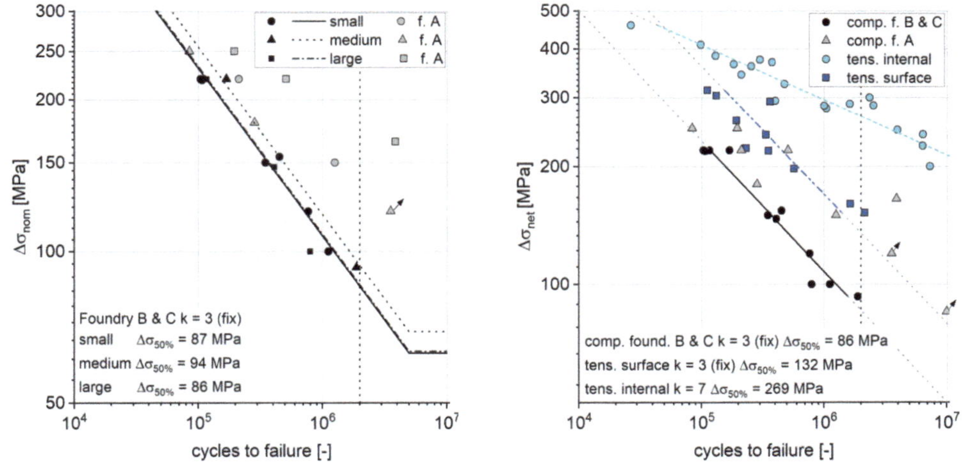


(a) (b)

**Figure 3.29:** a) Common presentation of all component tests and evaluations with k = 3.0 and a predefined $N_{endu} = 5 \cdot 10^6$ cycles according [156]; b) Comparison of the test results with the tensile specimens from Section 3.2

#### 3.3.2.2 Beach Marks

Figure 3.30a shows the fracture surface of the beach mark test GKC3 with $\Delta\sigma_{\mathrm{nom}} =$ 150 MPa. Up to the first marking, the specimen was loaded with 100,000 cycles. Between all other lines, 50,000 cycles were applied. The dominance of the crack in the lower part of the picture is clearly visible at the fracture surface. This resulted from geometrical deviations or the clamping in the testing machine. It was examined more closely, as shown in the microscope image in Figure 3.30b. All markers were clearly visible and were measured for further evaluation. Due to the lateral extension of the artificial defect, there were two competing stress hotspots per defect at each of the two interfaces in the specimen. In the present example, the crack started from the left side of the upper edge and from the right side of the lower edge and only united later to form a common crack front. This could be recognized by the diagonally running fragment on the fracture surface and explained by the crack front, which was not perfectly elliptical. In addition, it could be seen from the course of the residual fracture surface that an additional moment led to higher stresses on the left side of the specimen.

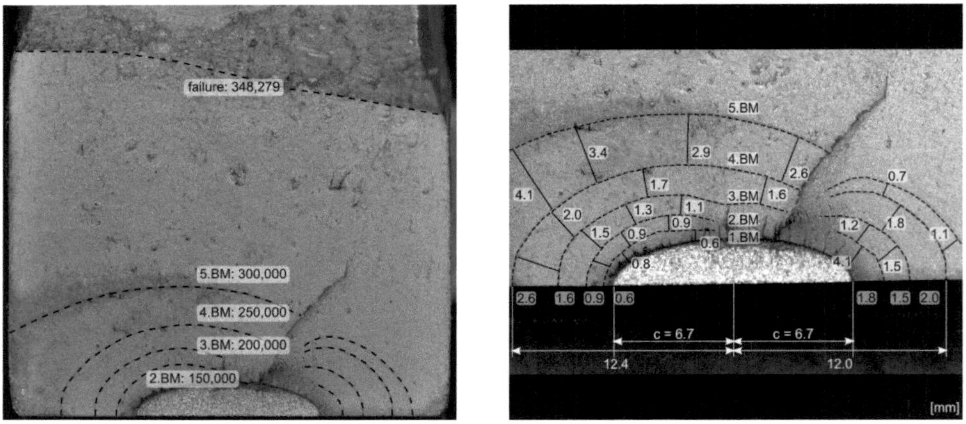


(a) (b)

**Figure 3.30:** Evaluation of the beach mark test GKC3: a) Full scale; b) Detailed view

The figure illustrates that the predominant part of the service life is attributable to short crack states. Crack growth up to a doubling of the initial crack width from c = 6.7 mm to c = 13.4 mm was considered. This corresponds to half of the minimum distance that must be maintained between two adjacent defects according to Section 5.2.6 and FKM-Rili-BM:2009 [161]. As marked in Figure 3.30b, this condition lies between the fourth and fifth marking lines. In the specific case, 250,000/348,279 $\approx$ 70 % to 300,000/348,279 $\approx$ 85 % of the total lifetime was passed until this state was reached.

A closer look at the shape of the crack front reveals a uniform expansion along the crack contour at the beginning. The interaction of the stress distribution caused by the notch with its maximum at the component surface and the additional moment led to comparable crack tip stresses along the crack contour. In the case of a purely semi-elliptical surface crack on a flat plate under tensile stress, a stress hotspot is present in the apex area. With increasing crack size in these real components, this behavior changed and the increments at the surface exceeded those in the apex area by a factor of approximately 1.5. The reason for this was the increase in additional moments resulting from a displacement of the surface center of gravity and the interaction with the component's edge.

### 3.3.2.3 Evaluation of the Strain Measurement

Figure 3.31a shows a typical strain history over the test duration. The labels of the strain gauges are defined in Figure D.1. Some of the curves shown in detail in Annex D contain interferences. A few measurements had to be aborted early due to technical problems. However, the basic statements could be derived in most cases and were not influenced by this evaluation. Figure 3.31a clearly and quantifiably shows the biaxial bending stress that is assumed when considering the residual fracture surfaces. In the present example, the "south" defect experienced higher stresses from the beginning. In addition, a small difference between "east" and "west" can be seen. A recalculation of these moments to substitute loadings and local stresses was done with the help of a numerical model given in Section 4.2. A straightforward approach to the calculation was not possible due to the superimposed notch effect. In addition to the strain ranges, Figure 3.31b considers changes with respect to the initial value.

In both diagrams, the qualitative crack propagation could be derived from the sequence of the failing strain gauges. The strains of the more-stressed crack increased continuously until failure. On the side of the less-stressed crack, the strains decreased continuously until just before failure. As soon as the strains of the side facing the crack started to grow disproportionately, a disproportionate decrease occurred on the side of the

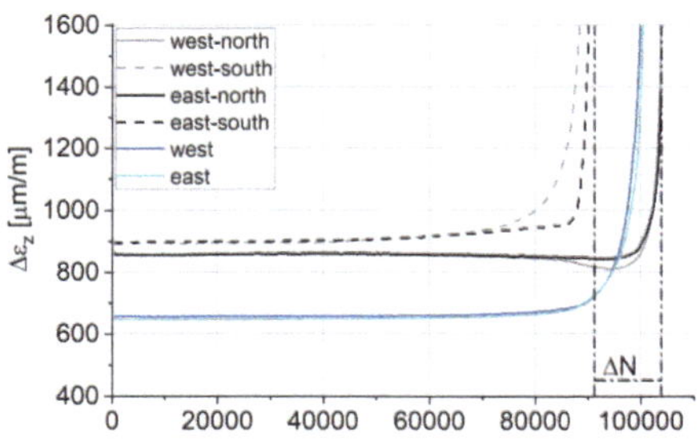


(a)

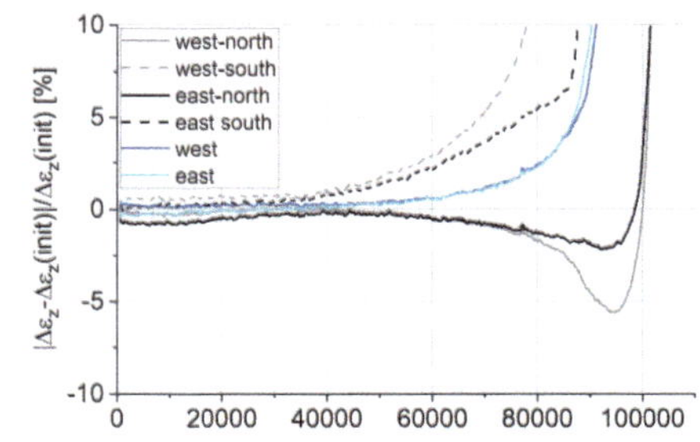


(b)

**Figure 3.31:** a) Strain range history of component tests; b) Definition of stain ranges from the initial values less-stressed crack. The strain gauges located in the center, labeled "east" and "west", did not experience this decrease and remained mostly unaffected by crack growth. In [121], the strain states were assigned to crack sizes by means of numerical crack propagation calculations. The share of large-area crack growth in the total service life can be evaluated based on the strain data. In the evaluation of the beach marks, the decisive part of the service life could be attributed to the area until twice the initial crack size was reached.

For simplicity's sake, $\Delta N$ was defined as the cycles between the first strain gauge tear-off and component failure. This was approximately the range between mark five and failure in the beach mark test shown in Figure 3.30. Figure 3.32a relates $\Delta N$ to the total fatigue life N determined in the test. This is a measure of the proportion of long-range crack growth that is clearly visible from the outside. The results are shown separately according to component size and foundry, analogous to the previous illustrations. The marked outlier is specimen GGB2. Here, the crack originated at the edge of the component and can therefore not be compared with the other components. In all other components, $\Delta N$ accounted for less than 12 % of the total service life. The proportions decrease as the load is applied (i.e., the number of cycles increases). However, this only occurs to a very small extent. A much more pronounced trend can be seen between the different component sizes. The proportion is considerably higher for the small samples than for the large ones. The effects are examined in more detail in Section 4.2.2 in connection with the crack tip stresses that depend on the component size. The size dependence is disturbed for the "large" components from

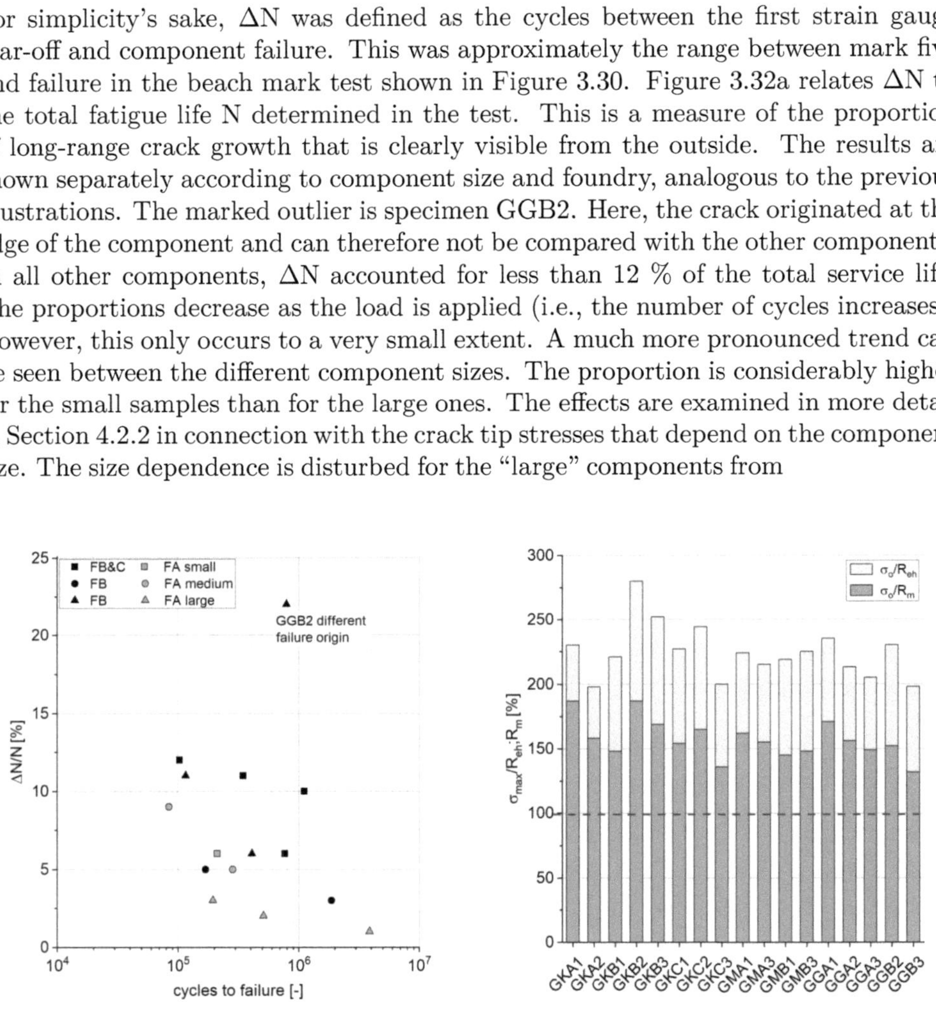


(a) (b)

**Figure 3.32:** a) Proportion of cycles between the first strain gauge tear-off and failure, expressed as a proportion of the total life; b) $F_O$ related to the residual fracture surface foundry B. As shown in Section 4.2.2, additional moments from geometric deviations acting perpendicular to the other cases were present here. The lower level of the samples from foundry A underlines the assumption of a delayed crack initiation (e.g., due to residual stresses).

#### 3.3.2.4 Residual Fracture

Similar to the evaluation of the tensile specimens in Figure 3.32b, the upper loads are related to the residual cross section and are compared to the nominal yield strength or tensile strength in Figure 3.32b. As observed and explained in Section 3.2.2.8 for material 1.6220, the stresses in the residual section also exceed both the yield strength and the tensile strength. The fracture surfaces show a honeycomb-like structure and reinforce the ductile residual fracture behavior. The present magnitude is considerably larger than that determined for the tensile specimen. A differently pronounced multiaxial stress state at higher stress triaxialities is assumed as the underlying cause.

### 3.3.3 Summary

These experimental investigations were performed on specimens with wall thicknesses between 35 mm and 100 mm made of the material G20Mn5 and conditions representative for real components. The investigations aimed at an integral consideration of the technological thickness effect as well as statements on the interaction between geometrically induced stress increases (notches) and casting defects. In addition, these investigations served to validate the design concept from Chapter 5. The specimens were designed with two symmetrically located notches (SCF $\approx$ 2.0) and artificial semi-elliptical surface defects were introduced into each notch root. The dimensions of the artificial defects were defined independently of the component dimensions in the size of the UT acceptance limit.

Due to deviations from straightness, additional bending moments acted in the test cross section. As a result, one of the two artificial defects was subjected to higher stresses, resulting in asymmetric crack growth. The stress conditions were mapped by strain measurements and the strain changes were evaluated over the test duration. This confirmed that only a small proportion of the total service life is due to large-scale crack growth. The regression curves of the component test showed a gradient k $\approx$ 3.0 comparable with the SN curves of the surface defects from Section 3.2. The fatigue resistance related to the nominal stress range at the reference value $2 \cdot 10^6$ cycles was $\Delta\sigma_{50\%} = 86$ MPa. The reduction compared to the results from Section 3.2 was attributed to the local stress increase caused by the geometric notches. The influence of component thickness on fatigue resistance was not significant. Residual fracture behavior was classified as ductile and was consistent with the tensile specimens of G20Mn5 material from Section 3.2. In one series of specimens, beneficial compressive residual stresses were present, so these results were excluded from further considerations.

# 4 Numerical Analyses

This chapter covers an extended evaluation of the experiments from Chapter 3 based on numerical simulations. The aim of these investigations was to generate a comprehensive understanding of the damage process caused by real casting defects. For this purpose, the tensile specimens from Section 3.2 were subjected to high-resolution crack propagation calculations. The second objective was to quantify the actual stress states present in the component tests from Section 3.3 (e.g., due to geometric deviations) and thus enable a more profound evaluation of these tests. The calculation approaches adopted within this chapter provide information essential to the simplified design concept developed in Chapter 5, but can also be used for a simulation-based component design. In all simulations, the commercial FEM software Ansys Workbench 2019 R2 [3] was used.

## 4.1 Tensile Specimens

In Section 3.2.2, a significant difference in the fatigue behavior between internal and surface defects was observed. At internal defects, a pronounced crack initiation phase was found, whereas surface defects seemed to behave as initial cracks. This hypothesis was based on the gradient of the SN curves and was underlined by simple analytical approaches. Within this chapter, these effects are confirmed and quantified. In doing so, different modeling depths and analysis strategies are compared. Any effects of eccentric defect locations for both internal and surface defects were directly taken into account. The focus was placed on material 1.6220 and selectively transferred to material 1.6760.

Following the assumptions of Section 2.1, these simulations purely focused on the phase of stable crack growth. The remaining part compared to the experimentally determined service life was assigned to a crack incubation phase. The research partners in [116] carried out further attempts to map the crack initiation using a combination of several online NDT methods [69] or numerical by means of damage models.

Figure 4.1a illustrates the basic procedure of these simulations. An FEM model was generated, evaluated and validated for each specimen considered. The casting defects were introduced into the FEM model at different modeling depths, and the crack tip stresses present in each crack growth increment were calculated.

Separated from this stress analysis, crack growth was evaluated. For this purpose, crack geometries and number of cycles were determined, and failure criteria were checked in newly developed Matlab [60] routines. After passing through this incremental calculation procedure, the results of the individual specimens were considered and evaluated in the overall context. The essential influences, modeling approaches, calculation strategies and results of these three blocks are described in detail in the following subsections.

### 4.1.1 Modeling

The Finite Element simulation was only used to determine the crack tip stresses at every individual crack growth increment. Its presentation focuses on the fundamental parameters, such as the boundary conditions, material models and especially the modeling of the casting defects.

#### 4.1.1.1 Overall Model and Boundary Conditions

The calculation was performed on the model of the tensile specimens shown in Figure 4.1b. Modeling the entire specimen was necessary to represent any additional stresses caused by eccentricities of the defects. The boundary conditions and stiffness ratios were taken into account according to the test fixture schematically shown in the same figure. The loading was force controlled and corresponded to the axial load range $\Delta F_{x,FEM}$ applied in the experimental investigation.

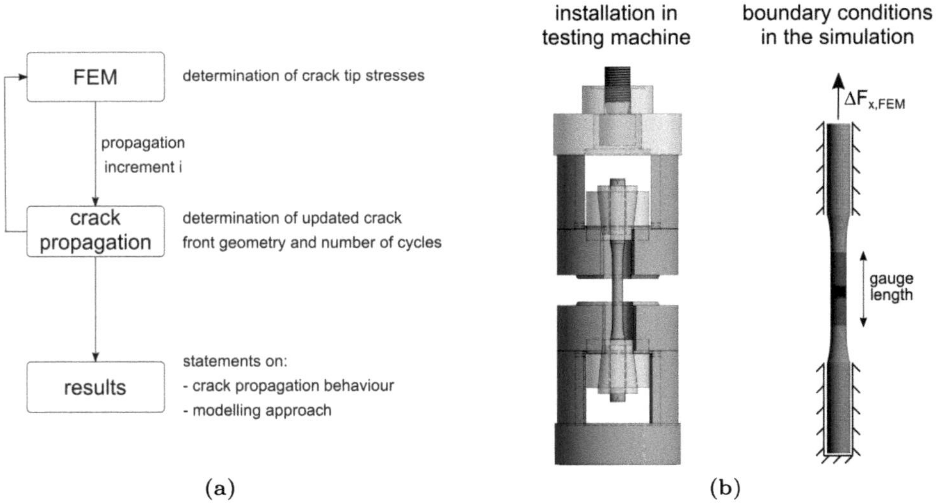


**Figure 4.1:** a) Basic procedure of numerical crack growth calculations; b) Specimen geometry, boundary conditions and installation of the specimen in the testing machine

#### 4.1.1.2 Modeling the Casting Defects

Since the numerical models were based on the CT records from Section 3.2.1, only specimens corresponding to failure cases I and II defined in Figure 3.6, could be considered. The three modeling variants shown in Figure 4.2 were evaluated and compared. The highest modeling depth was achieved by considering the defect volume and a planar initial crack surrounding the defect in the failure plane. Analogous to [103], a constant expansion increment of 0.1 mm starting from the surface of the defect volume was applied. In this way, potential crack-like fragments in the dissolved microstructure were considered conservatively. In a second model, an identical initial crack front was assumed, but the defect volume was neglected. By comparing these approaches, the influence of the 3D stress field caused by the casting defects was investigated. The application of this model was limited to a small number of selected specimens with internal defects at different load levels. The most far-reaching simplification was achieved through the approach of an elliptical substitute crack with an identical surface area in the failure plane. This aimed at a simplified and more cost-efficient consideration and should correspond to the elliptical failure state observed in the fracture surface analysis in Section 3.2.2.4.

Figure 4.2 compares the stress intensity factors of all three modeling approaches in the specimen U20O12 exemplar. Regions with no crack growth due to $\Delta K_I < \Delta K_{th}$ are marked in dark blue. The comparison of the results of models 1 and 2 illustrates shadowing effects caused by the defect volume, since high crack tip stresses were reduced to fewer regions of the crack front. Second, the stress trajectories are deflected by the defect in such a way that they acted as relief notches and reduced the crack tip stresses at identical locations. The maximum stress intensity factors of the elliptical crack front were significantly lower compared to the first two models. The reason for this was the lack of curvature in the undercut areas. These local effects thus dominated the shielding effects from the volume defects.

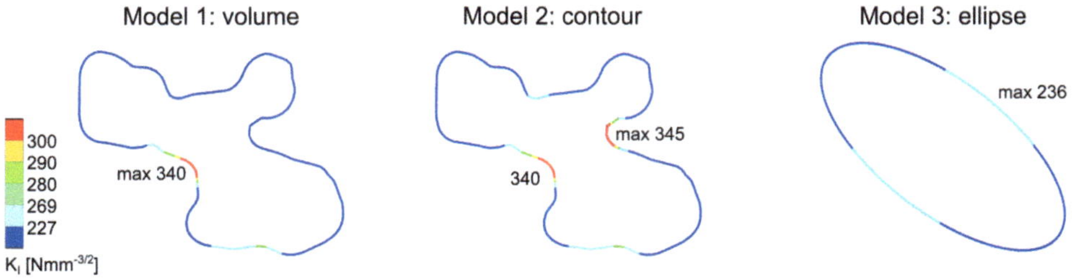


**Figure 4.2:** Modeling approaches for crack growth at casting defects

*Model 1 – Volumetric defect:*
Figure 4.3 shows the applied steps for the generation of an FEM model for the determination of the crack tip stresses based on CT data. These records were transformed within the pore analysis into surface models as described in Section 3.2.1.2. The resulting STL

file data was not suitable for a direct FEM meshing, as some triangles were inverted and the surface showed multiple indentations that would cause stress singularities. In particular, a defined planar initial crack as a prerequisite for these simulations was not included in this data.

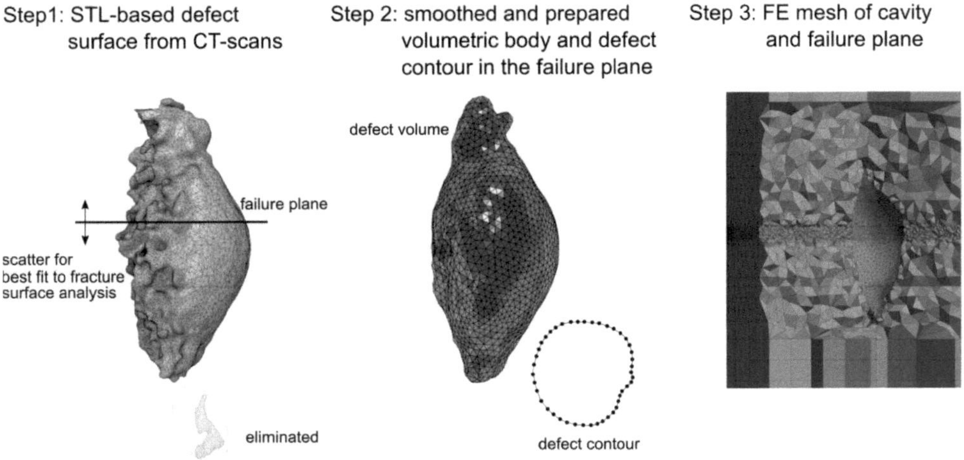


**Figure 4.3:** Necessary steps from CT records to the FE model

In a first step, the failure plane of each individual specimen determined within the experimental investigations was transferred to CT records. To eliminate measurement inaccuracies, the resulting defect contour in the section plane was compared with the fracture surface analysis. By iteratively adopting the axial location of the failure plane in the CT data, the best match between the observed failure and the simulation model was ensured. Additionally, pores aside the cavity that caused failure or defects that did not significantly affect the local crack tip stresses were removed.

In a second step, the remaining parts were smoothened and re-meshed in such a way that the characteristic effects of the pore on the stress field remained but the local singularities vanished. These data were transferred to volumetric bodies and exported as IGS files. Additionally, the defect contour in the failure plane was extracted as a list of nodes. The defined contour line was expanded by a uniform increment of 0.1 mm using the procedure described in Figure 4.6. The resulting contour line formed the initial crack front.

In the last step, both the volume and the crack front were induced into the FEM model. The pore volume was subtracted from the geometry of the tensile specimen shown in Figure 4.1b at its precise location. Afterwards, the geometry of the initial crack front was imprinted on both sides of the failure plane. Specimens with several failure planes could not be precisely reproduced with this approach, so a dominant plane had to be selected. Finally, the upper and lower parts of the specimen were merged together in all regions outside the crack front.

*Model 2 – Crack contour:*
The crack contour of this model corresponded to that of model 1; only the defect volume was deactivated for these calculations.

*Model 3 – Substitute ellipse:*
Figure 4.4a shows an example of a substitute ellipse. This was determined in such a way that it corresponded to the area weakened by the induced crack from model 1. The remaining geometry parameters were set in such a way that:

a) The crack centroid corresponded to that of the defect.

b) The point furthest from the centroid defined the orientation $\alpha$ and the length of the first half axis $h_1$.

c) The second half axis $h_2$ resulted from the predefined condition of equal surface areas.

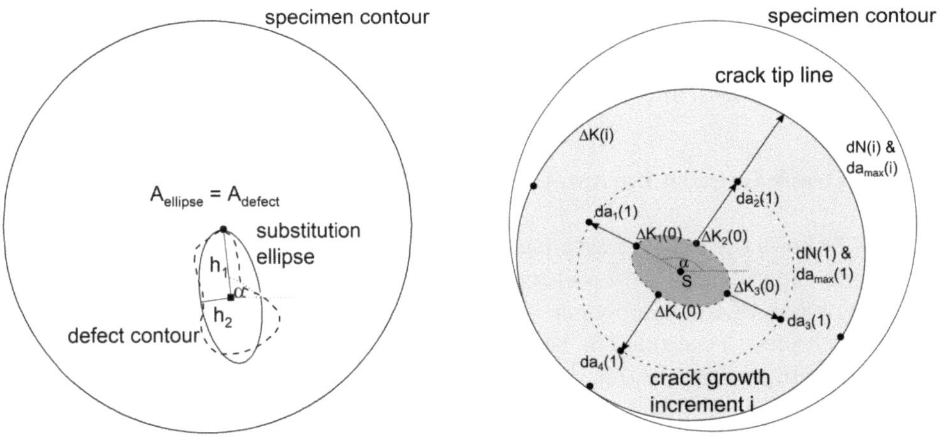


(a) (b)

**Figure 4.4:** Substitute ellipse: a) Definition of geometric properties; b) Incremental crack growth

### 4.1.1.3 Material Model, Meshing and Determination of $K_I$

The analysis was intended to cover stable crack growth only. To apply Linear Elastic Fracture Mechanics (LEFM), an isotropic linear elastic material model where E = 210,000 MPa and $\nu = 0.3$ was chosen analogous to [103]. The meshing was carried out using volume elements with a quadratic approach and an element size of 0.1 to 0.2 mm in the failure plane. The stress intensity factors were determined by Ansys [3] in a post-processing routine by the "interaction integral method" [4] according to [106].

#### 4.1.1.4 Validation

The general applicability of the procedure was validated through comparative calculations of simpler geometries such as those presented in Annex B.1.4. For the extension to more complex conditions, the different crack modeling approaches, which were implemented in Ansys were applied and compared. The procedure was confirmed by obtaining results that were close enough according to engineering judgment.

During the actual simulations, the convergence behavior of six evaluated integration paths was monitored for all specimens in all crack growth increments. Only solutions that converged completely along the entire crack path were transferred to the propagation calculation. The comparison of the results with previous and subsequent crack configurations as well as generally valid criteria (e.g., for element distortion) formed the final quality assurance step of the simulations.

### 4.1.2 Crack Growth

The description of the crack growth process in this section is divided into the material data, the crack growth approach and the applied failure criterion.

#### 4.1.2.1 Crack Growth Parameters

Figure 4.5 shows the crack growth parameters from Section 3.1.2 supplemented by the underlying experimental data and separated by material for a stress ratio of $R = 0.1$. Since any effect of the component thickness has no statistically significant influence on the crack growth parameters, the parameters listed in Table 3.2 were used in these simulations. In addition to this classical approach, a two-stage growth law was derived in accordance with the basic concept presented in BS 7910:2013 [138]. In particular, the crack propagation rate, which is considerably lower in reality when $\Delta K_I$ is close to $\Delta K_{th}$, was to be represented. The parameters used in the crack propagation calculation are summarized in Table 4.1. The values for "stage B" corresponded to the parameters of the classical Paris law. The cusp in the experimental data is discussed in the evaluation and derivation of the crack propagation parameters in [116]. Here, this effect is attributed to roughness-related crack closing effects.

**Table 4.1:** Summary of two-stage crack growth parameters

| | stage A | | stage B | | threshold region | |
|---|---|---|---|---|---|---|
| Mat. [-] | C [-] | m [-] | C [-] | m [-] | $\Delta K_{th}$ [$Nmm^{-3/2}$] | trans. [$Nmm^{-3/2}$] |
| 1.6220 | $8.33 \cdot 10^{-50}$ | 17.85 | $9.97 \cdot 10^{-14}$ | 3.00 | 227 | 269 |
| 1.6760 | $1.13 \cdot 10^{-48}$ | 17.37 | $2.37 \cdot 10^{-13}$ | 2.89 | 221 | 275 |

for da/dN in mm/cycle and $\Delta K$ in $Nmm^{-3/2}$

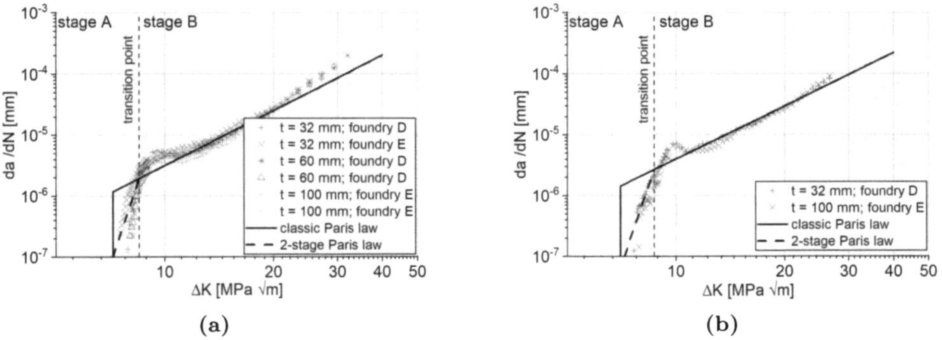


**Figure 4.5:** Test results of (one-stage) crack growth parameters determined according to ASTM E647:2015 [132] for R = 0.1 from [116] and a two-stage approach for material a) 1.6220 and b) 1.6760

### 4.1.2.2 Growth Procedures

A preceding analysis showed that $\Delta K_{II}$ and $\Delta K_{III}$ attributed to the remaining crack opening modes [28] were negligibly low, and thus a planar consideration of the crack propagation based on $\Delta K_I$ was considered sufficiently accurate. In the area of stable crack growth, this coincided with the findings of the fracture surface analysis in Section 3.2.2.4. The proposed procedures are presented separately according to the modeling depths. The approaches for model 1 and model 2 correspond to each other.

*Model 1 and 2 – Crack Contour:*
Figure 4.6 summarizes all the relevant steps of the applied crack growth procedure. The starting point of each increment $i$ was a disordered set of $m$ points describing the crack front. In the first increment ($i = 0$), this was derived from the CT data as described in Section 4.1.1. In all further steps ($i > 0$), it resulted from the nodal coordinates of the FEM simulation. For each propagation increment ($i > 0$), the following calculation process was applied:

Sorting: A sorting algorithm developed within this work organized the node set in such a way that a closed contour line was formed in ascending order. For this purpose, the neighboring node with the minimum distance in each case was determined beginning from a random starting point. When selecting the output data, it was important to ensure that the mesh size of the FEM model sufficiently resolved the geometries of all existing outgrowths in the crack front. In special cases, manual adjustment was necessary.

Crack Growth: The updated contour line was derived in three stages.

> Criterion and Intensity: The local crack growth increment $da(n)$ was determined as a function of the local $\Delta K(n)$ according to the Paris law Eq. (2.10). For this purpose, an individual $da_{max}(i)$ was applied for the respective crack growth

increment $i$. First, the number of cycles of the currently considered increment $i$ was determined by

$$\mathrm{dN}(i) = \frac{\mathrm{da_{max}}(i)}{\mathrm{C} \cdot \Delta\mathrm{K_{max}}(i)^{\mathrm{m}}} \tag{4.1}$$

and stored for the calculation of the total number of cycles. Then, a local crack growth increment was determined for each node $n$ along the contour line by

$$\mathrm{da}(n) = \mathrm{dN}(i) \cdot \mathrm{C} \cdot \Delta\mathrm{K}(n)^{\mathrm{m}}. \tag{4.2}$$

For $\Delta\mathrm{K}(n) < \Delta\mathrm{K_{th}}$, the local increment was set to $\mathrm{da}(n) = 0$ by default. If the two-stage crack growth law was applied, the Paris parameters C and m were selected according to the assignment of $\Delta\mathrm{K}(n)$ to stage A or stage B, according to Table 4.1. To create the initial crack front at $i = 0$, a constant value of $\mathrm{da} = 0.1$ mm was applied at all nodes. In the course of the calculation or with an increasing approximation of a circular shape, $\mathrm{da_{max}}(i)$ tended to increase. If local peaks in the stress intensity factors occurred along the contour line, large increments $\mathrm{da_{max}}(i)$ led to a breakthrough of the crack front and energetically impossible states.

Direction: Each local crack growth increment $\mathrm{da}(n)$ was applied in the direction of the bisector between the connection vectors of three adjacent nodes $\vec{\mathrm{da}}(n)$.

Orientation: The sign of each expansion vector $\vec{\mathrm{da}}(n)$ was ultimately determined using calculus of variations to maximize the area covered by the crack.

The procedure developed within these investigations allowed the consideration of arbitrarily shaped crack contours, undercuts, convex zones and concave zones in the defect contour.

FEM: The updated crack contour acted as the starting point for the numerical determination of the stress intensity factor for the next increment $i + 1$ using the FEM model according to Section 4.1.1.

After the failure criterion was reached, the total number of cycles $\mathrm{N_{tot}}$ was calculated by

$$\mathrm{N_{tot}} = \sum_{i=0}^{\mathrm{i_{tot}}} \mathrm{dN}(i). \tag{4.3}$$

In the actual calculations, typically 25 to 45 crack growth increments ($\mathrm{i_{tot}}$) were applied. As the changes between $\Delta\mathrm{K(n)}^{(i)}$ and $\Delta\mathrm{K(n)}^{(i+1)}$ were negligible and the new crack contour was unknown before applying the propagation procedure, iteration approach 1 according to Figure 2.4 was applied in these simulations.

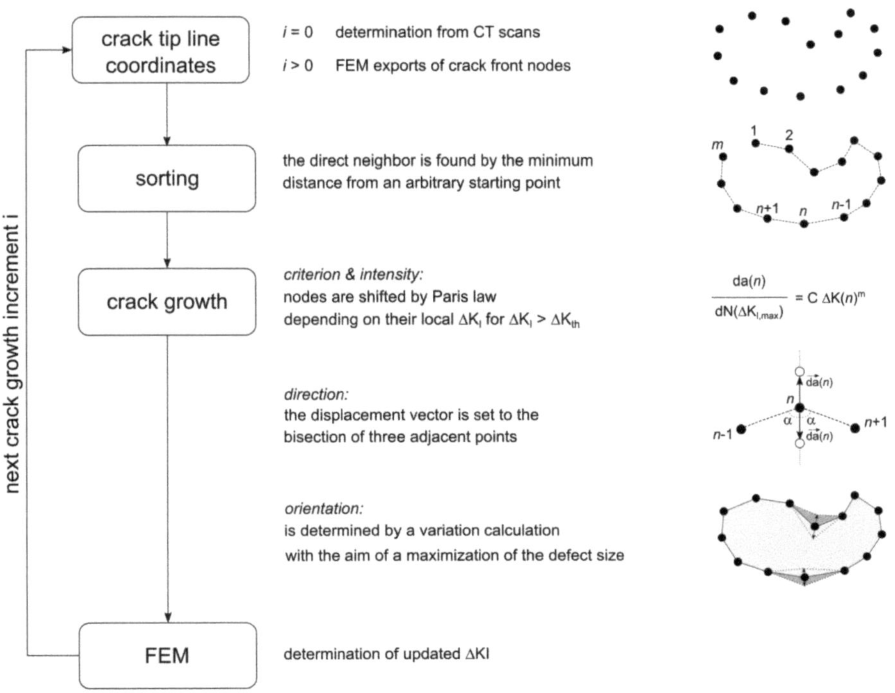


**Figure 4.6:** Steps in each crack growth increment for arbitrary contours

In the case of surface defects, an analogous procedure was performed. Additionally, the intersection between the crack contour and the specimen contour was specifically determined and shifted so that the intersection was perpendicular to the surface. This was necessary to eliminate numerical problems that occur at other angles, as discussed in [14, 77], and that are not yet solved in general.

*Model 3 – Substitute Ellipse:*
The fundamental idea of deriving the local crack growth increments $\mathrm{da}(n)$ from the local $\Delta\mathrm{K}(n)$ using the Paris law was the same as in the previous procedure. However, in this approach, the crack contour was not described by multiple points but by an ellipse equation. Its characterization was done using the orientation $\alpha$, the two main axes $\mathrm{h}_1$, $\mathrm{h}_2$ and the coordinates S of the centroid.

The local crack expansion $\mathrm{da}_{1-4}$ was determined as shown in Figure 4.4b by evaluating $\Delta\mathrm{K}_{1-4}$ at the four vertices orthogonal to the crack contour. From the coordinates of these newly determined vertices, updated ellipse shape parameters were calculated. Thus, eccentric growth of non-centrically arranged defects could be represented in a simplified way. The calculation of the stress intensity factors in the individual increments was carried out using the FEM model described in Section 4.1.1.

Due to the predefined crack contour having an elliptical shape, a successive outgrowth of the undercut areas was omitted. In this way, the stress intensity factors were more uniform along the crack contour, and larger crack growth increments could be selected. In the actual calculations, 5 to 10 increments had to be used. For the determination of the number of cycles $dN(i)$ associated with each increment, the application of iteration approach 3 according to Figure 2.4 was required. Surface defects were not considered in this approach.

#### 4.1.2.3 Failure State

The objective of the simulation was to map the experimental investigations. For this purpose the calculation was terminated when the area enclosed by the crack contour corresponded to that determined in the fracture surface analysis ($A_{sim} = A_{crack}$ according to Section 3.2.1.4). Additionally it was checked if $dN/da \rightarrow 0$ was valid. $K_o < K_{Ic}$ with $K_{Ic} = K_{Id}$ was checked for dynamic effects according to Section 3.2.2.8.

### 4.1.3 Results

#### 4.1.3.1 Comparison of the Calculated Crack Propagation with Fracture Surface Analysis

Figure 4.7 shows the fracture surfaces of three representative specimens. On these, the crack contours of the individual calculation increments according to model 1 using the single-stage Paris law are marked. Figure 4.7a shows U20O12, representing a specimen with small, branched, centrally arranged casting defects. Figure 4.7b illustrates a large, eccentrically arranged defect, and U20M10 in Figure 4.7c appears for specimens with large surface defects.

(a) (b) (c)

**Figure 4.7:** Comparison of the fracture surface analysis with the calculated crack contours for specimens a) U20O12, b) U20M7 and c) U20M10

In all cases, a good agreement was found between the calculated crack expansion and the area that was assigned to stable crack growth in the fracture surface analysis. The observations from Section 3.2.2 of an elliptical or approximately circular crack surface in the fracture state could be confirmed with these results. The largest stress intensity factors occurred in the indentations of the crack contour, provided that no shadowing effects were present. As a result, these grew out with a comparatively small increase in cracked area per growth increment, and a smooth crack contour was quickly obtained. Deep undercuts in the crack front led to local compressive stress states. This was caused by strain limitation and appeared both in model 1 and in model 2. As a result, negative K-factors could be observed locally. In these cases, subsequent crack configurations caused a closure of these undercuts by joining the edges. In these computationally determined areas, fragments of another fracture process could be spotted on the real fracture surface. The shifted center of gravity at eccentrically arranged defects led to additional stresses. They consequently grew toward the closest surface. Furthermore, the interaction (cf. Section 2.2.4.5) between crack tip stresses and the specimen surface intensified this behavior. In the case of surface defects, the propagation along the surface was dominant, and bending effects rapidly increased. The calculations on surface defects were limited to two examples. In all other cases, either failure case III according to Figure 3.6 was present or the defects were so small that the LEFM lost its validity.

#### 4.1.3.2 Comparison of the Calculation Approaches

Figure 4.8 shows the correlation between the crack propagation behavior and modeling depth exemplar for specimen U20O19. Considering model 1 (Figure 4.8a), the crack front grew eccentrically despite a centric arrangement. The area marked on the crack surface was in a plane with a longitudinal offset of approximately 1 mm. The calculation showed that this section of the crack front was subject to lower stress intensity factors in the considered failure plane. The reasons were shadowing effects that caused crack growth in another parallel plane. The unification of both cracks in the fracture state could be attributed to burrs on the fracture surfaces. In addition, deviations between the initial defects based on the CT data and the resolved microstructure observed on the fracture surface could be detected.

In the case of the contour model (Figure 4.8b), an almost concentric crack growth was observed. Undercuts quickly grew into an elliptical crack shape, which then expanded concentrically. For most specimens, the agreement between model 2 and the fracture surface was considerably poorer than the results of model 1. The third model based on the approach of a substitute ellipse is shown in Figure 4.8c. Its behavior was similar to that of model 2. In cases of eccentrically arranged defects, a clear directivity to the component surface was detected. This was caused by interaction effects between the crack tip and the specimen surface as well as additional bending moments that resulted from a shift of the centroid.

Figure 4.9a shows the evolution of crack growth for specimen U20O19 discussed in Figure 4.8. Herein, $\sqrt{A_{crack}}$ is plotted as a function of the calculated number of cycles. This operation eliminates the almost quadratic increase of the crack surface resulting from the elliptical shape. The resulting representation corresponds to the more familiar da/dN evaluation. The points in the graphs indicate the individual increments and are connected to each other by straight lines. The calculation according to model 1 led to significantly higher lifetimes compared to the two planar considerations. The behavior of the elliptical substitute defect and model 2 corresponded very well both qualitatively, as shown in Figure 4.8, and in absolute values Figure 4.9b. Due to the simple geometric description, the non-existent local peaks in the stress intensity factors and the considerably lower number of iteration steps, the calculation effort of the substitute ellipse was a fraction of that of the highly resolved crack contour. Independent of the calculation model, it became obvious that the major part of the calculated lifetime could be assigned to crack configurations close to the initial state. Besides the growth paths of the three calculation models, the limit state $A_{crack}$ derived from the fracture surface analysis is marked in the plot. The second horizontal line corresponds to the size of a crack surface in which the nominal stress in the remaining cross section reaches the yield strength $\sigma_y$ at $F_o$. Generally, the differences in the calculated service life were less than 5 %. So, besides the findings on the residual fracture in Section 3.2.2.4, $\sigma_y$ acts as a useful and not overly conservative criterion (e.g., in the evaluation in the FAD). The negligible differences result from the increase in ΔK and its potentiation in the Paris equation.

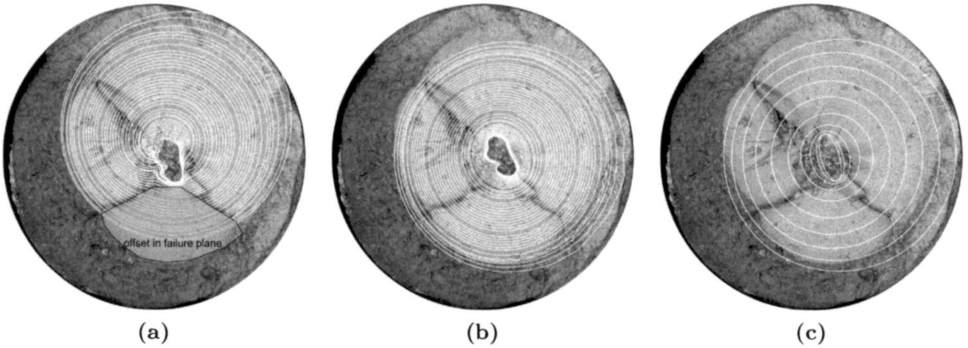


(a) (b) (c)

**Figure 4.8:** Comparison of the crack propagation approach for specimen U20O19: a) Volume; b) Contour; c) Ellipse

#### 4.1.3.3 Comparison of the Calculated Lifetimes with those Determined in the Experiments

Figure 4.9b compares the service life calculated based on crack growth with the load cycles determined in the experiments. The different modeling variants of the considered specimens of material 1.6220 are grouped for better clarity. Specimens marked with

"(S)" had surface defects. For comparison, the regression curve of the considered internal defects is shown in addition for the test results and the results of model 1. All other calculation approaches are supplemented and led to lower calculated service lives. On the other hand, the calculation of specimens with surface defects, regardless of stress level, corresponded very well with the service life observed in the experimental investigations. This underlines the hypothesis from Section 3.2.2 that as long as surface defects meet the minimum dimensions for fracture mechanics consideration, the damage behavior can be described by crack growth. For smaller defects, the approaches established in [66] apply. Since permissible defects of this size cannot be detected by common NDT on untreated cast surfaces, these approaches lose their practical relevance for application and are not examined further. In the case of low load amplitudes, stress intensity factors between $\Delta K_{th}$ and the transition point specified in Table 4.1 occurred locally so that higher lifetimes resulted from the two-stage crack growth law.

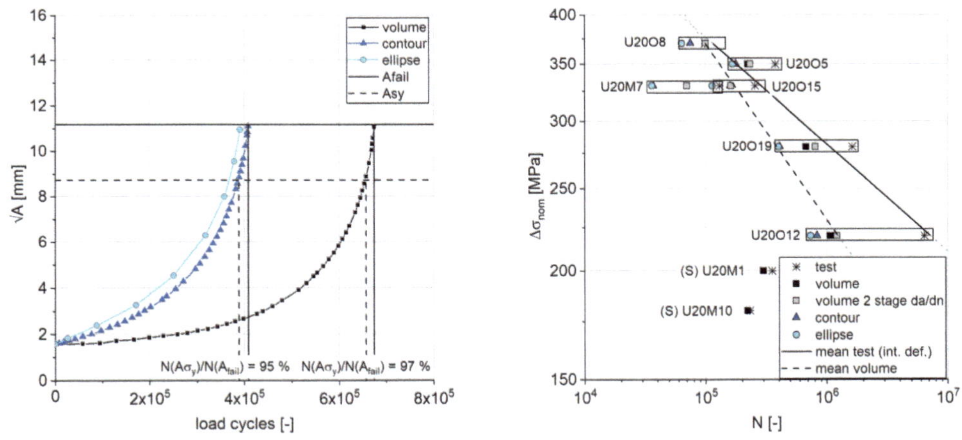


(a) (b)

**Figure 4.9:** a) Evaluation of crack size development over the service life; b) Comparison of simulated and experimentally determined SN data

Figure 4.10a illustrates the dependence of the proportion $N_{sim}/N_{test}$ of numerically determined stable crack growth $N_{sim}$ on the cycles to failure detected in the experimental investigations $N_{test}$ as a function of the experimentally determined service life. Additionally, the behavior is compared with that of material 1.6760. In the case of high loads, the stable crack growth determined by model 1 could represent almost the total service life determined in the experiments. At the lowest considered load levels, this dropped to about 20 %. Thus, a design model purely based on crack growth is increasingly conservative for lower loads and the same initial defect size. Despite the sharp-edged crack contour, the behavior indicated a pronounced crack initiation phase corresponding to the observations in [42]. Specimen U20M7 was outside the trend of the other tests, which resulted from a large, eccentrically located casting defect. In this case, the crack propagated exclusively in the direction of the nearest surface, causing additional moments and pronounced interaction effects. For statements on the effect of the material, three specimens with internal defects of different sizes and stress levels were examined. The calculations were limited to model 1 using the one-stage Paris law. The proportions of the total service life were between 20 % and 30 % and thus were considerably lower than those of material 1.6220. The tendency that high stresses lead to a large proportion of crack growth during the total service life was still present, but the expression was significantly lower. Otherwise, no further differences between the two materials under consideration could be determined in these calculations.

### 4.1.3.4 Quantification of the Calculation Models

Figure 4.10b provides a more detailed analysis of the effects of the calculation model. For this purpose, the calculated cycles were placed in relation to model 1 with a single-stage crack growth law. The results are illustrated as a function of the number of cycles determined in the experiments and indirectly relate to the applied load level. The approach of a planar crack, regardless of whether it followed the exact crack contour or was represented by a substitute ellipse, provided between 50 % and 80 % of the service life calculated with model 1. A dependence on the load level could not be found. In addition, the influence of the two-stage crack growth model did not depend on the load level in the considered examples.

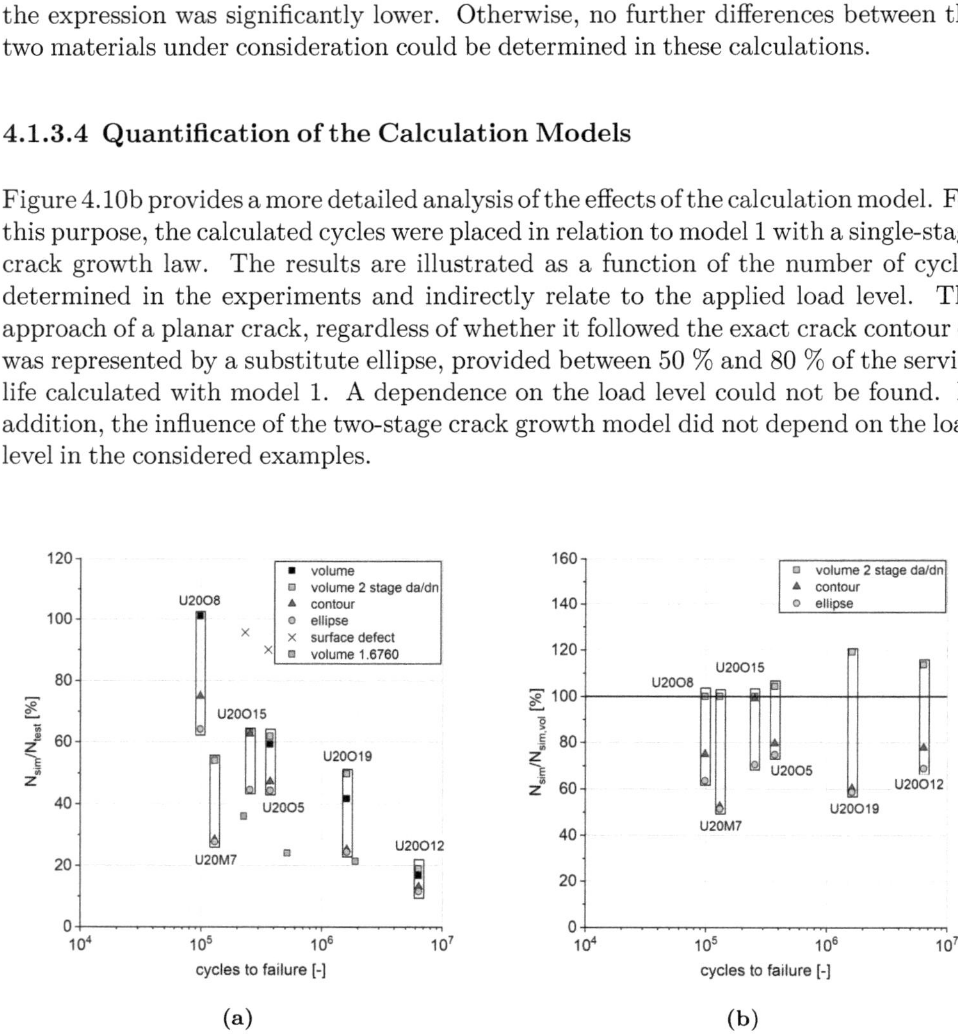


**Figure 4.10:** a) Share of the calculated number of cycles on the experimentally determined service life; b) Comparison of the calculation models

Since the maximum $\Delta K$ in the first crack growth increment was already above the transition point between stage A and stage B, the two-stage law had only an indirect influence on the service life by a slightly modified crack shape. The resulting gain in the calculated service life was in the range of 5 % to 20 %. In this case, the characteristics of the local crack contour and the shadowing effects of the 3D volume defect were the main factors. In later crack growth states, $\Delta K_I$ exceeded the transition point along the entire crack contour. Therefore, the two-stage law was no longer decisive.

### 4.1.4 Summary

The numerical crack propagation calculations aimed at a deeper understanding of the dominant damage mechanisms and formed the basis of a numerical component design. For this purpose, algorithms for crack propagation calculations of the branched casting defects were developed. The computational models were derived from the CT records and different modeling depths were considered.

The calculations confirmed the different effects of internal and surface defects. In the case of the internal defects, a pronounced crack initiation phase was met, especially at lower load amplitudes. The portions ascribed to crack growth seem to be small and overly conservative at first glance for a design model. A consideration in a double logarithmic diagram and the transfer to tolerable stress amplitudes reduces the effects to an order of magnitude that lies within the scatter known for tests on welded specimens.

Any initial crack geometry grew over an ellipse into an approximately circular state. The volumetric characteristic of the defects or the resulting 3D stress field influenced the crack growth in form, direction and simulated load cycles. A planar consideration was always conservative and the simplified consideration of a substitute ellipse was sufficient. With the methodology shown and the approach of an elliptical plane initial defect, a component design in the HCF range could be carried out numerically. Compared to the method derived in Chapter 5, load transfers and actual stress distributions could be considered with this method. For material 1.6760, the share of crack growth within the total service life was lower, so a design approach based on fracture mechanics is more conservative. However, this statement does not include the residual fracture behavior discussed in Section 3.2.2.8 or the brittle fracture behavior investigated in [103]. Furthermore, the simulation results suggested that the disintegrated microstructure surrounding the casting defects did not act as a sharp crack. If this was the case, the crack growth would have occupied a much larger proportion of the total service life.

## 4.2 Component Tests

In the component tests discussed in Section 3.3, non-uniform crack growth was detected. Nevertheless, the previous evaluations were based exclusively on the nominal stress ranges. Additional stresses resulted from geometric deviations such as non-straightness, the position of the holes in the components, but also from a not entirely centric clamping in the testing machines. The aim of these investigations was to derive the actual stress situation for the initial crack geometry at the beginning of the fatigue tests. Their representation was done using substitute moments, which led to a condition comparable with the measurements. This allowed the derivation of SN curves based on the maximum stress range. Furthermore, the effects of geometric deviations could be quantified. Finally, the main findings on the crack propagation behavior from [121] based on these models and presented in this section are shown.

### 4.2.1 Modeling

The calculations were performed on the model cutout shown in Figure 4.11 for the "small"-type specimen. This contained the geometric notch, significant influences from the transition areas and the artificially introduced surface defects. The definition of the boundary conditions, designations, the selection of the coordinate system and the transfer to other specimen types are shown in Figure D.1. After a preliminary study, the dimensions and boundary conditions were selected in a way that the resulting stress state corresponded to that of the entire component. The meshing was carried out by volume elements with a quadratic approach (Solid 186 & 187 according to [3]), whereby the element sizes were set in such a way that a converged mesh was obtained in the notch area.

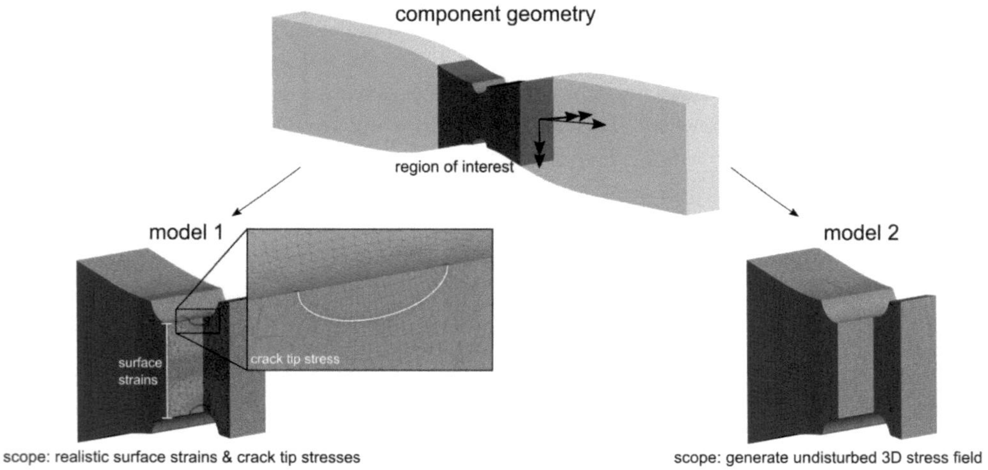


**Figure 4.11:** Section and model overview – component tests

For the stress and strain analyses, two models with different objectives were set up. In model 1, the surface defects were introduced as semi-elliptical cracks to derive the surface strains with consideration of the defect size. In addition, model 1 was used to determine the stress intensity factors. In model 2, the initial defects were not considered so that a regular mesh in the investigation cross section could be achieved. The aim of this model was to generate a homogeneous equivalent stress distribution. From this, the maximum stress ranges were determined. This approach corresponds to a typical model used in the design phase.

The loading was force controlled. All calculations were made on the nominal geometry, and the axial load was scaled by a ratio of the actual and nominal cross-sectional areas. Exemplary investigations on the effect of the actual geometry in [121] justified this procedure and showed that the local geometrical deviations of the notch and connection areas had only a negligible effect on the stresses in the investigated cross section. Young's modulus was varied in Section 4.2.2 for the comparison of the measured and simulated strain states within the range discussed in Section 3.1.

### 4.2.2 Calculation of Substitute Loads and Actual Stresses

The calculation of the substitute loads was used to quantify the notch stresses present in the investigated cross section. This enabled the determination of SN curves based on $\Delta\sigma_{\mathrm{max}}$, corresponding to the approach in a design process and served to validate the design concept in Chapter 5. In addition, the substitute loads could quantify the effects of geometric deviations. On the one hand, the mechanical effects of the investigated specimens could be classified according to ISO 8062-3:2008 [157]. The relevance of the consideration of geometric deviations in a design concept was underlined. Finally, the substitute loads were used as a starting point for crack propagation analyses in [121].

#### 4.2.2.1 Calculation Procedure

The procedure is summarized in Figure 4.12. The starting point was the strain measurements from Section 3.3. The sequential application of both FEM models from Figure 4.11 was necessary for the reconstruction of the stress states. In the first step, a test section was selected in the form of a cycle range in the measured strains. In this selection, the initial settlement must have subsided; the values of all measuring channels had to be as constant as possible in this cutout. To compensate for fluctuations and peaks, the strain ranges in this interval were averaged and form the reference value for the calibration of substitute stresses. An initial value of the substitute stress $\mathrm{M_x}$ and the final value of the load $\Delta\mathrm{F_{x,FEM}}$ were determined directly from these values using the actual dimensions by

$$\Delta\mathrm{F_{x,FEM}} = \Delta\mathrm{F_{test}}\frac{\mathrm{A_{nom}}}{\mathrm{A_{act}}}. \tag{4.4}$$

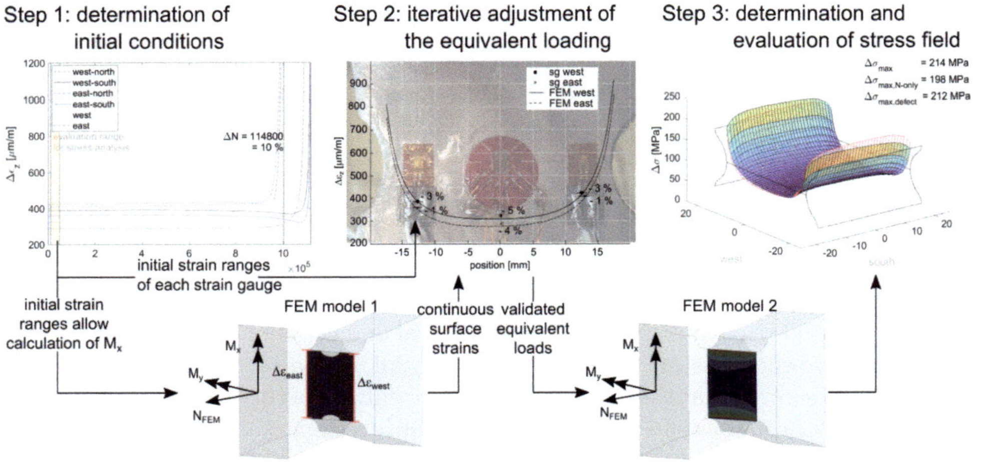


**Figure 4.12:** Procedure for determining the substitute loads and actual stress distributions

The determination of $M_y$ had to be done iteratively. A direct calculation was not possible due to the superposition with the notch stress being coupled with the disproportionate effects of small positional deviations of the strain gauges in the high-gradient areas. Therefore, $M_y$ was adjusted, and the calculated distribution of the surface strains was compared with the measurements. In order to minimize the errors of the strain gauges in the high-gradient areas, the comparison was carried out with mean values from the FEM. The calculated strains were averaged over a length corresponding to the strain gauge width. The respective positions of the strain gauges were taken into account and documented in Annex D. The substitute loads $M_x$ and $M_y$ that led to the smallest deviation between the calculated and measured values were used for further calculations.

In a third step, the substitute loads were applied to model 2. From this, the stress field in the undisturbed investigation cross section was determined and compared to a state without substitute loads. The parameters $\Delta\sigma_{\text{max}}$, $\Delta\sigma_{\text{max,N-only}}$, $\Delta\sigma_{\text{max,defect}}$ and SCF were determined according to the following definitions.

| | | |
|---|---|---|
| $\Delta\sigma_{\text{max}}$ | = | Maximum stress range under substitute loads. |
| $\Delta\sigma_{\text{max,N-only}}$ | = | Maximum stress range without substitute loads $M_x$ and $M_y$. This value is calculated in a component design. (In addition, the deviations in wall thickness according to ISO 8062-3:2008 [157] must be taken into account.). |
| $\Delta\sigma_{\text{max,defect}}$ | = | Maximum stress range under substitute stress in the area of the artificially introduced surface defects. Due to $M_x$, $\Delta\sigma_{\text{max,defect}}$ is generally not equal to $\Delta\sigma_{\text{max}}$. |
| SCF | = | $\Delta\sigma_{\text{max}}/\Delta\sigma_{\text{nom}}$ |
| $\text{SCF}_{\text{max,defect}}$ | = | $\Delta\sigma_{\text{max,defect}}/\Delta\sigma_{\text{nom}}$ |

The results of steps 2 and 3, including the substitute loads and parameters, are listed for all test specimens in Annex D. When recalculating the strains in step 2, it was found that a choosing of Young's modulus E = 200,000 MPa resulted in the smallest deviation between the calculated and measured strains. The deviations in the strains with E = 210,000 MPa were about 5 % higher. Since the loading was force controlled both in the test and in the simulation, there was no significant effect of Young's modulus on the calculated stresses or on $\Delta K$.

#### 4.2.2.2 Results

Figure 4.13 compares representative stress states for the components "small," "medium" and "large – Foundry B." The colored plane represents the actual stress state, whereas the red mesh represents a state without any additional bending moment. In the case of the "small" and "medium" components, the effects of $M_y$ dominated, whereas $M_x$ played only a minor role. One artificial defect was loaded more strongly compared to the stress distribution without an additional $M_y$ loading. As can be seen from the fracture surfaces in Section 3.3.1.4, the crack subjected to higher stress propagated in an almost straight line toward the opposite side. In the case of the "large – Foundry B" component, deviations from straightness led to a dominance of $M_x$. In this case, the defects were located in the neutral fiber related to $M_x$ and were only influenced during propagation, resulting in asymmetrical crack propagation. The situation for the "large – Foundry A" specimen corresponded to those of the "small" and "medium" specimens.

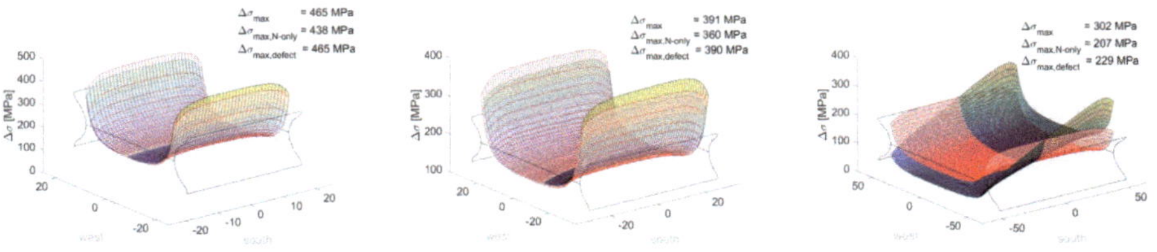


(a) (b) (c)

**Figure 4.13:** Typical stress distributions for a) small, b) medium and c) large components

An individual evaluation of the SCFs and $SCF_{max,def}$ for all specimens is shown in Figure 4.14a. The differences between SCF and $SCF_{max,def}$ were mostly negligible with the exception of GGB1, GGB2 and GGB3. In these cases, SCF was higher by a factor of up to 1.3. In GGB2, this led to the starting point of the crack growth at the component edge where a grinding groove was located. $SCF_{max,def}$ was relevant for both crack tip stress at the artificially introduced defects and crack growth and was thus considered exclusively. This was in the range of 2.1 to 2.3 and thus slightly above the design objective of 2.0 due to the additional stresses.

Figure 4.15a shows $\Delta\sigma_{max,defect}$ as a function of the number of cycles to failure determined in the experiments. Additionally, the corresponding values for $\Delta\sigma_{nom}$ were added to this plot. The regression curve of both parameters according to EN 1993-1-9:2010 [156]

was limited to the samples from Foundry B and C for identical reasons, as discussed in Section 3.3.2. The difference between $\Delta\sigma_{50\%}$ corresponded approximately to the mean value of the $\Delta\sigma_{\text{max,defect}}$. Gradients and standard deviations were not significantly influenced by this consideration.

The comparison of $\Delta\sigma_{\text{max,defect}}$ with the test results of the tensile specimens with surface defects from Section 3.2 in Figure 4.15b shows that $\Delta\sigma_{\text{max,defect}}$ led to higher stress ranges. The $\Delta\sigma_{50\%}$ value of the notched specimen was about 50 % greater than the $\Delta\sigma_{\text{net}}$ of the tensile specimens. This was due to the supporting effect of the geometric notches and illustrates the potential of considering the notch effect in the design and not performing the check purely based on maximum values.

Figure 4.14b compares the effect of the additional bending moment $M_x$ and $M_y$ with the effects of deviations from straightness according to ISO 8062-3:2008 [157]. For this purpose, the axial stresses $\sigma_{x,Mx}$, $\sigma_{x,My}$ resulting from the substitute moments were related to the nominal stress caused by the pure tensile force $\sigma_{x,N}$ by

$$\frac{\sigma_{x,M}}{\sigma_{x,N}} = \frac{M/W}{N/A}. \tag{4.5}$$

This indicates the extent to which the normal stress was increased by the additional bending moment on a square reference section. In this simplified approach, the notch effect was not taken into account. The different behavior of the component types is also visible in the graph. In the case of the "small," "medium" and "large – Foundry A" components, the shares of $M_y$ dominated and led to a related additional stress $\frac{\sigma_{x,M}}{\sigma_{x,N}}$ of

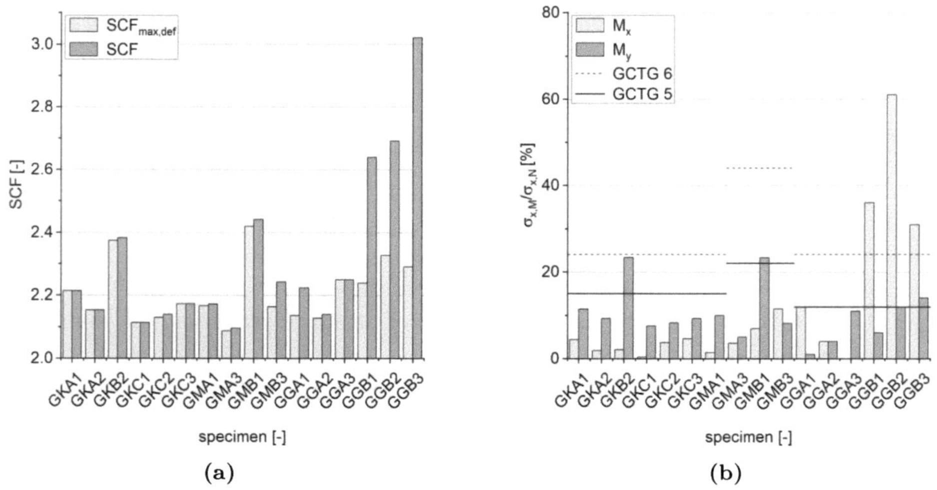


**Figure 4.14:** a) Resulting SCFs; b) Determination of additional stress caused by the substitute moments and their comparison with the straightness tolerances according to ISO 8062-3:2008 [157]

5 % to 10 % and in exceptional cases, up to 23 %. The effects of $M_x$ were in the order of 1 % to 5 % and reached 10 % in exceptional cases. In the case of the "large – Foundry B" component, the fractions from $M_x$ dominated and considerably exceeded all other comparison values with a related additional stress $\frac{\sigma_{x,M}}{\sigma_{x,N}}$ of 30 % to 60 %. The effects of $M_y$ were in the same order of magnitude as the other specimens.

Figure 4.14b also contains the limit values for the straightness tolerance group GCTG 6 according to ISO 8062-3:2008 [157] for the respective component dimensions. GCTG 6 is the most stringent requirement that can be economically applied to steel castings produced by hand molding [157]. The length of the component determines which deviation from straightness e is permissible. ("small," l < 300 mm: e = 1.4 mm; "medium," l = 1350 mm: e = 2 mm; "large" l = 1500 mm: e = 2 mm) The underlying background and conversions are described in detail in Section 5.2.7. As an additional comparative value, GCTG 5, the highest group for steel casting at mechanically supplemented sand forming is given. Most of the components were well below the GCTG 6 tolerance limits and were in the range of GCTG 5; only GGB(1-3) exceeded the values significantly. In summary, the large components were representative components. However, the geometric imperfection or effect of eccentric clamping in the testing machine was below the admissible limits according to ISO 8062-3:2008 [157]. Therefore, the test results cannot be used directly for component design, and the approaches in Section 5.2.7 are essential to a generally applicable design method.

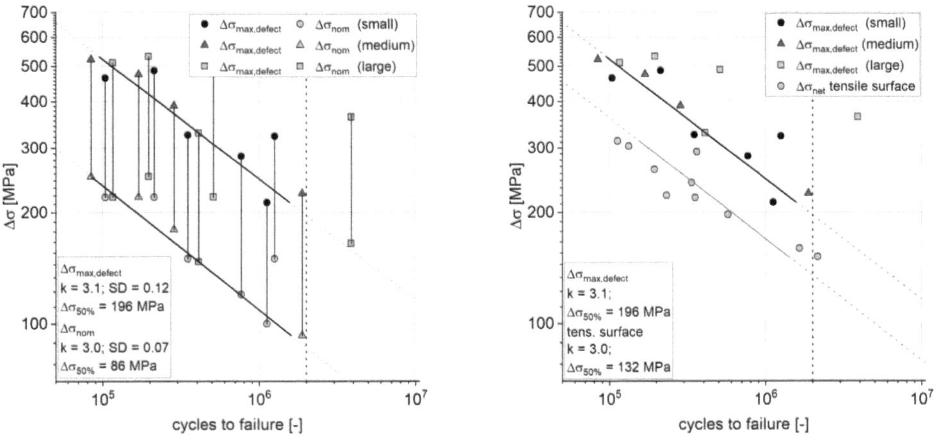


(a) (b)

**Figure 4.15:** a) Comparison of nominal stress range and maximum stress range in artificial defects; b) Comparison with tensile specimens with surface defects

### 4.2.2.3 Influence of Component Size on Crack Tip Stress

The small scatter between the lifetimes of different component sizes (see Figure 3.29) and the wall thickness independence of the Paris parameters shown in Section 3.1.2 are leading to the assumption that the influence of wall thickness plays a minor role at surface defects and that the absolute defect size dominates the effects on service life.

As the artificial defects have been defined according to UT detection limits, they had identical dimensions for all component sizes. Secondly, the components were designed such that the stress concentrations and gradients were consistent. This resulted in the situation shown in Figure 4.16a. The defects extend into the interior of the component with different percentages of component thickness. Therefore, the crack vertex of the small component was located in areas of lower stress compared to the larger components. The calculated stress increase at this point in a defect-free component would correspond to $SCF_{defect,tip,small} = 1.3$. For the large components, the substitute defects were almost completely located in the highly stressed area and resulted in $SCF_{defect,tip,large} = 1.8$.

The consideration carried out in Figure 4.16a for the nominal dimensions without considering the additional bending moment showed that up to 40 % higher stresses occurred at the crack front of "large" components. These differences should be reflected in $\Delta K_{defect,tip}$ due to the linear relationship between $\sigma$ and K according to Eq. (2.5).

Figure 4.16b shows that these effects were not fully transferred to the actual maximum stress intensity factors. For this purpose, the maximum stress intensity factors occurring under the respective substitute loads were numerically evaluated. Despite the superposition of notch effects and bending stresses, the location of the highest

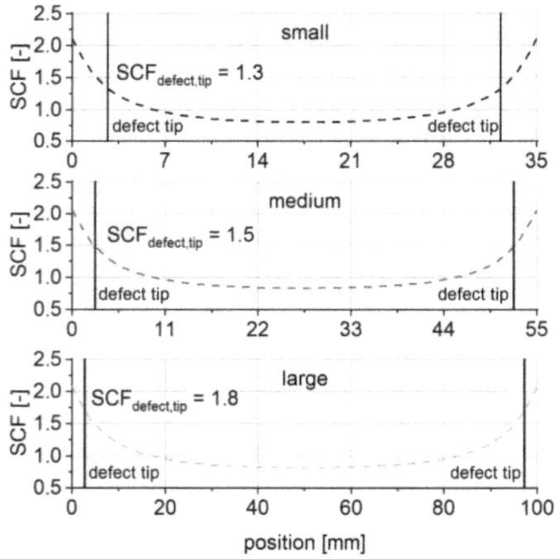


(a)

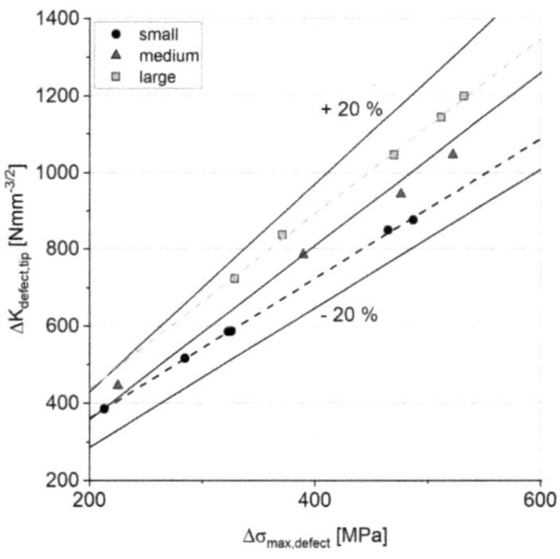


(b)

**Figure 4.16:** a) Comparison of the nominal stress range and the maximum stress range in artificial defects; b) Effect of crack tip stress in the initial state stress intensity factors in all specimens was at the vertex inside the components and not on their surfaces. The stress intensity factors determined in this way are shown as a function of $\Delta\sigma_{max,defect}$ and thus of a virtual stress on the component surface. The regression curve of all data points as well as a scattering of ± 20 % complete the representation and serve to classify the $SCF_{large}$ and $SCF_{small}$, which differ by 40 %.

It can be seen that the "large" components occupied the top line of the scatter band and the small ones occupied the bottom line. Nevertheless, the difference between the regression curves for "large" and "small," with a deviation of < 25 % in the parameter range under consideration, was smaller than the stress analysis would suggest. An analytical estimation according to Annex B.1.1 showed that the effect of component width in the considered parameter range on $\Delta K_{defect,tip}$ was less than 1 % and could not be the reason for the contrary effect.

If the major part of the service life were assigned to crack growth in the region after the initial failure, the crack tip stresses could be qualitatively transferred to the total service life. According to Eq. (5.9) and $m = 3$, the service life of "small" components would be more than twice as long as that of "large" components.

The observation of the beach mark test in Figure 3.30 and the results from [121] confirm the dominance of crack growth near the initial defect in the total service life. Nevertheless, a change in the crack shape was visible, which resulted in a disproportionate increase of crack width c caused by the component width (interaction with the surface of the finite component width see influence of c/w in Appendix ?? & stress gradient). Thus, the small differences between the component sizes in the total service life could explain and indicate an independence of the dimensions of the surface defects from the wall thickness in practical application.

### 4.2.3 Crack Propagation

Within the framework of these investigations, crack propagation calculations were carried out on the component tests in [121]. The calculations were based on the models and boundary conditions presented in Section 4.2.1. Modeling approaches as well as calculation procedures correspond to the data given herein. The simulations were limited to stable crack growth; crack initiation and plastic effects were not considered. The aim of these investigations was to quantify crack propagation behavior and evaluate essential influencing factors.

This presentation is limited to the findings relevant to this work, for a comprehensive study of the background the reader is referred to [121]. It was shown that the crack propagation of the large component tests could be recorded both qualitatively and quantitatively by calculation. The growth behavior was divided into different areas. At the beginning, an elliptical crack growth was considered to start from the artificially introduced initial defects. When the crack flanks reached the lateral surfaces of the component, a crack front propagating as a straight line was assumed. Since only a small part of the total lifetime was attributed to the transition period between these two states, this transition period was neglected.

A closer examination of the two competing initial defects in Figure 3.27b showed that the load on and consequently the growth rate of one of the two crack was reduced by the additional moment and came to a standstill in the course of growth. The highly loaded crack grew much faster from the beginning, so the center of gravity moved toward the less-loaded crack. The crack tip stresses that occurred in relation to the more heavily loaded crack resulted in increasingly smaller local crack growth increments with subsequent calculations. Finally, a compressive stress state acted on that crack and stopped its propagation. The behavior calculated by [121] corresponds well with the observations on the crack surfaces. A comparison of the calculated and experimentally determined surface strains (cf. Figure 3.31a) showed that the simulations qualitatively represented the behavior very well. Effects such as the cusp in the strain curve on the side of the lower-stressed defects were present. However, the substitute moments acting constantly during the entire crack growth led to increasing deviations from the measured data. This resulted in greater calculated stress intensity factors and consequently lower calculated lifetimes.

The sensitivity analyses in [121] showed that the selected crack propagation parameters dominated. On the other hand, the deviations between the actual and nominal geometry had no decisive influence on crack tip stress or crack growth. The formulation of the stress state at the beginning of the test by substitute moments for a crack propagation calculation was less suitable due to the facts described above.

For an actual component design, it can be derived from [121] that the mathematical evaluation of surface defects in components with stress gradients can be conservatively represented in the HCF range by a crack growth calculation. In most cases, a focus on crack growth in the immediate vicinity of the initial defect is sufficient. Its numerical representation is comparatively simple using tools such as “semi-elliptical cracks” in Ansys [3].

### 4.2.4 Summary

The aim of these investigations was to quantify the notch effect and the additional bending moments resulting from deviations from straightness from Section 3.3. The calculations were performed on representative model sections. The stress states that occurred in the test sections of the individual specimens were determined iteratively based on the strain measurements and the actual dimensions. For a more accurate assessment of damage mechanism and crack propagation, the artificially introduced defects were subjected to numerical fracture mechanics evaluation.

The stress states derived from the strain measurements were converted into substitute moments. The evaluation of the stress states confirmed the targeted SCF $\approx 2.0$. The evaluation of the fatigue resistance using local maximum stress ranges under consideration of the substitute moments, showed $\Delta\sigma_{50\%} = 196$ MPa. This fatigue resistance was a factor of 1.5 higher than the value for tensile specimens with surface defects and approximately constant stress distribution from Section 3.2. This illustrated the effect of the stress gradient on the crack propagation and consequently on the fatigue resistance.

A comparison of the effects of the additional bending moments with the tolerance groups for permissible deviations from straightness according to ISO 8062-3:2008 [157] showed that the specimens were considerably below the GCTG 6 quality requirements, which are the highest for hand-molding processes. The substitute moment My resulted in one of the two artificial defects experiencing higher stresses. Compared to a condition without $M_y$, this usually resulted in a local stress increase of 5 % to 10 % and in one exceptional case up to 23 %. Only one series (GKB1-3) showed significant deviations. In contrast to the other specimens, in these cases $M_x$ dominated and led to a stress increase on the undisturbed sides of the components(no notch and no artificial defects). Their magnitude exceeded the threshold values according to GCTG 6 and in one case led to a failure that did not originate from one of the artificial defects.

The substitute defect sizes selected independently of the wall thickness resulted in their vertices being stressed to different extents depending on the component type. Thus, the stresses in the region of the defect vertex of the "large" specimens were 40 % higher than in the case of the "small" specimens. As the calculations showed, this increase was not completely transferred to $\Delta K$. The deviations of the crack tip stresses between the two component types amounted to values $< 25$ %. From this approach, the wall thickness dependent residual cross-section between the opposing defects and interaction effects between defects and free surfaces, an explanation could be found for the insignificant influence of wall thickness on load cycles observed in the tests for wall thickness independent crack propagation parameters.

# 5 Design Concept

Within this chapter, the previous findings on defects' effects on fatigue behavior are transferred into a design concept. The objective is a concept that is easy to apply for the special requirements of the construction industry as defined in Section 1.2. The following summary of the dominating features provide the starting point for a generalized concept:

- Surface defects have a significantly stronger effect on fatigue strength than internal defects.
- Crack growth was identified as the dominant damage mechanism for surface defects, regardless of their shape. Both voluminous as well as crack-like defects act as initial cracks.
- The expression of the dominating damage mechanism in internal defects strongly depends on their shape. Shrinkage cavities were considered in the present experimental investigations. They were classified as sharp, but not crack-like, defects in [42], which was confirmed in Section 4.1 by a simulated crack initiation phase. In addition, the derivation in the gradient of the experimentally determined SN curves in Section 3.2 from the Paris exponent supports this statement. Unless it cannot be clearly ensured that a specific internal defect is a pure volume defect, the worst case of a crack should be assumed in this generalization. This corresponds to the approaches in [35, 96, 103].
- The crack tip stresses of a crack that surrounds a volume defect are lower than those of a pure planar crack of the same shape due to changed stress trajectories, as shown in [103] and Section 4.1.
- Geometric tolerances of a component cause additional non-negligible stresses, as shown in Section 4.2.

The derivation of a design concept based on the quality grades according to EN 1559-2:2014 [152] is not target oriented for the following reasons:

- As discussed in Section 2.1, different acceptance limits for casting defect sizes of the same quality grade are allowed depending on the NDT method. Since the defect size is the decisive influencing factor apart from its location, a generalization is not possible.

- The current edition of EN 12680-1:2003 [144] (UT) does not define any minimum spacing between multiple defects. From a technical point of view, this is a step backwards from past codes. In Section 2.2.4.5, it was shown that the interaction of adjacent defects could lead to increased crack tip stresses. Thus, additional requirements like those given in DIN 1690-2:1985 [139], FKM-Rili-BM:2009 [161] and BS 7910:2013 [138] are indispensable for a stress analysis.

- The comparison in Figure 5.1 clarifies the differences in the permissible defect size and NDT evaluation. A semi-elliptical surface crack is compared to an elliptical internal crack. Both, the ratio a/c = 0.4 as well as boundary conditions concur. The illustrated defect sizes were derived from a limitation of $\Delta K_{I,max}$ to $\Delta K_{th}$ at $\Delta\sigma = 60$ MPa. In other words, they represent the maximum defect sizes for a calculated endurance strength of $\Delta\sigma_{endu} = 60$ MPa. In the specific example, the internal defect may cover three times the area of the surface defect and would be assigned to the UT2 quality class according to EN 12680-1:2003 [144]. The surface defect can be assigned to AM5 according to EN 1369:2013 [148] or UT1 according to EN 12680-1:2003 [144]. From a technical point of view, the evaluation of the surface defect according to UT is questionable, but it is permissible in terms of the standards.

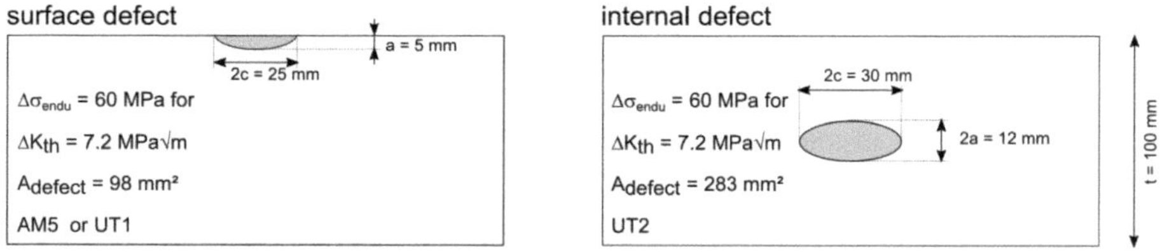


**Figure 5.1:** Comparison of the maximum permissible single defect sizes on the surface and inside the component with $\Delta\sigma = 60$ MP and $\Delta K = \Delta K_{th}$

## 5.1 Basic Idea

The design concept is outlined by the following principles:

- The design resistance as well as the verification of the quality at the cast component is based on newly defined Resistance Categories (RCs).

- Internal and surface defects are represented by initial cracks, so only crack growth is considered. This covers the worst case and is very conservative, especially for lower stress ranges in internal defects, but still allows for large defect sizes and economic requirements as proven in Section 5.4.

- The design is carried out by comparing local or nominal stresses with resistance curves using a stress approach similar to that described in EN 1993-1-9:2010 [156] for welded connections.

Figure 5.2 provides an overview of the core issues and quality requirements the designer needs to give to the foundry. The RCs act as a link and communication medium between the designer thinking in fatigue classes (FAT) and the foundry providing evidence with respect to quality requirements and defect properties.

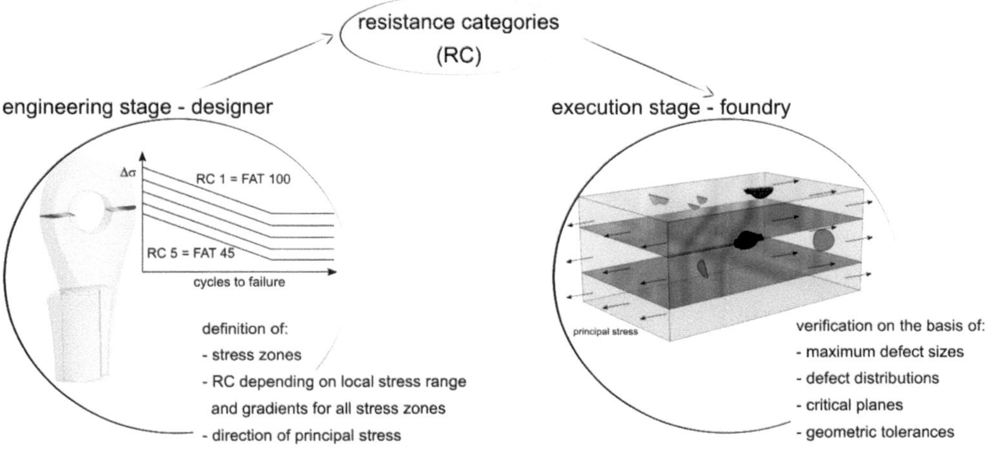


**Figure 5.2:** Resistance categories linking engineering stage with execution stage

In an application of this concept, the following additional points must be considered:

- In order to use the full potential, the designer should make statements on the RCs for separate zones of different stress utilization.
- Information on the direction of maximum principal stresses enables a focused NDT testing adapted to the intended use of the component.
- Failure due to brittle fracture under static loading is covered by this approach when certain additional requirements are fulfilled.
- If stress gradients occur, an additional subdivision of the RCs into the core and rim zones (definition of core and rim zone acc. to [144]) can be useful.
- The requirements for geometric tolerances must meet DCTG 11 for component thickness and GCTG 6 for straightness according to ISO 8062-3:2008 [157]. Additionally, more rigorous requirements can be defined. Their positive effects can be taken into account directly when determining the stresses.
- The specified permissible defect sizes determined in this thesis are not NDT indications. A translation of the defect sizes into NDT display characteristics is still necessary. The special requirements of cast components need to be considered, as it is the scope of the DeQaGuss research project [115].

## 5.2 Development of the Concept

The design concept is developed according to the steps illustrated in Figure 5.3. Each of these topics is discussed in an individual section. Useful FAT classes are selected from EN 1993-1-9:2010 [156] based on engineering requirements. Minimum material properties are discussed in the context of the present investigations and the literature. Following the concept of [103], simplified reference cases are specified and form the basis of all investigations. These cases are used to generate the link between fatigue resistance and tolerable defect sizes. In further steps, influencing factors such as additional ultimate limit state (ULS) loading, stress gradients, the effect of multiple defects and geometric deviations are investigated. Finally, the design concept and derived defect sizes are specified.

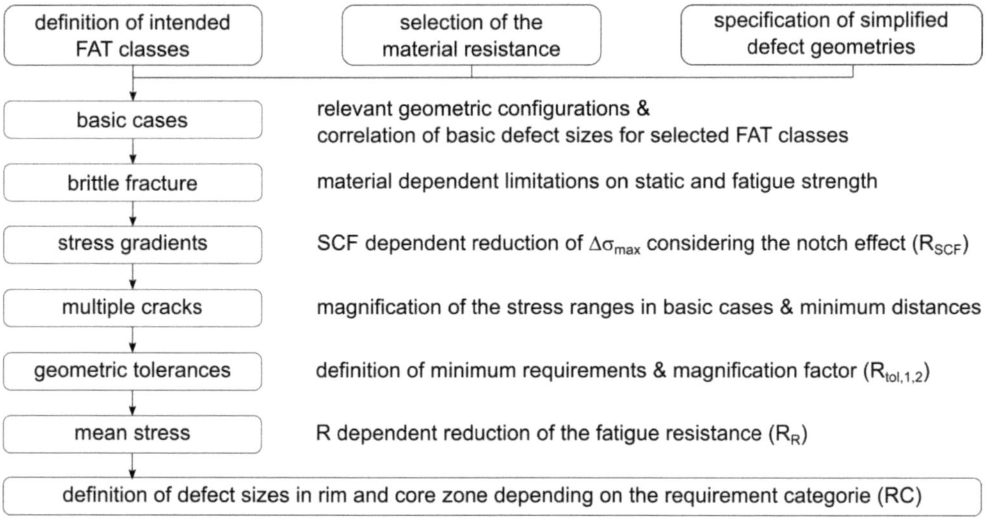


**Figure 5.3:** Procedure for the development of the design concept

### 5.2.1 Definition of the Loading Classes

EN 1993-1-9:2010 [156] provides 14 FAT classes downwards with a spacing of $\frac{\sqrt{5}}{2}$. The derivation of quality requirements for each individual class is not feasible, but Table 5.1 summarizes a well-founded selection.

Table 5.1: Selection of FAT classes

| FAT | reasons for the selection |
|---|---|
| 112 | Corresponds to the maximum FAT from [78] (k = 5; bending stress of butt weld joint of circular hollow section, with backing strip). Cannot be proven by fracture mechanics. |
| 100 | Corresponds to the FAT class from [78] (k = 5; bending stress of butt weld joint of circular hollow section, without backing strip). Presumably, the highest value that can be verified by fracture mechanics when taking the UT and MT detection limits into account. |
| 90 | FAT class for fully penetrated butt welds of circular hollow sections with $t > 8$ mm according to EN 1993-1-9:2010 [156] (table 8.6 detail 3). In [35], 87 MPa was assumed decisive for the welded joint. |
| 71 | FAT class for fully penetrated butt welds of circular hollow sections with $t \leq 8$ mm according to EN 1993-1-9:2010 [156] (table 8.6 detail 3). |
| 56 | Low FAT class to allow larger defects in low-stress areas. |
| 45 | Low FAT class to allow larger defects in low-stress areas. |
| 36 | Not suitable, as this leads to excessive defect sizes. |

### 5.2.2 Input Parameters

The material parameters to be used for the development are summarized in Table 5.2. These correspond to the minimum requirements for material 1.6220 according to EN 10340:2008 [142]. Parameters that are not directly regulated are discussed and chosen conservatively or are limited via substitute parameters. The Paris coefficients, material strength and the results of the fatigue tests in Chapter 3 exceeded these minimum requirements by far and underline the conservatism. The ductile failure mode observed in Section 3.2.2.8 is not taken into account, as the minimum required KV value could cause brittle fracture as identified in [103]. Material 1.6760 according to SEW 520:2017 [164] needs to fulfill higher values and thus is covered by the following considerations. Only in the case of brittle fracture tests under ULS loads, special considerations are necessary for this material. This is caused by higher permissible stresses under static loading.

A generally applicable selection of the crack propagation parameters or a coupling to existing minimum requirements according to the technical delivery conditions is not possible. The aim of the following considerations is to select an approach based on historical data that is as universal as possible and not overly conservative. A mandatory determination of the Paris parameters in each individual case is not considered an objective and contradicts this general approach.

The results from ErStaGu [116] (no. 1 in Table 3.2), a simplified approach according to BS 7910:2013 [138] (no. 16 in Table 3.2) and the approach for a more specified use according to BS 7910:2013 [138] (no. 12 in Table 3.2) are compared. Both approaches in BS 7910:2013 [138] are not explicitly derived for cast steel, but rather aim at a much wider range of applications.

**Table 5.2:** Summary of material parameters (minimum requirements, characteristic or design values)

| parameter | explanation for the selection |
|---|---|
| $\sigma_y = 300\,\mathrm{MPa}$ | Minimum requirements according to EN 10340:2008 [142]. The additional requirement $\sigma_y = 360\,\mathrm{MPa}$ is useful for static load cases, but its effects on fatigue resistance are insignificant according to the findings in Section 2.3. |
| $K_{Ic} = 4200\,\mathrm{Nmm}^{-3/2}$ | Calculated from the minimum required values in a Charpy test for $T_{Ed}$ = -30 °C, as done in [103] according to the master curve concept of [161]. As discussed in [103], other concepts (EN 1993-1-9:2010 [156] or SINTAP [6]) result in higher values. The specification in [35] of $K_{Ic} = 2400\,\mathrm{Nmm}^{-3/2}$ according to [88] was assessed as very conservative in [35]. |
| $\Delta K_{th} =$ <br> $149\,\mathrm{Nmm}^{-3/2}$ (R = 0.1) <br> $63\,\mathrm{Nmm}^{-3/2}$ (R > 0.5) | According to the estimation $\Delta K_{th} = 170 - 214 \cdot R$ for $0 \leq R \leq 0.5$ from BS 7910:2013 [138]. The test results are thus represented conservatively. In the case of surface cracks $\Delta K_{th} > 63\,\mathrm{Nmm}^{-3/2}$ is only valid for a > 1 mm |
| | (in this context understood as a characteristic value) |
| m = 2.88 <br> $C = 3.98 \cdot 10^{-13}$ | As a result of the following discussion. For $\Delta K$ in $\mathrm{Nmm}^{-3/2}$ and da in mm/cycle (mean value acc. to BS 7910:2013 [138]) |

For a detailed comparison, the fracture mechanics-based SN curves resulting from the considered parameters are extracted from Figure 3.2b and shown in Figure 5.4a. In addition, selected reference values are presented and discussed. Figure 5.4b shows the maximum permissible defect sizes in order to achieve $2 \cdot 10^6$ cycles to failure at $\Delta\sigma = 71\,\mathrm{MPa}$. The defect sizes are based on a crack propagation calculation. In all cases, semi-elliptical surface cracks are considered according to the situation in Figure 3.2b. For a classification into quality grades according to EN 1369:2013 [148] (MT), references are supplemented in the diagram.

The characteristic values for the fracture mechanics parameter (m, C(R), $\Delta K_{th}(R)$) determined in the experimental investigations of Section 3.1.2 lead to significantly longer service lives and considerably larger permissible initial defects compared to the values from BS 7910:2013 [138]. In thicknesses t > 30 mm [87, 155], these correspond to AM 5–6 according to EN 1369:2013 [148]. For a generally applicable design concept, the number of samples for the da/dN tests carried out in ErStaGu [116] is too small for a profound statistical evaluation, and a generally applicable conservative approach cannot be derived.

The simplified approach according to BS 7910:2013 [138] leads to extremely low resistances and is strong overly conservative compared to all other literature data. According to BS 7910:2013 [138], the objective of this definition was to provide "preliminary screening assessments" for a general initial assessment with low requirements on the material properties. The permissible initial defect sizes derived from this cannot be implemented economically and contradict the experiences summarized in Section 2.3 and Chapter 3.

Table 3.2 shows identical values according to HSE:1995 [166] and BS 7910:2013 [138] at different stress ratios. The values according to HSE:1995 [166] for R = 0.1 correspond to the values according to BS 7910:2013 [138] for R < 0.5. Compared with the results from

ErStaGu [116], this corresponds to the behavior at R = 0.5. Furthermore, the values according to HSE:1995 [166] for R = 0.5 are assigned to R > 0.5 in BS 7910:2013 [138]. The results from the experimental investigations of [44] for R = 0.1 tend to behave as determined in ErStaGu [116]. The approach according to BS 7910:2013 [138] for R < 0.5 seems to be calibrated conservatively by results for R = 0.5. The derivation of the design value using a 2SD shifting is less than the difference between the test results for R = 0.1 and the mean value according to BS 7910:2013 [138] for R = 0.5.

For further calculations, the following approach is used as indicated in Table 5.2, using the mean value according to BS 7910:2013 [138] for R < 0.5 for an assessment at R = 0.1. According to Figure 3.2b, good agreement is achieved with the assumption in [35]. An additional validation via $\Delta K_{th}$ of these characteristic values is done in Section 5.2.3.

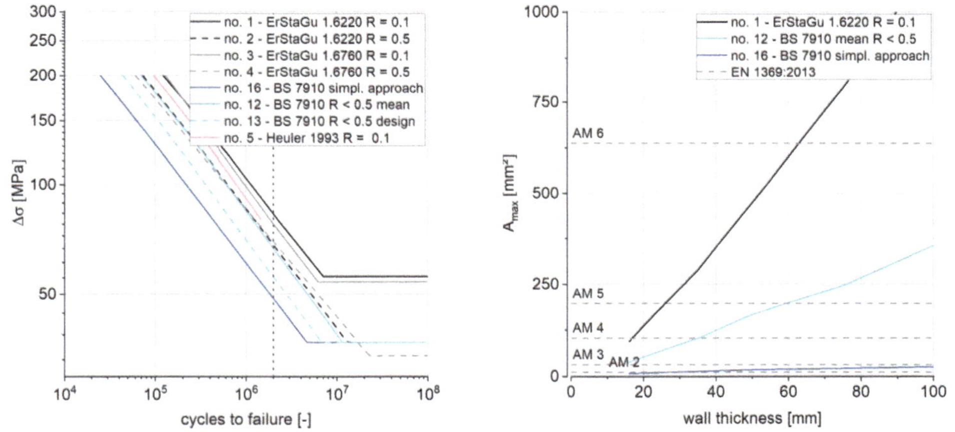


(a) (b)

**Figure 5.4:** a) Comparison of Paris parameters using fracture mechanics-based SN curves; b) Permissible defect sizes for surface defects at $\Delta\sigma = 71$ MPa and $2 \cdot 10^6$ cycles to failure from crack growth calculations

## 5.2.3 Derivation of Basic Resistance Values and Defect Sizes

Figure 5.5 illustrates the applied procedure to link fatigue resistance curves to maximum tolerable defect sizes. Analogous to [103], the basic investigation cases shown in Figure 5.6a and Figure 5.9a form the foundation of the following procedures. Due to the different effects of internal and surface defects, both are studied separately using analytical crack growth calculations. Loads causing nominal stress ranges according to FAT 100, 71 and 45 at $2 \cdot 10^6$ cycles are applied to the basic investigation cases with variable wall thicknesses of 16 mm, 35 mm, 50 mm, 75 mm and 100 mm. The initial crack geometry is selected to represent the most unfavorable state with respect to crack tip stresses by a/c = 0.4 [28]. The maximum initial defect sizes $A_{cg}$ result from crack growth calculations and are defined by a resulting lifetime of $2 \cdot 10^6$ cycles to failure at $\Delta\sigma = \Delta\sigma_c(\text{FAT})$. Based on investigations on endurance behavior, the maximum initial defect sizes are updated to $A_{Kth}$. Finally, the resulting defect sizes are compared to quality requirements according to EN 12680-1:2003 [144] and EN 1369:2013 [148].

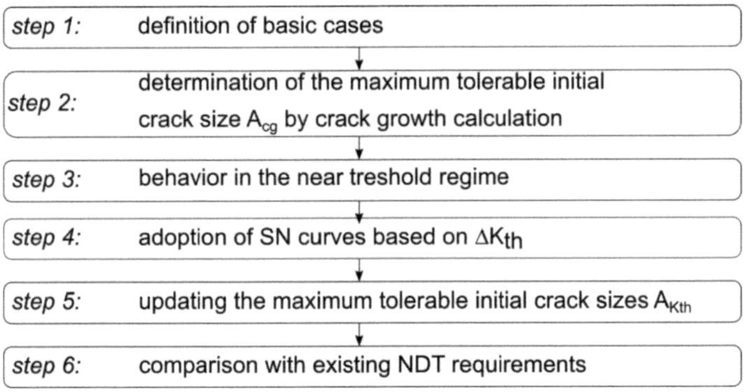


**Figure 5.5:** Procedure to link fatigue resistance to defect sizes for basic cases

These investigations purely focus on single defects and are based on the material parameters for R = 0.1 according to Table 5.2. Any effects of stress concentration caused by additional bending moments or geometric shapes are not considered within these investigations. A generalization follows in the subsequent sections.

#### 5.2.3.1 Surface Defects

*Step 1: Definition of basic cases*
Figure 5.6a illustrates the basic situation for the investigations of surface defects. Stress intensity factors are calculated according to Annex B.1. The plate geometry defined by the ratio $\frac{2w}{t} = 7.5$ is chosen according to [103] and [54]. The crack growth calculation follows the concept described in Section 2.2.4 using the stress intensity factors at the surface $\Delta K_{I,c}$ and the internal apex $\Delta K_{I,a}$. The step size is set to an increment of $da_{max} = 0.01$ mm at the point of maximum crack tip stress. The remaining increment is calculated according to Eq. (2.10). Both the exceedance of the limit curve in a FAD and the wall breakthrough of the crack define failure.

*Step 2: Determination of the maximum tolerable initial crack size $A_{cg}$ using crack growth*
Figure 5.6b summarizes the resulting defect sizes that cause failure after $2 \cdot 10^6$ cycles under an applied nominal stress rage of 100, 71 and 45 MPa. The calculations are marked as dots and are interpolated linearly. As a reference, the quality requirements according to EN 12680-1:2003 [144] (MT) for surface defects and according to EN 1369:2013 [148] (UT) for subsurface defects in the rim zone are illustrated. In both cases, the NDT displays are transformed into defect geometries using the equal surface concept according to FKM-Rili-BM:2009 [161].

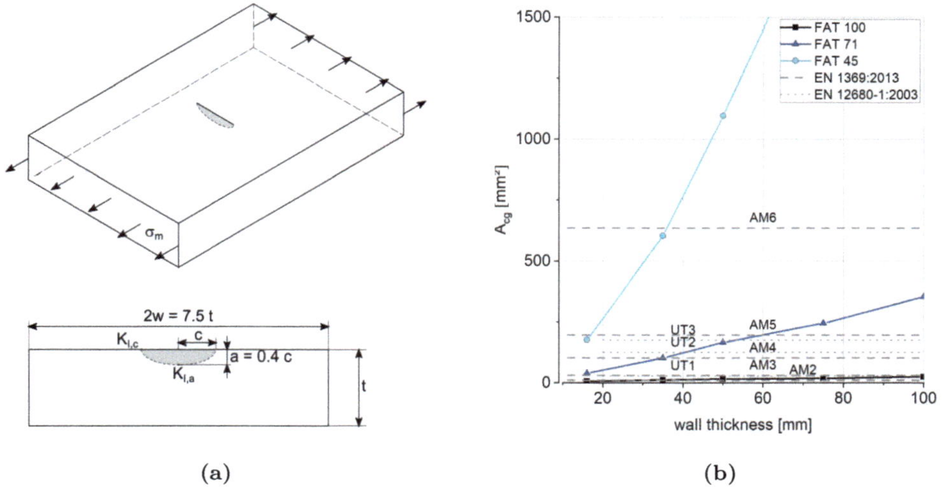


**Figure 5.6:** Surface defects: a) basic investigation case; b) maximum defect surface sizes determined by crack growth simulations

In contrast to the requirements of the NDT standards, the calculations show a dependency on wall thickness. The defect sizes that result for a nominal stress range of 100 MPa are in the range of the highest MT requirements for an untreated casting surface. For lower FAT classes, the resulting crack sizes distinctively exceed the NDT requirements, which is in line with the findings of [35].

*Step 3: Behavior in the near threshold regime*
In this fracture mechanics approach, endurance is not defined a priori by a fixed number of cycles but by the loading that causes $\Delta K \leq \Delta K_{th}$. To set the cycles to failure $N_{endu}$ at the endurance limit and check its conformity with EN 1993-1-9:2010 [156], the behavior in the threshold region is investigated more thoroughly. $N_{endu}$ is calculated according to the procedure discussed in Section 2.2.4.4 to determine fracture mechanics-based SN curves. This was done for all wall thicknesses, FAT-classes and the associated maximum defect sizes $A_{cg}(FAT)$. In all cases, $N_{endu}$ varies from $4 \cdot 10^6$ to $3.5 \cdot 10^7$ cycles and thus is higher than defined in EN 1993-1-9:2010 [156] in most cases. The resulting $\Delta\sigma$ where $\Delta K = \Delta K_{th}$ are shown in Figure 5.7a. Additionally, the stress ranges that are associated with the considered FAT classes at $5 \cdot 10^6$ and $1 \cdot 10^7$ cycles to failure are shown. These are determined by extrapolation according to Eq. (5.1) and $m = 2.88$.

$$\Delta\sigma_a = \Delta\sigma_c \cdot \left(\frac{2 \cdot 10^6}{N_a}\right)^{\frac{1}{m}} \tag{5.1}$$

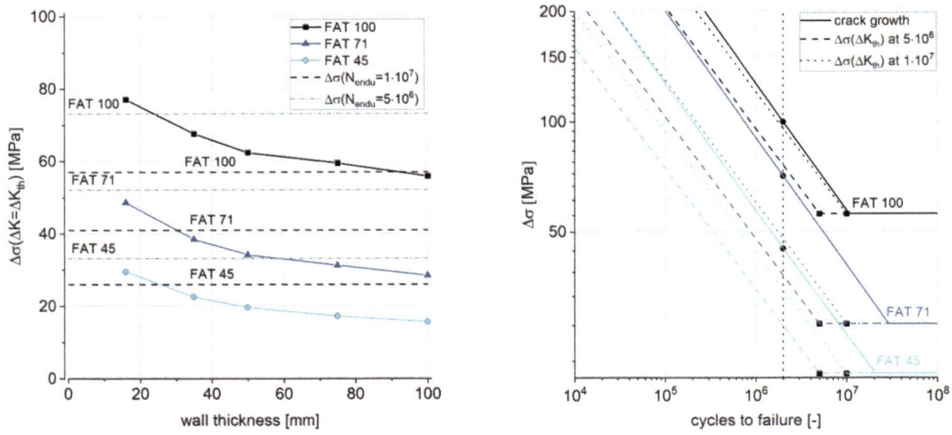


(a) (b)

**Figure 5.7:** a) Nominal stress ranges at $\Delta K_{max} = \Delta K_{th}$ and $A_{cg}$; b) SN curves with $\Delta\sigma(\Delta K_{th})$ at different reference points compared to the results of crack growth simulations (t = 50 mm)

The comparison of the wall thickness and defect dependent values with $\Delta\sigma$ at $N_{endu} = 5 \cdot 10^6$ deviates from the FAT class and shows that cracks still propagate at stress ranges that would be considered endurant according to EN 1993-1-9:2010 [156]. This means that the maximal tolerable initial crack sizes derived from the crack growth simulation and the definition of $N_{endu} = 5 \cdot 10^6$ according to EN 1993-1-9:2010 [156] underestimates the resistance for low stress ranges. It is pointed out that $\Delta K_{th}$ and the considered Paris parameters have a different background. This is assumed to be the reason for these differences but adds robustness as the worst of two approaches is considered. Choosing $N_{endu} = 1 \cdot 10^7$ is in line with the concept of [96] and most of the studies discussed in Section 2.3.1. With this adaption, theoretically unsafe design scenarios are limited to wall thicknesses of less than $\approx$ 30 mm. This condition is addressed in Section 5.4.3.

*Step 4: Adoption of SN curves based on $\Delta K_{th}$*

To eliminate these uncertainties, a $\Delta K_{th}$-based approach to derive the maximum initial defect sizes is applied. At first, the endurance limit is defined according to the results of step 3 at $1 \cdot 10^7$. Then, $\Delta\sigma(\Delta K = \Delta K_{th})$ is calculated and provides an anchor point for the following SN curves. The gradient of the so-defined SN curves is set to m = 3 in accordance with EN 1993-1-9:2010 [156]. Exemplary results for defect sizes $A_{cg}$(FAT 100), $A_{cg}$(FAT 71) and $A_{cg}$(FAT 45) for t = 50 mm are presented in Figure 5.7b. These curves are compared to the fracture mechanics-based curves from step 2. To illustrate the effects of the definition of $N_{Kth}$, the state for $N_{Kth} = 5 \cdot 10^6$ cycles is also shown. These $\Delta K_{th}$-based curves provide increasingly conservative results with an increasing initial defect size (i.e., with decreasing FAT class). An approach based on $N_{Kth} = 5 \cdot 10^6$ cycles provides overly conservative results. As a result of these investigations, $1 \cdot 10^7$ is defined as the number of cycles from which the endurance limit is reached. Further service life curves are derived with the gradient m = 3 based on $\Delta\sigma(\Delta K_{th})$.

*Step 5: Update of the maximum initial defect sizes based on* $\Delta\sigma(\Delta K_{th})$
To derive maximum defect sizes associated with the FAT classes that are targeted in Section 5.2.1, the stress ranges $\Delta\sigma_{endu}$ are evaluated by

$$\Delta\sigma_{endu} = \Delta\sigma_c \cdot \left(\frac{2 \cdot 10^6}{1 \cdot 10^7}\right)^{\frac{1}{3}} \tag{5.2}$$

and form the load cases for a static calculation of the stress intensity factors. Iteratively, the defect sizes $A_{Kth}$ are determined for all wall thicknesses in such a way that $\Delta K_{max} = \Delta K_{th}$ for the individual $\Delta\sigma_{endu}$. The results are shown in Figure 5.8.

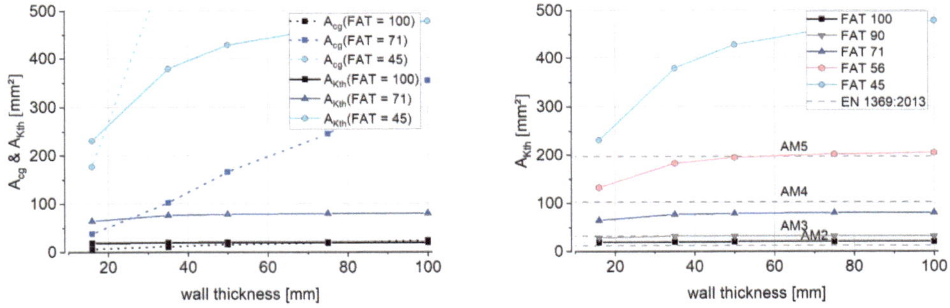


(a) (b)

**Figure 5.8:** a) Comparison of $A_{cg}$ and $A_{Kth}$; b) Classification of $A_{Kth}$ into NDT requirements

The comparison in Figure 5.8a between $A_{Kth}$ and $A_{cg}$ shows that initial defect sizes defined by $\Delta K_{th}$ are smaller than those defined by crack growth for wall thicknesses larger than 30 mm. The design approach of [103] generally requires the highest quality grades for wall thicknesses t < 30 mm and quotes [87, 155] for approval. The present approach tries to avoid this limitation, so a combined approach is used for t < 30 mm, as discussed in Section 5.4.3.

*Step 6: Comparison with existing NDT requirements*
In Figure 5.8b, a comparison with the quality requirements according to the existing NDT standards is made analogous to Figure 5.7a. In addition, FAT 90 and FAT 56 are added. The newly defined maximum initial defect sizes $A_{Kth}$ are of equal dimensions to common manufacturing qualities. The permissible defect sizes for RC 5 with FAT 45 are classified as large for designs subject to fatigue loading and are not extended to lower FAT classes and larger permissible initial defects.

#### 5.2.3.2 Internal Defects

Internal defects are considered according to the geometrical situation shown in Figure 5.9a. The procedure and essential statements are identical to those for the surface defects. The representation is reduced to the comparison to existing quality requirements analogous to Figure 5.8b. Intermediate steps discussed in detail in the context of surface defects are given in Annex E without further comments.

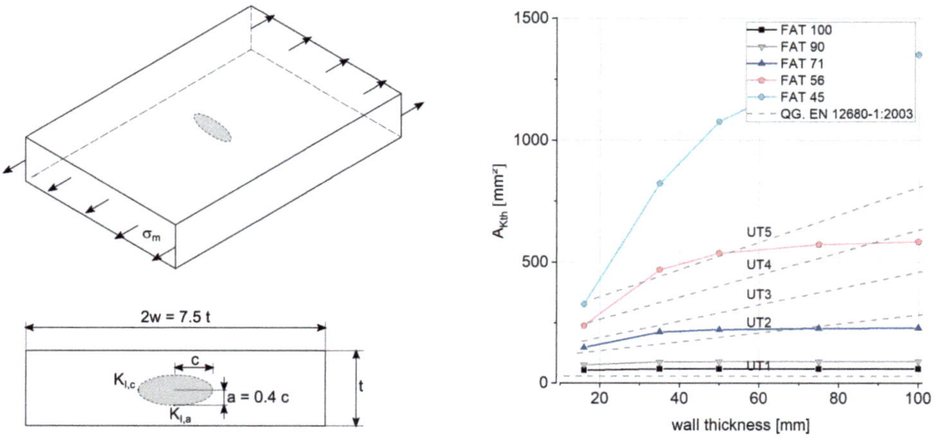


(a) (b)

**Figure 5.9:** Internal defects: a) Basic investigation case; b) Classification of $A_{Kth}$ into the requirements of NDT standards

### 5.2.4 Interaction of Static and Cyclic Loading

The approach developed herein is based on the assumption of a crack-like initial defect. This defect grows in the course of the calculated service life until a critical state is reached. In common structural applications, static forces act in addition to the fatigue loading. These can occur during the entire service life as mean stress (see Section 5.2.8) more or less frequently as well as in exceptional design situations only. The most unfavorable condition is shown in the lower part of Figure 5.10 and is reached when the design load case for brittle fracture occurs at the end of the calculated service life. In this concept, the maximum allowable defect size required for the brittle fracture analysis (e.g., according to [103]) must be considered as the final state of crack growth from fatigue loading. This approach was also implemented in [35]. In addition, Figure 5.10 shows an example of a component that is exclusively subjected to fatigue loads. In this case, the state in the FAD at $\sigma_o$ or the wall breakthrough defines the failure state.

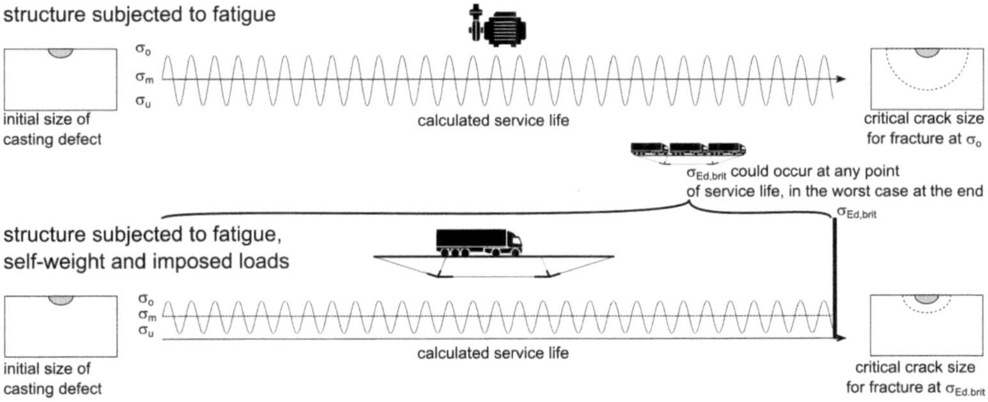


**Figure 5.10:** Influence of the load situations on the critical crack size relevant for failure

According to EN 1993-1-10:2010 [154], the actions for brittle fracture design are to be determined as an exceptional load case. Thus, $\sigma_{\mathrm{Ed,brit}} < 0.75\,\sigma_{\mathrm{y}}$ is generally applicable. Both EN 1993-1-10:2010 [154] and [103] consider the brittle fracture resistances in clusters of $\sigma_{\mathrm{Ed,brit}} = 0.50$ to $0.75\ \sigma_{\mathrm{y}}$, $0.25$ to $0.50\ \sigma_{\mathrm{y}}$ and $0$ to $0.25\ \sigma_{\mathrm{y}}$. The condition $\sigma_{\mathrm{o}} = \sigma_{\mathrm{Ed}}$ corresponds to pure fatigue loading and values of $\sigma_{\mathrm{o}} > \sigma_{\mathrm{Ed}}$ is not considered. For each FAT class and material to be tested, the scenarios shown in Table 5.3 cover the entire parameter range.

For each scenario, crack growth was calculated according to the procedure on the reference components defined in Section 5.2.3. The calculations were carried out on components with a representative wall thickness of 50 mm, and the maximum permissible initial defect sizes $\mathrm{A_{Kth}}$ for RC 1, RC 2 and RC 5 deviated from $\Delta\mathrm{K_{th}}$, as discussed in Section 5.2.3. The results of these calculations were load cycles $\mathrm{N_{brit}}$ at which the crack reached the critical state according to the FAD under brittle fracture stress $\sigma_{\mathrm{Ed,brit}}$ or wall breakthrough. To determine the comparison value $\mathrm{N_{fat}}$, the calculation was carried out until the critical state was reached under the effect of $\sigma_{\mathrm{o}}$. $\mathrm{K_{Ic}}$ was selected according to Table 5.2 and considers a design temperature of $\mathrm{T_{Ed}}$ = -30 °C.

Figure 5.11a shows a representative FAD for both fatigue loading and $\sigma_{\mathrm{Ed,brit}}$. In the present example, the calculated failure state at $\sigma_o$ did not occur due to plastic or brittle fracture but due to a wall breakthrough as shown in Figure 5.11b. Herein, both $\mathrm{N_{brit}}$ and $\mathrm{N_{fat}}$ are marked, and the critical crack dimension for a failure at $\sigma_{\mathrm{Ed,brit}}$ can be determined by the dimensions a and c.

**Table 5.3:** Summary of the cases under investigation

| static load | | $\sigma_{\mathrm{Ed,brit}} = 0.75\,\sigma_{\mathrm{y}}$ | $\sigma_{\mathrm{Ed,brit}} = 0.50\,\sigma_{\mathrm{y}}$ | $\sigma_{\mathrm{Ed,brit}} = 0.25\,\sigma_{\mathrm{y}}$ |
|---|---|---|---|---|
| | case 1 | $\Delta\sigma = 0.75(1-R)\sigma_{\mathrm{y}}$ | $\Delta\sigma = 0.5(1-R)\sigma_{\mathrm{y}}$ | $\Delta\sigma = 0.25(1-R)\sigma_{\mathrm{y}}$ |
| fatigue | case 2 | $\Delta\sigma = 0.9\,\Delta\sigma_{\mathrm{case\ 1}}$ | $\Delta\sigma = 0.9\,\Delta\sigma_{\mathrm{case\ 1}}$ | $\Delta\sigma = 0.9\,\Delta\sigma_{\mathrm{case\ 1}}$ |
| load | case 3 | $\Delta\sigma = \Delta\sigma_{\mathrm{FAT}}$ | $\Delta\sigma = \Delta\sigma_{\mathrm{FAT}}$ | $\Delta\sigma = \Delta\sigma_{\mathrm{FAT}}$ |
| | case 4 | $\Delta\sigma = \Delta\sigma_{\mathrm{endu}}$ | $\Delta\sigma = \Delta\sigma_{\mathrm{endu}}$ | $\Delta\sigma = \Delta\sigma_{\mathrm{endu}}$ |

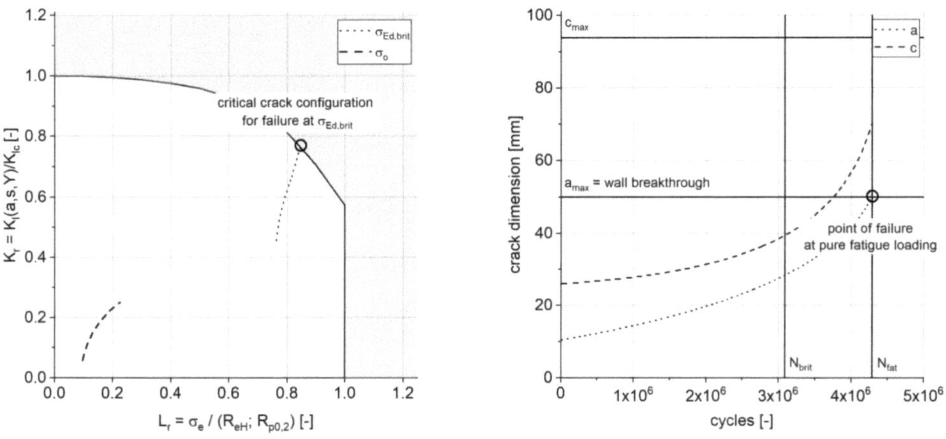


(a) (b)

**Figure 5.11:** Failure criterion for crack growth simulation: a) FAD for fatigue and ULS load, b) Crack dimensions depending on the load cycles

### 5.2.4.1 Surface Defects

These investigations were carried out for both materials 1.6220 and 1.6760 with $\sigma_y$, according to the technical delivery conditions and similar crack growth parameters used in Section 5.2.2.

Figure 5.12 shows the ratio $N_{brit}/N_{fat}$ depending on the utilization clusters $\sigma_{Ed,brit}/\sigma_y$ and on the ratio $\sigma_o/\sigma_{Ed,brit}$ for surface defects and both materials separately. In the cases of $\sigma_o/\sigma_{Ed,brit} = 1.0$, $N_{brit}/N_{fat} = 100\ \%$ follows by definition. Values of 100 % for other $\sigma_o/\sigma_{Ed,brit}$ are possible and occur when the fracture state is not defined by the FAD but is caused by a wall breakthrough. At a low utilization of $\sigma_{Ed,brit}/\sigma_y = 0.25$, both materials behave similarly, and their service lives are not influenced by $\sigma_{Ed,brit}$ as the geometric boundaries of the crack are dominant in the failure state. For material 1.6220, this is also valid for $\sigma_{Ed,brit}/\sigma_y = 0.5$. Only for the stress cluster $\sigma_{Ed,brit}/\sigma_y = 0.75$ larger deviations in the service life are noted and more pronounced with an increasing initial defect size. For defects with an initial size that corresponds to RC 1, failure due to $\sigma_{Ed,brit}$ occurs at 93 % of the service life for pure fatigue loading. For an initial defect according to RC 5, this value decreases to 72 %.

The results for material 1.6760 in Figure 5.12b are more pronounced. In several cases, the differences between $N_{brit}$ and $N_{fat}$ are large, but $N_{brit}$ exceeded $1 \cdot 10^7$ cycles. These cases occur for low $\sigma_o/\sigma_{Ed,brit}$ ratios and are marked in the diagram. Similar to material 1.6220, larger initial defects cause a larger influence on $N_{brit}/N_{fat}$, especially for $\sigma_{Ed,brit}/\sigma_y = 0.75$ and RC 5. In the worst scenario, only 20 % of the fatigue life is reached.

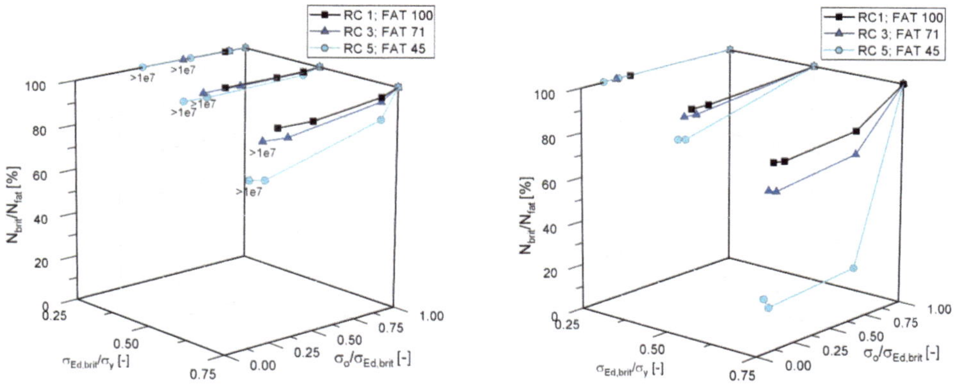


(a) (b)

**Figure 5.12:** Influence of $\sigma_{\mathrm{Ed,brit}}$ on the service life of the basic investigation case: surface defects for material a) 1.6220 and b) 1.6760

For a more precise observation, the cycles $\mathrm{N_{brit}}$ and $\mathrm{N_{fat}}$ resulting from the crack growth simulations are shown in Figure 5.13 separately by material for all three initial defect sizes. As the effects are the most pronounced for $\sigma_{\mathrm{Ed,brit}}/\sigma_{\mathrm{y}} = 0.75$, this case is studied exclusively. In addition, the SN curve derived from $\Delta\mathrm{K_{th}}$ without any consideration of $\sigma_{\mathrm{Ed,brit}}$ in Section 5.2.3 is added. As discussed in Section 5.2.3, the design curves based on $\Delta\mathrm{K_{th}}$ are more conservative from an increasing initial defect size compared to the results of the crack growth simulation. At the same time, the effects of $\sigma_{\mathrm{Ed,brit}}$ increase with increasing RC respectively increasing defect size. For material 1.6220, the design SN curves overcompensate for the negative effects of $\sigma_{\mathrm{Ed,brit}}$ in the entire parameter range. As shown in Figure 5.13b for material 1.6760, the differences are more significant and the design SN curves are no longer conservative for large initial defects.

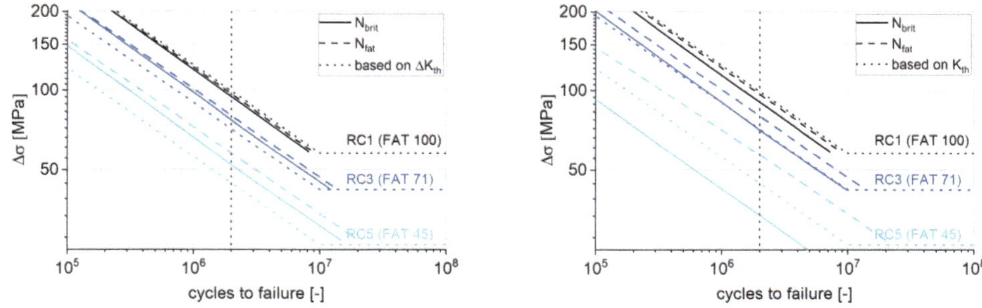


(a) (b)

**Figure 5.13:** Comparison of $\mathrm{N_{brit}}$ and $\mathrm{N_{fat}}$ with the design SN curves from $\Delta\mathrm{K_{th}}$ for surface defects at t = 50 mm and $\sigma_{\mathrm{Ed,brit}}/\sigma_{\mathrm{y}} = 0.75$ for material a) 1.6220 and b) 1.6760

### 5.2.4.2 Internal Defects

Analogous to the investigations on surface defects, the results of the studies on internal defects are shown in Figure 5.14. For the material 1.6220, no negative effects of $\sigma_{\mathrm{Ed,brit}}$ on the service life were found. In the entire parameter range, the wall breakthrough occurs before a crack tip stress-induced failure or exceeding of the yield strength is detected. For material 1.6760, the effects are much more critical than for surface defects. At $\sigma_{\mathrm{Ed,brit}}/\sigma_{\mathrm{y}} = 0.75$, only 5 % of $\mathrm{N_{fat}}$ is activated for an initial defect according to RC 5. A comparison of the service life curves is omitted for the internal defects, since $\mathrm{N_{brit}}$ and $\mathrm{N_{fat}}$ are almost identical throughout the parameter range.

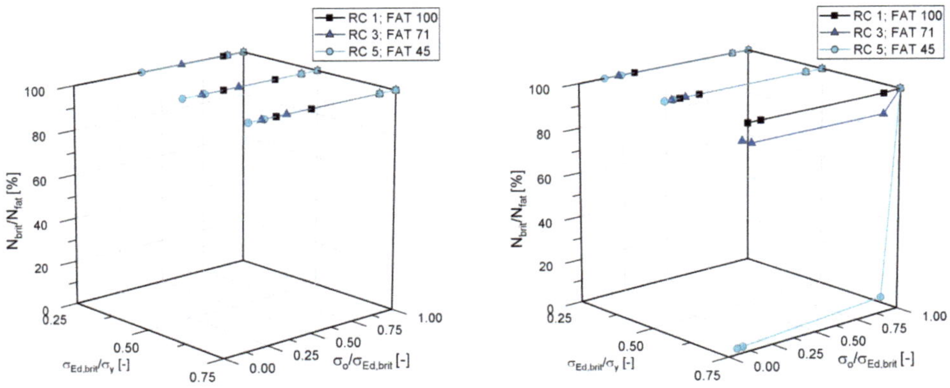


(a) (b)

**Figure 5.14:** Influence of $\sigma_{\mathrm{Ed,brit}}$ on the service life of the basic investigation case: internal defects for material a) 1.6220 and b) 1.6760

### 5.2.4.3 Design Approach

The following conclusions for the design procedure are deduced:

- For material 1.6220, no additional requirements are necessary regarding the combination of ULS and fatigue, as long as $\sigma_{\mathrm{Ed,brit}}/\sigma_{\mathrm{y}} = 0.75$ is met.
- If materials with a higher yield strength than $\sigma_{\mathrm{y}} = 300$ MPa are used and stresses of more than 50 % of $\sigma_{\mathrm{y}}$ occur under static loading, at least RC 3 must be selected regardless of lower stress ranges.

### 5.2.5 Effects of Stress Gradients

Stress peaks caused by geometric notches or by the superposition of normal and bending stresses and their effects on crack growth are evaluated separately for internal and surface defects. According to BS 7910:2013 [138], a constant stress curve corresponding to the maximum stresses represents the worst case and can always be considered as conservative. In FKM-Rili:2012 [162], the influence of the stress gradient on crack growth behavior is taken into account in fatigue calculations on stress levels by the fracture mechanics support index $n_{bm}$. This depends both on the statistical and stress mechanics support factor considering the related stress gradient. Although the determination of $n_{bm}$ with the help of the FEM is no restriction, this approach is not used in this context. The reason for this is that a generalization of the resistance values based on crack propagation is no longer possible, and the effects taken into account (macroscopic support effect and statistical defect distribution) do not affect the present worst-case consideration.

The linearization of the stress curve leads to a conservative estimation of the lifetime, but still considers stress variation. In BS 7910:2013 [138], instructions for action are given separately for static and cyclic loads as well as for internal and surface defects. Their application, however, requires an individual crack propagation calculation for each case of practice. In order to avoid these problems, a separate, more conservative and general linearization approach is chosen.

Figures 5.15a–e show stress curves for representative scenarios. The given SCFs are exemplary and apply to the specific geometric conditions. In addition to the stress curves, linearization according to BS 7910:2013 [138] and the new approach is shown. In contrast to BS 7910:2013 [138], the nominal stress $\Delta\sigma_{\mathrm{nom}}$ is selected as the second reference point in addition to the maximum stress. The definition is shown in Figure 5.15f. With:

$$\Delta\sigma_{\mathrm{max}} = \Delta\sigma_{\mathrm{N}} + \Delta\sigma_{\mathrm{M}}; \qquad \Delta\sigma_{\mathrm{nom}} = \Delta\sigma_{\mathrm{N}} - \Delta\sigma_{\mathrm{M}}; \qquad \mathrm{SCF} = \Delta\sigma_{\mathrm{max}} - \Delta\sigma_{\mathrm{nom}}$$

the division into a substitute bending stress $\Delta\sigma_{\mathrm{M}}$ and a substitute tensile stress $\Delta\sigma_{\mathrm{N}}$ results in

$$\Delta\sigma_{\mathrm{M}} = \Delta\sigma_{\mathrm{max}} \cdot \frac{\mathrm{SCF} - 1}{2\,\mathrm{SCF}}; \qquad \Delta\sigma_{\mathrm{N}} = \Delta\sigma_{\mathrm{max}} - \Delta\sigma_{\mathrm{M}}. \tag{5.3}$$

This allows a consideration purely based on the maximum stress range and the SCF with the basic models from Section 5.2.3 and their extensions to bending loads according to Annex B.1. As it becomes apparent from the examples, this analysis always leads to a conservative approach for applications relevant to cast steel components. This is due to the higher degree of filling in the equivalent stress curve for the same maximum stress in the region of the assumed defect.

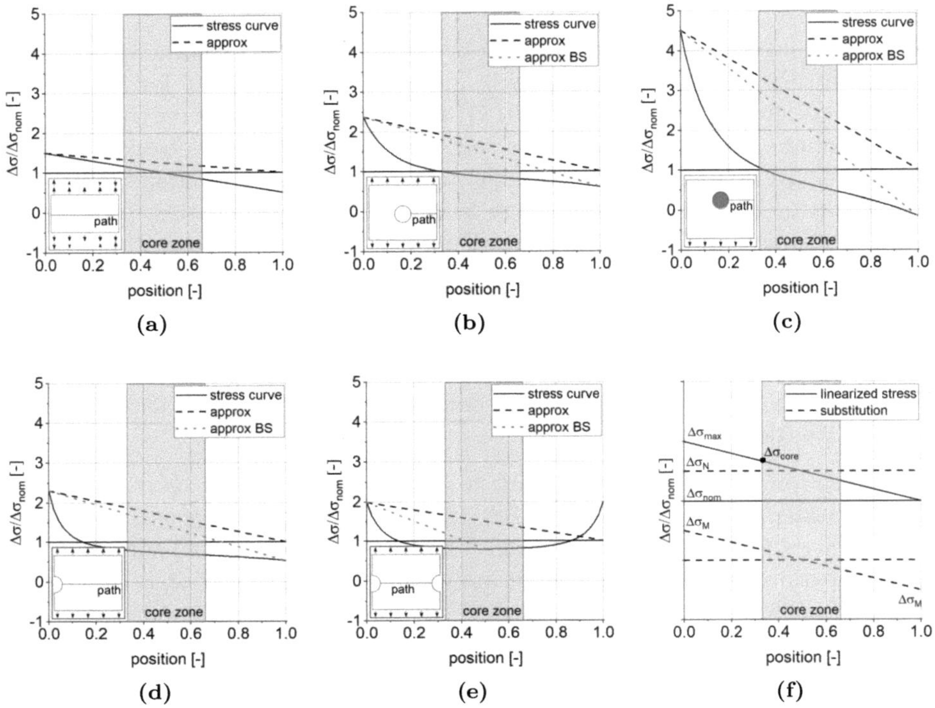


**Figure 5.15:** Exemplary stress distributions and approximations for plates with a) bending and tensile loading, b) an internal hole, c) a bolt load in a hole, d) a surface notch and e) two surface notches; f) definitions for equivalent loading

### 5.2.5.1 Surface Defects

Based on this new linearization, the example geometries summarized in Table 5.4 were investigated. The parameter range was $1.0 \leq \text{SCF} \leq 4.5$. The crack propagation calculations were carried out at two arbitrary stress levels: $\Delta\sigma_1 = 71$ MPa and $\Delta\sigma_2 = 110$ MPa, for different initial defect sizes ($A_{cg}$) according to Section 5.2.3.1.

**Table 5.4:** Summary of the cases under investigation

| case [-] | t [mm] | w [-] | a(FAT 45) [mm] | a(FAT 71) [mm] | a(FAT 100) [mm] |
|---|---|---|---|---|---|
| 1 | 16 | 3 t | 6.7 | 3.1 | 1.3 |
| 3 | 50 | 3 t | 16.7 | 6.5 | 2.0 |
| 5 | 100 | 3 t | 28 | 9.5 | 2.5 |

Figure 5.16a shows the resulting fatigue lives for the two limit cases, t = 16 mm and t = 100 mm, for initial defects according to FAT 71, and their extrapolation to the relevant load cycle range. By definition, the SN curves for SCF = 1.0 correspond

to the design fatigue curve selected in Section 5.2.3. Note, that the results compare identical maximum stress ranges and not nominal ranges. Figure 5.16a shows that the consideration of the linearized stress gradient in the considered parameter range results in up to 65 % longer service lives compared to SCF = 1.0 and an identical $\Delta\sigma_{\text{max}}$. For the evaluation in Figure 5.16b, the applied stress ranges $\Delta\sigma_{\text{max}}$ were determined in such a way that they led to the same number of load cycles as SCF = 1.0. In the design, the resistances and quality requirements according to Section 5.2.3 can be selected and the actions in Eq. (5.14) can be approximated conservatively by an empirical approach

$$\mathrm{R_{SCF}} = 1 - \frac{(\mathrm{SCF}-1)^{0.4}}{15}\left(\frac{\mathrm{t}}{16\,\mathrm{mm}}\right)^{-0.3}. \tag{5.4}$$

$\mathrm{R_{SCF}}$ results from the curve fit shown in Figure 5.16b in the parameter range $1.0 \leq \mathrm{SCF} \leq 4.5$ for the worst case of the smallest defect sizes based on FAT 100. The reduction was more pronounced for larger initial defect sizes as shown in Figures 5.16c and d, as the defect tips reach into areas of lower stresses. This effect is not considered in Eq. (5.4).

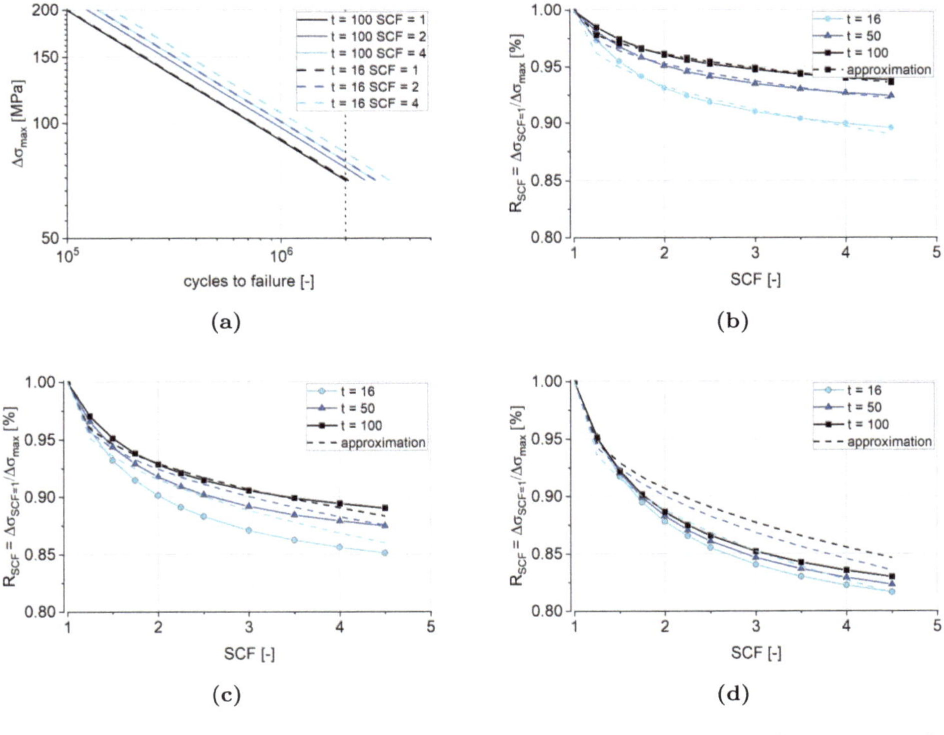


**Figure 5.16:** a) Influence of the SCF on fatigue life (exemplary for FAT 71); Deviation of the maximal allowable stress range resulting in similar cycles to failure depending on the SCF including approximation for initial defects according to b) FAT 100; c) FAT 71; d) FAT 45

If sufficient fatigue resistance could not be demonstrated according to Eq. (5.16) for lower FAT-classes, Eq. (5.5) offers an opportunity to consider these effects. This is only valid for FAT < 100 MPa and might require an iterative procedure. As shown in Figures 5.16c and d this approach is increasingly conservative for decreasing FAT classes.

$$\mathrm{R_{SCF,mod}} = 1 - \frac{(\mathrm{SCF}-1)^{0.4}}{0.11\cdot\Delta\sigma_\mathrm{c}/\mathrm{MPa}+4}\left(\frac{\mathrm{t}}{16\,\mathrm{mm}}\right)^{-0.1}. \tag{5.5}$$

### 5.2.5.2 Internal Defects

In the examples in Figure 5.15, the core zones are marked according to the definition of EN 12680-1:2003 [144], which is adopted in this concept. In addition, Figure 5.17a shows the results of a parameter study for the example of an external notch according to Figures 5.15d and e. The detailed situation for a single surface notch is given in Figure 5.17b. The region where stresses exceed $b_{\sigma,\mathrm{nom}}$ was determined for different geometrical situations, related to a parts width b as a function of SCF. It is noted that the stress ranges in the core zone are lower than $\Delta\sigma_\mathrm{nom}$ for the relevant geometric conditions. Thus, the check for internal defects can be carried out at nominal stress level. The exceptional case represents a component stressed in bending with and without normal stress. In this case, the maximum stress at the edge of the core zone

$$\Delta\sigma_\mathrm{core} = \Delta\sigma_\mathrm{nom} + \frac{1}{3}(\Delta\sigma_\mathrm{max} - \Delta\sigma_\mathrm{nom}) \tag{5.6}$$

according to Figure 5.15f must be taken into account. The maximum stress can also be applied as a constant value for simplification in any case. This approach is only necessary when defining different RCs in core and rim zone.

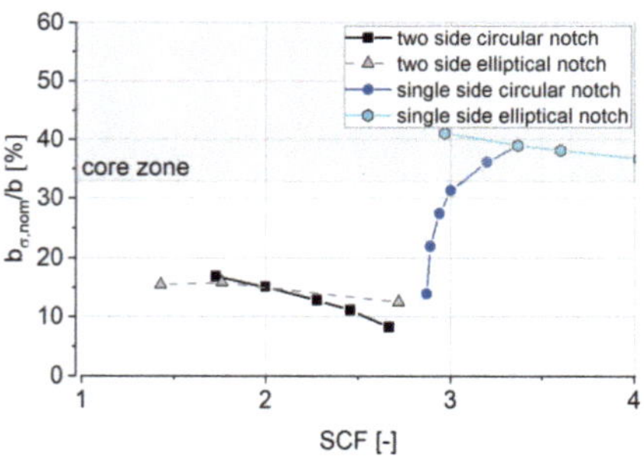


(a)

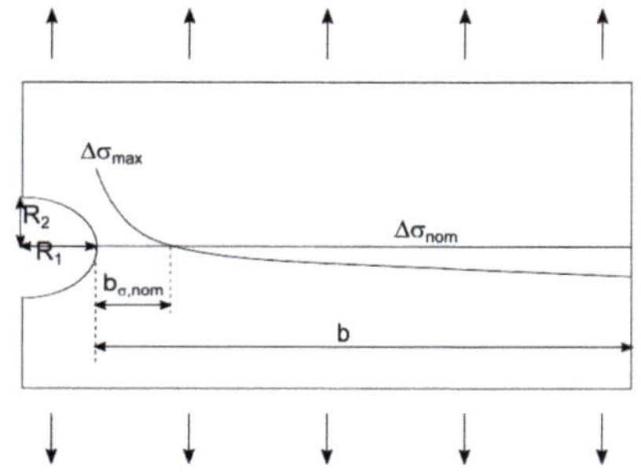


(b)

**Figure 5.17:** a) Range related to wall thickness in which $\Delta\sigma > \Delta\sigma_\mathrm{nom}$ for different example geometries; b) Definition of $\mathrm{b}_{\sigma\mathrm{nom}}$ using the example of a plate with a notch on one side

### 5.2.6 Effects of Multiple Casting Defects

Generally, casting defects do not occur as individual flaws but as an accumulation of different types and sizes at various positions. For a comprehensive design concept, the previously derived maximum single defect sizes must be supplemented by additional rules for the minimum distances between defects and total defect sizes based on the behavior of two interacting defects. Cracks in a cross section perpendicular to the direction of the maximum principal stresses represent the most unfavorable case. The starting point of these investigations was a model to estimate the effect of adjacent defects on the service life using crack propagation calculations. Thereby, the basic concepts of crack interaction from Section 2.2.4.5 were used, and the procedures from Section 5.2.3 were adapted. The calculations were based on the following assumptions and represent a worst case scenario.

- Consideration of two semi-elliptical surface cracks with a ratio $a/c = 0.4$ and identical sizes.
- The interaction of the cracks was limited to the K factors at the component surface ($\Delta K_{c,1}$ and $\Delta K_{c,2}$) and was described conservatively by the interaction relationships of two cracks of identical size in the 2D case according to Figure 2.9c. In accordance with [67], an increase in $\Delta K_S$ was not necessary. A 3D consideration could not be applied due to an insufficient data basis and few published situations. Against the background of a conservative design approach, a numerical calculation was not necessary.
- Phase 3 of the crack coalescence according to Figure 2.11 was neglected conservatively. According to [100], the crack growth rates occurring herein were up to three orders of magnitude above the da/dN occurring after unification.
- Material properties, failure conditions, crack propagation increments and Paris coefficients were selected analogous to Section 5.2.3.

The calculation was carried out according to the scheme in Figure 5.19 with the definitions from Figure 5.18.

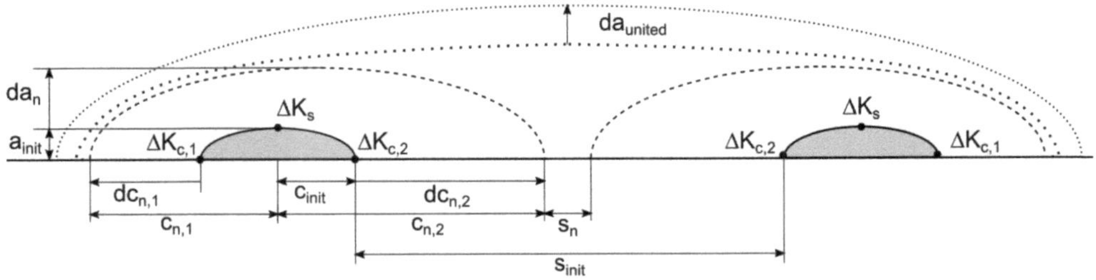


**Figure 5.18:** Geometric situation and definitions of two interacting semi-elliptical surface cracks

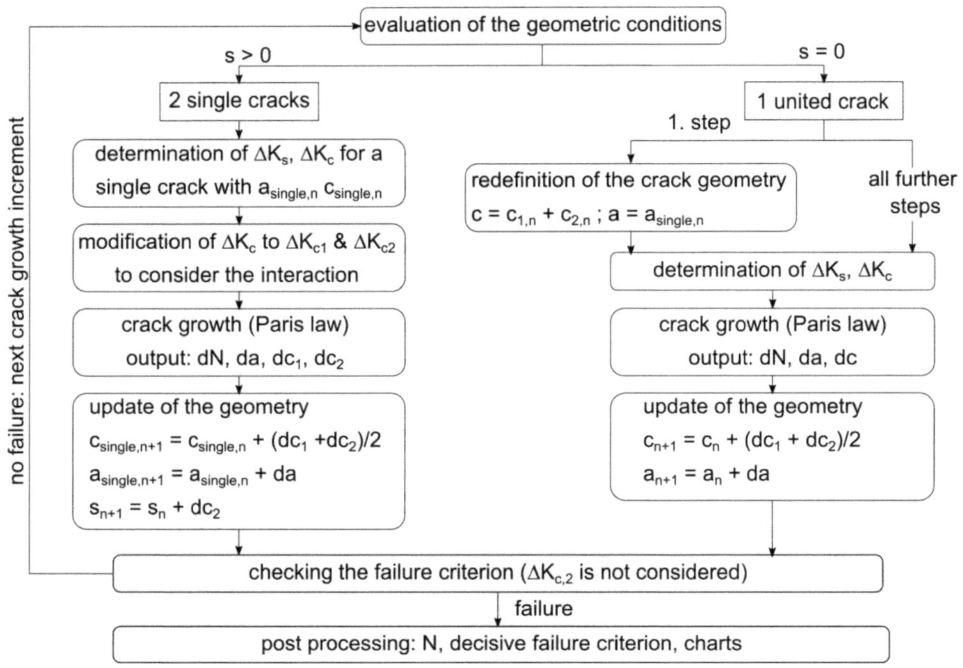


**Figure 5.19:** Flow chart of the crack propagation calculation for two interacting surface cracks

As an example, the propagation behavior is shown in Figure 5.20 for s = 2c, $\Delta\sigma = 71$ MPa, t = 50 mm and the maximum permissible individual defect $A_{Kth}$ according to RC 3. The parameters describing the crack geometry are shown in the upper diagram. The point in time when the cracks join and form a united crack can be clearly identified. The lower diagram shows the development of the stress intensity factors at the characteristic points. In accordance with the discussion in Section 5.2.3, $2 \cdot 10^6$ cycles to failure are exceeded despite the interaction of the cracks. Most of the service life could be assigned to phase 1 and 2 until unification without much mutual influence.

During the calculation, the parameters listed in Table 5.5 were varied in the range of the displayed values. In addition to the described interaction consideration (marked "Type B"), a comparative calculation was performed on a single crack ("Type A"). The difference between these two approaches quantifies the influence of the interaction on the lifetime. Analogous to the interaction conditions for static loads in BS 7910:2013 [138], an enclosing substitute crack ("Type D") was considered. Besides, a second substitute crack ("Type C") should allow a less conservative estimation of the total service life. In all calculations, the failure state was assigned to a wall breakthrough. In order to counteract premature failure by exceeding the validity range ($c_{max}$) of the formulas according to Annex B.1 and to realistically represent the considerably higher total crack widths = 4c + s, a plate width of 2w = 2 · 7.5t was chosen for these calculations. This approach was applied exclusively for the interaction calculations. When deriving the final admissibility limits in Section 5.4, additional requirements become decisive.

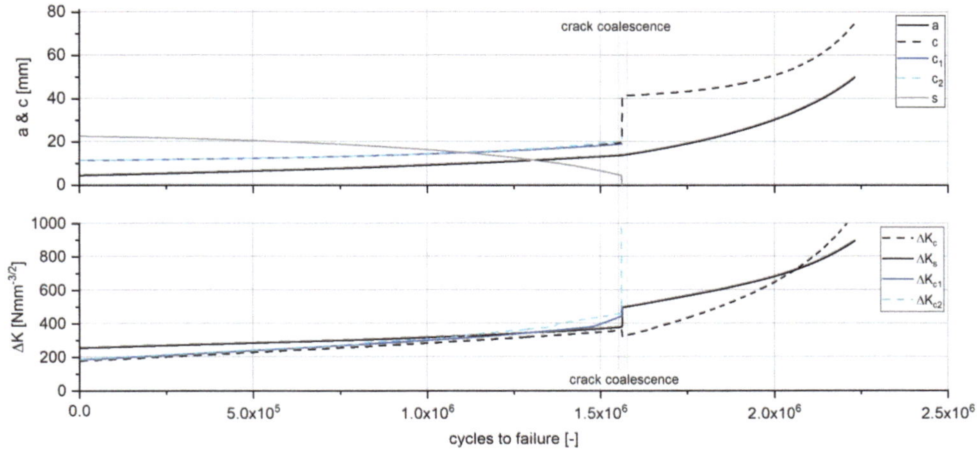


**Figure 5.20:** Course of crack geometry and stress intensity factors in two interacting cracks as a function of the load cycles

Figure 5.21 shows the results separately by loading according to FAT class and the corresponding initial crack sizes according to Section 5.2.3. The ratio $N_{int}/N_{sing}$ is the target function and describes the share of the number of cycles $N_{int}$ considering the interaction of two identical single defects related to the lifetime of a single defect $N_{sing}$ with the dimensions of one of the interacting defects. This corresponds to "Type A" according to Table 5.5. The results are shown both as a function of the related distance between the individual defects $\frac{s}{2c}$ and separately by component thickness. The results for s = 0 are based on a "Type C" calculation according to Table 5.5.

**Table 5.5:** Factors that are varied in the parameter study

| FAT [MPa] | t [mm] | Type [-] | s/(2c) [-] |
|---|---|---|---|
| 100 | 16 | A = single crack: $c_{sing}$ | 1/4 |
| 71 | 30 | B = two cracks: $c_{crack\text{-}1} = c_{crack\text{-}2} = c_{sing}$ | 1/2 |
| 45 | 50 | C = substitute crack $c_{rep} = 2\ c_{sing}$ | 1 |
| | 100 | D = substitute crack $c_{rep} = 2\ c_{sing} + s/2$ | 2 |

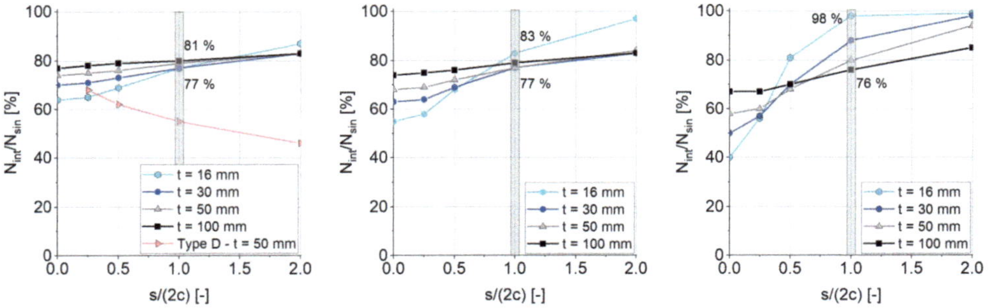


(a) (b) (c)

**Figure 5.21:** $N_{int}/N_{sing}$ evaluated for a loading and $A_{Kth}$ according to a) FAT 100, b) FAT 71 and c) FAT 45

The following findings can be extracted:

- From a distance $\frac{s}{2c} = 1.0$, stable behavior occurs. The exceptions are wall thicknesses less than 30 mm. Anyhow, these are not conservatively represented by the approach via $\Delta K_{th}$ to determine the maximum initial defect size.
- With an increasing initial defect size (i.e., with a decreasing FAT class), the dependence of $\frac{s}{2c}$ increases.
- In contrast to a large wall thickness, an increasing dependency between $N_{int}/N_{sing}$ and $\frac{s}{2c}$ can be seen for a small wall thickness t.
- The evaluation of the damage progressions, see Figure 5.20, shows a proportion of 50 % to 95 % of the calculated total service life until the two individual defects combine.
- Considering the two substitute cracks "Type C" respectively "Type D" s = 0 in Figure 5.21a as an example for t = 50 mm shows a more conservative behavior.

In order to enable a consistent evaluation for static check, a minimum distance between two individual defects of s = 2c is selected for further procedures. If the actual spacing falls below this minimum distance, the multiple defects must be evaluated as a substitute crack according to FKM-Rili-BM:2009 [161], BS 7910:2013 [138]. From these findings, this choice makes sense because the differences scattered significantly more at smaller ligaments. On the other hand, the definition of a larger minimum distance does not result in a significant increase in performance.

A closer look at the range $\frac{s}{2c} = 1$ in Figures 5.21a–c shows that, 76 % of the service life of a single crack can be achieved, in the worst case. For simplification, an equivalent lifetime-reducing effect on a single crack with the same dimensions as one of the multiple cracks can be represented by an increase of the imposed stress range. According to Eq. (5.9) and an assumed value of $f_N = 75$ %, a load increase $f_\sigma = 1.1$ follows. For a simplified consideration of the effects of multiple defects with distances $s \geq 2c$, the loadings for the determination of allowable single defect quantities shall be increased by a factor of 1.15 in accordance with the approach from Section 5.2.3. This increase by another 5 % compared to $f_\sigma$ shall ensure additional safety due to the strong reduction of the complex processes. According to [66], the effects of internal defects are even smaller, so the approach of an identical load increase leads to a conservative assessment. The effect of different sized initial defects is also smaller according to the illustrations in Section 2.2.4.5. In extreme cases, one or more of these defects is so small that it cannot grow. All of these cases lead to longer lifetimes and are thus represented by this approach.

$$\text{With} \quad dN = f_N \cdot dN_{red} \quad \text{and Eq. (2.10)} \tag{5.7}$$

$$\Leftrightarrow \quad \frac{f_N \cdot dN_{red}}{dN} = \frac{da \cdot C \cdot (\Delta K)^m}{da \cdot C \cdot (f_\sigma \cdot \Delta K)^m} \quad (f_\sigma \text{ superpos. acc. to Eq.(2.5)}) \tag{5.8}$$

$$\Leftrightarrow \quad f_\sigma = \sqrt[m]{\frac{1}{f_N}} \tag{5.9}$$

## 5.2.7 Effects of Dimensional Deviations

Geometric deviations of cast components are regulated in ISO 8062-3:2008 [157]. In the context of this design concept, two types of deviation – dimensional and straightness deviations – cause additional stresses and must be taken into account.

### 5.2.7.1 Dimensional Deviations

Dimensional deviations are divided into 16 tolerance groups (DCTG). For each of these groups, tolerance dimensions are specified in relation to the nominal dimension. Unless special agreements are made, half of the tolerance dimensions are regarded as positive deviations and half as negative deviations. Groups DCTG 12–15 are relevant for cast steel components with a wall thickness greater than 25 mm that are individually produced and small series components, both of which are typical for structural components. The requirements increase with a decreasing group index. In general, the deviation from the nominal wall thickness is to be classified one level lower. Additionally, components with lower wall thicknesses must meet requirements that are more stringent. For series components, or if the shell molding process is used, lower geometric deviations can be economically achieved. The reduced cross sections caused by wall thickness deviations are relevant for the consideration of tolerance dimensions in the design concept. Since the highest quality standards shall apply to components subject to fatigue, only DCTG 11 for wall thickness dependence is considered in the preparation of this concept. Table 5.6 summarizes the requirements in the relevant parameter range.

Table 5.6: Allowable thickness deviations for cast steel components according to ISO 8062-3:2008 [157]

| $t_{min}$ [mm] | $t_{max}$ [mm] | DCTG [-] | deviation [mm] | $A_{nom}/A_{real}$ [%] |
|---|---|---|---|---|
| | 10 | 8 | 1.0 | |
| 10 | 16 | 9 | 1.6 | 15 |
| 16 | 25 | 10 | 2.4 | 15 |
| 25 | 40 | 11 | 3.6 | 14 |
| 40 | 63 | 11 | 4.0 | 10 |
| 63 | 100 | 11 | 4.4 | 7 |
| 100 | 160 | 11 | 5.0 | 5 |
| 160 | 250 | 11 | 5.6 | 3 |

In addition, Table 5.6 considers a worst-case deviation from the tolerance dimensions. A square cross section with the minimum nominal dimensions of the respective cluster was assumed. For these cases, the maximum permissible deviation was applied as a thickness reduction in both directions. Finally, the resulting cross-sectional reduction was compared to the nominal dimensions. Depending on the wall thickness cluster, the resulting deviations were between 3 % and 14 %.

To account for this effect in the design process, the loads $\Delta\sigma_{Ek}$ need to be increased by a factor of $R_{tol,1}$. If the dimensional deviations are not examined for the specific geometric conditions or if requirements with regard to dimensional tolerance deviate from DCTG 11 are defined, $R_{tol,1}$ can be chosen conservatively according to Table 5.7. Otherwise, possible effects on $\Delta\sigma_{Ek}$ can be determined using one's own thoughts on possible cross-sectional reductions under the user-defined requirements. The effects under bending loads are not included in this table, but are greater due to the cubic dependence.

Table 5.7: Stress increase factors for dimensional deviations $R_{tol,1}$

| $t_{min}$ [mm] | $t_{max}$ [mm] | $R_{tol,1}$ [-] |
|---|---|---|
| 0 | 25 | 1.15 |
| 25 | 63 | 1.10 |
| 63 | 100 | 1.07 |
| 100 | - | 1.05 |

Table 5.8: Maximum permitted deviation from straightness for GCTG 6 of ISO 8062-3:2008 [157]

| $L_{min}$ [mm] | $L_{max}$ [mm] | e [mm] |
|---|---|---|
| 30 | 100 | 0.9 |
| 100 | 300 | 1.4 |
| 300 | 1000 | 2.0 |
| 1000 | 3000 | 4.0 |

#### 5.2.7.2 Straightness Deviations

According to ISO 8062-3:2008 [157], the straightness tolerance is divided into seven GCTGs. The most stringent requirements for cast steel components produced by hand molding are assigned to GCTG 6. The deviations in the relevant parameter range of component length $L_{min} \leq L \leq L_{max}$ are summarized in Table 5.8.

For components loaded with a normal force, straightness deviations result in an additional bending moments $M_{tol}$, causing additional axial surface stresses:

$$M_{tol} = e \cdot F_x. \tag{5.10}$$

Depending on the type of cross section and the actual dimensions, the fatigue stress $\Delta\sigma_{Ek}$ must be increased by the factor $R_{tol,2}$.

$$R_{tol,2} = 1 + \frac{\Delta\sigma_{x,M}}{\Delta\sigma_x} = 1 + \frac{e \cdot A}{W} \tag{5.11}$$

Figure 5.22a shows typical values of $\frac{\Delta\sigma_{x,M}}{\Delta\sigma_x}$ for rectangular and circular hollow sections. These deviations cause high additional stresses for compact sections with low wall thicknesses but are insignificant for hollow sections.

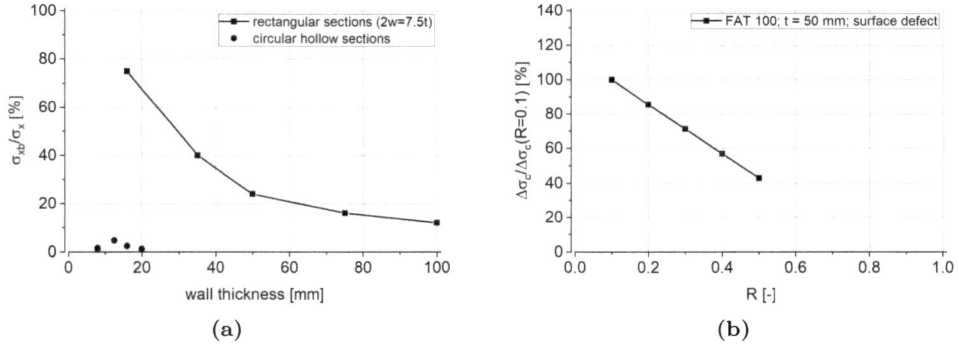


**Figure 5.22:** a) Influences of section type and dimensions on the additional stresses caused by straightness deviations $e = 2.0$ mm; b) Relation between $\Delta\sigma(\Delta K_{th})$ and R for different $\Delta K_{th}$, related to $R = 0.1$

## 5.2.8 Effects of Mean Stress

Due to their essential heat treatment, cast steel components are almost free from residual stresses [103]. In contrast to the fatigue analysis of welds (e.g., according to EN 1993-1-9:2010 [156]), the mean stress influence must be taken into account for cast steel components. The fatigue strength diagrams according to Smith or Haigh [83] show that fatigue resistance decreases with an increasing R. Thus, lower stress ratios can always be conservatively considered by using the resistances of a higher R. The resistance values and the RCs derived in Section 5.2.3 are based on $\Delta K_{th}(R = 0.1)$.

Analogous to the descriptions in Section 5.2.2, $\Delta K_{th}$ dependents on the stress ratio and can be estimated conservatively according to BS 7910:2013 [138] using Eq. (5.12).

$$\Delta K_{th} = 170 - 214R \qquad \text{for } 0 \leq R \leq 0.5 \tag{5.12}$$

Since the quality requirements shall only depend on the RC and the defect location, the maximum individual defect sizes for each RC determined in Section 5.2.3 are kept constant. Therefore, the loads or the design resistances need to be reduced for situations where $R > 0.1$. Doing so, new R-dependent values of $\Delta\sigma_c(R)$ are calculated for the defect sizes $A_{Kth}$ and scenarios conducted in Section 5.2.3. Irrespective of the geometric conditions (e.g., internal or surface defects, part width and thickness, defect size) this leads to comparable $\Delta\sigma_c/\Delta\sigma_c(R = 0.1)$ ratios in all cases. Due to LEFM, the linear dependency of Eq. (5.12) is transferred to the applicable stress ranges, as shown in Figure 5.22b for an exemplary situation. This relation is described by Eq. (5.13) and acts as a reduction factor of the fatigue resistance in Eq. (5.15).

$$R_R = \frac{\Delta\sigma_c}{\Delta\sigma_c(R = 0.1)} = 1.14 - 1.43R \qquad \text{for } 0 \leq R \leq 0.5 \tag{5.13}$$

Because this design concept considers only crack growth, the tensile part of the stress range can be exclusively considered as an action for stress ratios $R < 0$ according to Eq. (2.11) and Eq. (2.5). This approach is based entirely on fracture mechanics considerations and has not been verified experimentally within the context of these investigations. When applying this approach, it is also essential to ensure that local residual stresses resulting, for example, from plastic deformations or mechanics processing are excluded.

## 5.3 Summary of the Design Procedure

In Sections 5.2.1 through 5.2.8, the relevant influencing parameters of casting defects on fatigue resistance were examined, and their influences were represented by reduction factors. For the designing engineer, the main points can be summarized as follows.

This approach basis of numerous worst-case considerations summarized and discussed in Section 5.5.1. In this case, a complete statistical evaluation cannot be made and the actual safety level could not be determined precisely. However, the result of the discussion in Section 5.5.1 is that the considered assumptions and worst-case considerations are assumed to lead to a survival probability of at least 95 % can be considered as characteristic component resistances in the sense of EN 1993-1-9:2010 [156]. This concept is based on fracture mechanics considerations. The actual maximal permissible defect sizes relevant to the foundry are summarized in Section 5.4. With this approach, the designer does not need to consider these explicitly. The basic principles of structural design, action and stress analysis shall be based on the Eurocode concept. The design loading is determined by:

$$\Delta\sigma_{\mathrm{Ed}} = \gamma_{\mathrm{Ff}} \cdot \Delta\sigma_{\mathrm{max,Ek}} \cdot \mathrm{R}_{\mathrm{SCF}} \cdot \mathrm{R}_{\mathrm{tol,1}} \cdot \mathrm{R}_{\mathrm{tol,2}}. \tag{5.14}$$

With:

| | |
|---|---|
| $\gamma_{\mathrm{Ff}}$ | Load partial safety factor according to EN 1990:2010 [153]. |
| $\Delta\sigma_{\mathrm{max}}$ | Maximum principal stress range under consideration of local stress concentrations. |
| $\mathrm{R}_{\mathrm{SCF}}$ | Reduction factor to consider the positive influence of support effects according to Section 5.2.5. |
| $\mathrm{R}_{\mathrm{tol,1}}$ | Factor to consider the stress increase by dimensional deviations according to Section 5.2.7. |
| $\mathrm{R}_{\mathrm{tol,2}}$ | Factor to consider the stress increase by additional bending moments due to straightness deviations according to Section 5.2.7. |

The design resistance is determined by:

$$\Delta\sigma_{\mathrm{Rd}} = \frac{1}{\gamma_{\mathrm{Mf}}} \cdot \Delta\sigma_{\mathrm{c}} \cdot \mathrm{R}_{\mathrm{R}}. \tag{5.15}$$

With:

| | |
|---|---|
| $\gamma_{\mathrm{Mf}}$ | Resistance partial safety factor according to EN 1993-1-9:2010 Table 3.1 [156]. |
| $\Delta\sigma_{\mathrm{c}}$ | Reference value for fatigue strength for $\mathrm{N_c} = 2 \cdot 10^6$ load cycles according to Table 5.9 and Section 5.2.1. |
| $\mathrm{R}_{\mathrm{R}}$ | Reduction factor to consider stress ratios $\mathrm{R} > 0.1$ according to Section 5.2.8. |

**Table 5.9:** Assignment of the fatigue strengths ($\Delta\sigma_c$) to the resistance categories (RC)

| RC | [-] | 1 | 2 | 3 | 4 | 5 |
|---|---|---|---|---|---|---|
| $\Delta\sigma_c$ | [MPa] | 100* | 90 | 71 | 56 | 45 |

for $16 \geq t \geq 100$ mm; *only valid for $t \geq 30$ mm

The values of Table 5.9 are valid for stress ratios $R \leq 0.1$. The gradient of the fatigue strength curve is $m = 3$. The endurance strength $\Delta\sigma_{endu} = \Delta\sigma(N = 1 \cdot 10^7)$. Check of sufficient fatigue resistance is done in accordance with

$$\frac{\Delta\sigma_{Ed}}{\Delta\sigma_{Rd}} \leq 1.0. \tag{5.16}$$

Additionally, the following objectives must be considered:

- $\sigma_o \leq \sigma_y$ and $N > 1 \cdot 10^4$ load cycles.
- The number of load cycles is to be determined in accordance with the relevant technical standards (e.g., EN 1991).
- To consider spectrum loading, the counting methods for determining a damage equivalent according to EN 1993-1-9:2010 [156] as well as the effects on fatigue resistance should be applied. Note, this issue was not considered specifically within these investigations.
- If the component is stressed by live loads in addition to fatigue loading (incl. central loads from permanent actions), the additional requirements according to Section 5.2.4 must be considered.
- The stress state must be approximately uniaxial. The influence of complex multiaxial states were not investigated.

To use the full potential of the concept, the following additional points should be indicated in the drawings:

- the local resistance categories (RC)
- direction of the maximum principal stresses
- tolerance requirements other than DCTG 12 and GCTG 6 according to ISO 8062-3:2008 [157]

## 5.4 Resulting Quality Requirements

In this section, the influencing factors and defect sizes discussed in the previous sections are summarized and transferred to actual quality requirements dependent on the RCs. For this purpose, permissible single defect sizes $A_{sing}$, dimensions in thickness direction 2a, maximum permissible total defect sizes $A_{mult}$ for defined inspection cross sections $A_{proof}$ and evaluation rules are defined, depending on the defect location. Check is carried out in verification planes perpendicular to the maximum principal stress. Volumetric or inclined defects are projected onto these planes and must meet specified requirements. If the information of the direction of maximum principal stresses is not available, the check needs to be carried out in the planes of the most unfavorable defect arrangements. All mentioned defect sizes are referred to as the projection of the defect to the verification plane.

### 5.4.1 Requirements and Application Limits

Each component is subdivided into core and rim zones, as shown in Figure 5.27, analogous to EN 12680-1:2003 [144]. In addition to the fact that this definition is established and its application is common practice, the relevant mechanical effects are conservatively represented, as shown in Sections 5.2.5 and 5.2.6 assuming the surface as a possible interacting crack [67]. The maximum extension of a defect in the thickness direction of the component must not exceed the zone thickness or half the zone thickness for surface defects. This statement is based on the considerations of Section 5.2.6, is initially relevant for the derivation of the permissible defect sizes and does not represent a generally valid verification criterion. As a comparative value, EN 12680-1:2003 [144] states that internal defects may cover a maximum of 20 % of the wall thickness, and surface defects may cover 15 % of the rim zone thickness.

In Section 5.2.6 the effect of interacting multiple flaws was quantified, and minimum spacing was defined. The following definitions are available in the codes of practice for defects of different sizes, according to which adjacent defects must be combined into a substitute defect:

- EN 1369:2013 [144] (MT): The distance between two displays is less than the length of the longest defect in a row.
- DIN 1690-2:1985 [139] (UT): The distance between two displays is less than the largest extension of the larger of the two defects under consideration.
- FKM-Rili-BM:2013 [161] and BS 7910:2013 [138]: The evaluation is direction-related. Two defects must be considered as one substitute defect if the ligament is smaller than the dimensions of the smaller defect in the direction of observation.

The previous rules according to EN 1369:2013 [148] and the withdrawn DIN 1690-2:1985 [139] are slightly conservative compared to [161, 138] and the requirements defined in Section 5.2.6. As these are already established in the foundry industry, this formulation is used. The assignment of non-coplanar defects to a design level is carried out according to the definition in [161].

The determination of the permissible sizes of internal defects in Section 5.2.3 required a centric defect arrangement. In general practice, eccentric fault locations must be assumed. The greatest negative effects under the assumptions made for this design concept occur if the internal defect is located at the edge of the core zone. This effect is most significant with small defect sizes, since the potential for eccentricity is greatest here. An evaluation of the stress intensity factors for the worst case of initial defect $A_{Kth}$, according to FAT 100 in accordance with Section 7.4.2.1.7 of FKM-Rili-BM:2013 [161], showed an increase in the stress intensity factors of less than 1 % compared to the centric arrangement. This applied to a pure tensile stress and was independent of the wall thickness. Thus, the effects are compensated by the general increase of $\Delta\sigma$ by 15 % to account for interaction effects. A bending stress as less favorable case is already taken into account by the additional requirements formulated in Section 5.2.5 for existing stress gradients.

## 5.4.2 Derivation of Permissible Defect Sizes

The procedure for deriving the quality requirements is summarized in Figure 5.23. First, the maximum permissible single defect size is determined depending on the FAT class and the wall thickness. Certain additional requirements must be met. For a practice-oriented implementation, the results are linearized and presented as a formal relationship depending on the component thickness. The maximum permissible defect size and the maximum extension in the thickness direction of the component are the relevant parameters. In the case of internal defects, total defect size is defined based on the actual permissible defect sizes. Otherwise, in the case of small permissible defect sizes (i.e., high FAT classes and large wall thicknesses), critical defect arrangements can occur. In the case of surface defects, only the minimum distances between the individual defects and the sizes of both must be met.

### 5.4.2.1 Single Defect Size

In pre-trials for Section 5.2.3, an influence of the applied plate width on the service life was observed. Following [54, 103], this approach is technically reasonable and corresponds to a converged state. In this case, a cross-sectional weakening and stress increase due to multiple defects must be considered, such as by additional reduction factors as done in [103]. Note, that this is not the background of the stress increase factor discussed in Section 5.2.6. In order to take both influences into account, the allowable initial defect sizes are determined iteratively according to the approach for $A_{Kth}$ shown in Section 5.2.3, which considers increasing crack tip stresses at interacting cracks. In this final evaluation, the boundaries are set more tightly and the plate width is selected as $w = 2c$, thus corresponding to both the minimum distance to any adjacent defect and to the most unfavorable geometric conditions permitted according to Annex B.1.

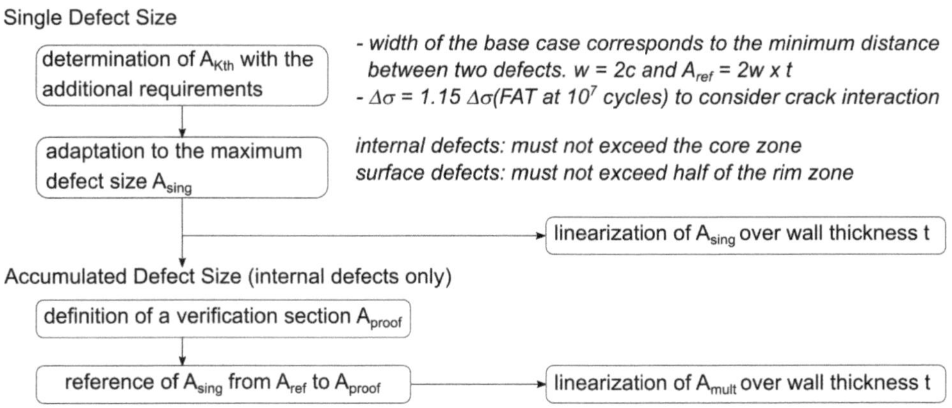


**Figure 5.23:** Procedure for determining quality requirements

At the same time, a stress increase caused by the reduction in cross section due to flaw accumulations can be avoided. The applied stress corresponds to the extrapolation $\Delta\sigma$(FAT at $1 \cdot 10^7$ cycles) and is increased by a factor of 1.15 to take into account the influences of interacting cracks on the service life described in Section 5.2.6.

Some of the defect sizes for RC 4 and RC 5 determined in this way exceed the requirements defined in Section 5.4.1 of the extension in the thickness direction. Internal defect shall not exceed the thickness of the core zone and a surface defect shall not exceed half the thickness of the rim zone. Formally, this limitation is achieved by

$$\mathrm{a_{sing}} = \begin{cases} \frac{\mathrm{t}}{6} & \text{if}\, \mathrm{a_{Kth}} > \frac{\mathrm{t}}{6} \\ \mathrm{a_{Kth}} & \text{otherwise} \end{cases} . \tag{5.17}$$

The maximum sizes of single defects determined in this way are referred to as $\mathrm{A_{sing}}$. They are summarized in a linearized form as a function of the wall thickness in the range 30 mm < t < 100 mm in Table 5.10. In addition to the permissible defect size, a/c = 0.4 is used to indicate a maximum permissible extension 2a of the defect in the thickness direction. This is necessary so that the wall crack identified in the crack propagation calculations as the decisive failure case occurs early in the case of narrow, deep cracks. For the surface defects (rim zone), the requirements according to EN 1369:2013 [148] have been supplemented for information. The resulting defect sizes thus continue to meet common quality requirements.

Figure 5.24 shows the calculated defect sizes without linearization for surface and internal defects separately. Again, the calculated cases are marked as points and connected by lines. The changes in defect sizes increase over the wall thickness with a decreasing FAT class. The identical defect sizes for t = 30 mm at RC 4 and RC 5 result from the limitation of elongation in the direction of the wall thickness. In addition, the differences for different wall thicknesses are more pronounced for internal defects and negligible for surface defects. This is why that the sizes of surface defects in Table 5.10 are covered by a single value corresponding to t = 30 mm, whereas internal defects are given in a linearized function of t. As the effects are only dominant for defects of RC 3, RC 4 and RC 5, $A_{sing}$ is defined for RC 1 and RC 2 at t = 30 mm only.

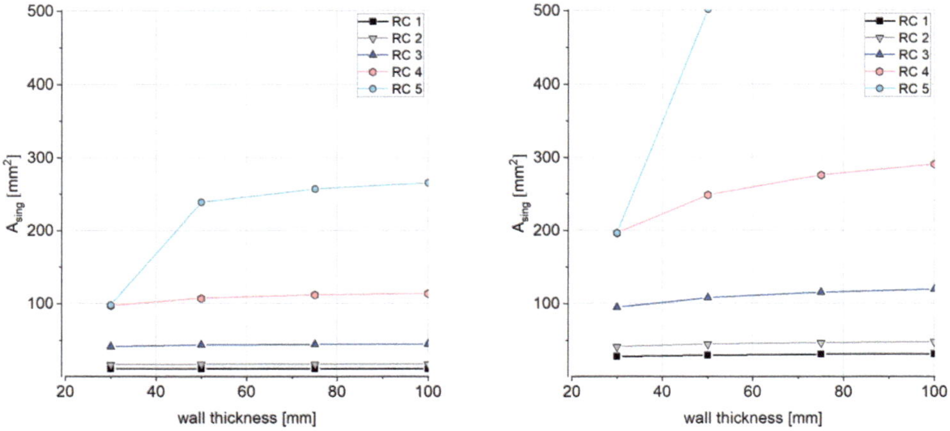


(a) (b)

**Figure 5.24:** Non-linearized defect sizes $A_{sing}$ for a) surface defects and b) internal defects

### 5.4.2.2 Accumulated Defect Size

If defects of different sizes occur, situations can appear that are not conservatively represented by this design scenario, despite its compliance with the minimum distances between the defects. In order to regulate these situations, a permissible total defect size $A_{mult}$ is defined. For this purpose, the determined single defect size in a basic case according to Section 5.2.3 with the cross section $A_{ref} = t \cdot 4c$ is transferred to an RC-independent reference cross section $A_{proof}$ and thus allows a unified assessment. This reference cross section is based on the largest permissible defect size with $a_{sing} = \frac{t}{6}$ according to

$$A_{proof} = t \cdot 4c = t \cdot 4 \cdot a/0.4 = t \cdot 4 \cdot (t/6)/0.4 = 1.67t^2 \tag{5.18}$$

and the relationships in Figure 5.25a. For simplification in further context, this is simplified to $A_{proof} = 1.5t^2$ The total defect size $A_{mult}$ is defined by

$$A_{mult} = A_{sing} \cdot A_{proof}/A_{ref}. \tag{5.19}$$

Figure 5.25b shows an example of the resulting defect sizes $A_{mult}$ at $A_{proof}$ and $A_{ref}$. Analogous to the single defect quantity, $A_{mult}$ is summarized in a linearized form in Table 5.10.

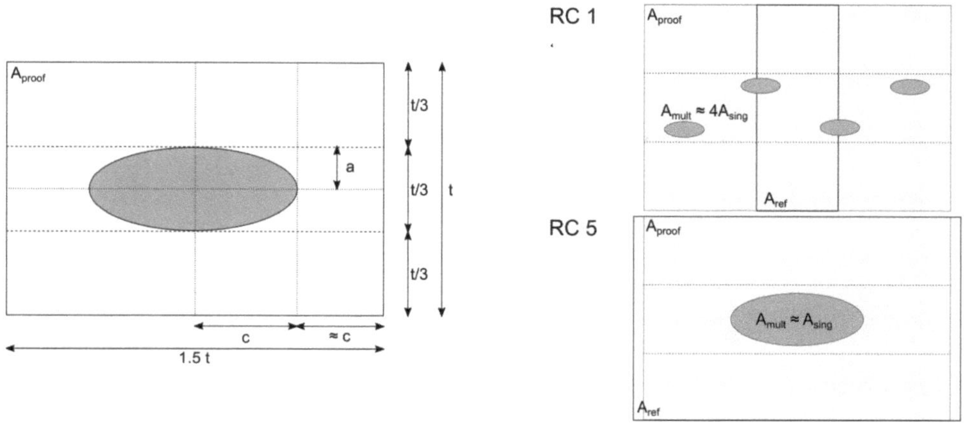


(a) (b)

**Figure 5.25:** a) Geometric relations to determine $A_{proof}$; b) Exemplary defect sizes $A_{mult}$ for RC 1 and RC 5 for t = 50 mm

### 5.4.3 Additions to Wall Thicknesses t = 16 mm to 30 mm

As discussed in Section 5.2.3, $A_{Kth}$ leads to conservative results compared to $A_{cg}$ only for $t \geq 30$ mm. According to [155, 87], the highest quality grade must generally be used in this range of wall thickness independent of the loading. In order to enable an evaluation in this parameter range, requirements are defined according to FAT class. For this purpose, the increase factor $\Delta\sigma$ is increased from 1.15 to 1.20, and both $A_{cg}$ and $A_{Kth}$ are determined analogous to the previous treated wall thickness range. In addition, the requirements that a surface defect must not exceed half of the rim zone thickness and that an internal defect must have a maximum extension in the thickness direction equal to the core zone thickness apply. The resulting defect sizes are termed as $A_{sing}$. In this narrow parameter range, wall thickness-dependent defect sizes are not considered suitable for practical use, especially since this would provide only negligible benefit. The values determined for t = 16 mm correspond to the worst-case scenario and are applied to the entire wall thickness range up to 16 mm < t < 30 mm. Figure 5.26 compares these defect sizes separately for surface and internal defects and the FAT class of both approaches.

It becomes obvious that, especially with low RCs, $A_{cg}$ leads to considerably lower values compared to $A_{Kth}$ and is therefore relevant for the design. In Table 5.11, $A_{sing}$ implements these results including the additional requirements. All previously defined additional requirements (e.g., the minimum distances between adjacent defects) must still be observed. To what extent the defined requirements for RC 1 can be verified with state of the art NDT for cast steel components is doubtful. Except for further findings (e.g., from [115]), a design FAT 100 at t < 30 mm must be omitted.

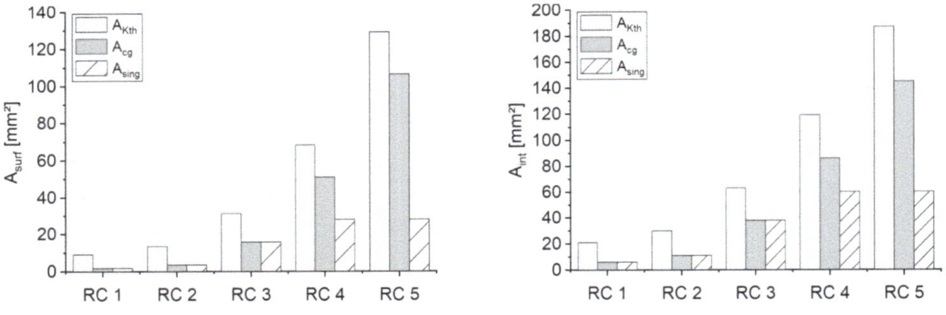


(a) (b)

**Figure 5.26:** Comparison of the different approaches to determine the maximum permissible defect size for t = 16 mm for a) surface defects and b) internal defects

### 5.4.4 Summary

Figure 5.27 summarizes the principles of this concept. It shows a section of a component with defects and is divided into a core and a rim zone. The requirements for surface defects are not only applied to actual surface defects but also to near-surface defects that are almost as critical under mechanical aspects. For this purpose, the specific maximum permissible defect sizes are transferred to the entire rim zone. For the core zone, the requirements defined for internal defects apply.

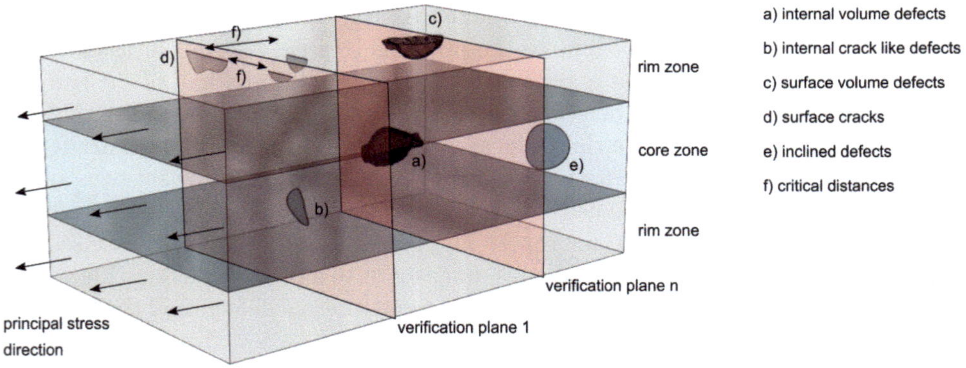


**Figure 5.27:** Exemplary display of different defect types, verification planes and evaluation zones

The principal stress direction is indicated, and possible levels of verification are shown. Irrespective of whether the defects are volumetric or planar, the projected surface is decisive for the verification plane. Distances between several flaws in one detection plane are given in the evaluation as single flaws. Furthermore, it defines in which cases two defects with an offset in the direction of the principal stress can be related to a common detection plane or be evaluated separately. Maximum allowable defect sizes are summarized in Table 5.10 and Table 5.11.

**Table 5.10:** RC-dependent quality requirements for wall thicknesses 30 mm $\leq$ t $\leq$ 100 mm. For a well-arranged layout the formulas are not dimensional-true. t needs to be inserted in mm as common in casing applications.

| | | core zone | | | rim zone | | |
|---|---|---|---|---|---|---|---|
| RC<br>[-] | FAT<br>[MPa] | $A_{sing}$<br>[$mm^2$] | 2a<br>[mm] | $A_{mult}$<br>[$mm^2$] | 2c<br>[mm] | $A_{sing}$<br>[$mm^2$] | LM/AM<br>[−] |
| 1 | 100 | 25 | 4 | 2.4t - 6 | 8 | 11 | 2 |
| 2 | 90 | 40 | 5 | 3.0t - 9 | 10 | 17 | 3 |
| 3 | 71 | 0.4t + 85 | 7 | 4.8t - 20 | 16 | 41 | 4 |
| 4 | 56 | 1.4t + 155 | 11 | 7.7t - 55 | 25 | 98 | 5 |
| 5 | 45 | 6.3t + 7 | 0.1t+7 | 13.0t - 200 | 25 | 98 | 5 |

**Table 5.11:** RC-dependent quality requirements for wall thicknesses 16 mm $\leq$ t $<$ 30 mm

| | | core zone | | | rim zone | | |
|---|---|---|---|---|---|---|---|
| RC<br>[-] | FAT<br>[MPa] | $A_{sing}$<br>[$mm^2$] | 2a<br>[mm] | $A_{mult}$<br>[mm] | 2c<br>[mm] | $A_{sing}$<br>[$mm^2$] | LM/AM<br>[−] |
| 1 | 100 | – | – | – | – | – | – |
| 2 | 90 | 11 | 2.4 | – | 4 | 3 | 2 |
| 3 | 71 | 38 | 4.4 | – | 10 | 15 | 3 |
| 4 | 56 | 55 | 5.0 | – | 13 | 28 | 3 |
| 5 | 45 | 55 | 5.0 | – | 13 | 28 | 3 |

The following additional requirements must be met:

- The verification area $A_{proof}$, which is decisive for an evaluation of the total defect area, must be selected as a function of the wall thickness as $A_{proof} = 1.5t^2$.
- The verification planes must be set perpendicular to the direction of the maximum principal stresses. If the designing engineer does not specify this direction, the evaluation must be carried out for the most unfavorable direction for the respective check.
- Non-coplanar defects are combined into one detection level according to [161]. Projection onto a common plane must be done if the distance perpendicular to the plane of observation is smaller than the sum of half the defect lengths. This must also be taken into account for the interaction of surface and internal defects. The resulting flaw area must be determined according to [161].

- In the case of multiple flaws (coplanar in the verification plane), the evaluation criteria according to EN 1369:2013 [148] or DIN 1690-2:1985 [139] must be followed when determining the existing defect sizes, meaning that the distance between two adjacent defects must be greater than the largest dimension of the larger defect. If this is not the case, the defect area must be determined from the sum of the individual defect areas including the ligament.
- If defects occur in both the core and rim zones in one verification plane, the maximum defect area in the rim zone shall not be exceeded by the sum. Moreover, the total defect size defined for the core zone must not be exceeded in the entire cross section of the detection plane.
- Defects that extend beyond the zone boundary must be evaluated as surface defects.
- Near-surface defects where h < a (cf. Figure 5.28) must be evaluated as substitute defects according to the requirements of [161].

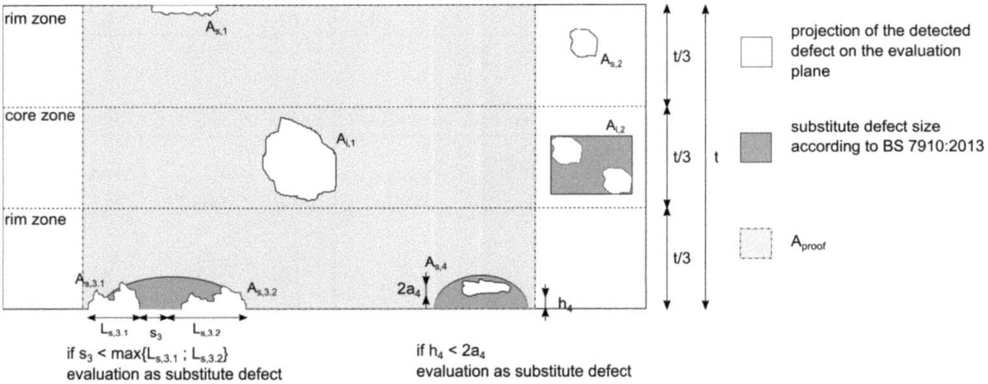


**Figure 5.28:** Definitions and properties of potential defects and their arrangements

# 5.5 Validation

In this section, all aspects of the worst-case consideration are summarized and the design procedure is compared with experimental investigations as well as literature data. It is shown that the calculated component resistances are a conservative reflection of these experiences. Furthermore, a principle conformity with existing design approaches is shown.

## 5.5.1 Assessment of the Safety Level

The present design approach is based on EN 1993-1-9:2010 [156] that requires characteristic values for actions and resistances. The resistance values of EN 1993-1-9:2010 [156] are determined by a complete statistical evaluation as shown in [22]. Characteristic fatigue resistance is defined by a survival probability of at least 95 %. This attempt could not be implemented directly in the present case as the correlation of defect sizes to component resistances needed to be carried out by worst-case considerations. This was necessary as the occurrence of the defects do not only follow a statistical distribution but are affected by the individual casting system. Unfavorable configurations are tolerated by the quality grades and limited by distinct constraints in an extensive NDT. To ensure an easy application of the design approach, further generalizations were necessary. For instance, a separate consideration of voluminous, internal defects would require additional gradients of the design SN curves and separate calculations in the design process.

Due to numerous worst-case approaches and generalizations, an estimation of the overall safety level was impossible. Due to the sum of the worst-case estimates, it was assumed that a survival probability of at least 95 % is maintained. This justifies the use of $\Delta\sigma_c$ in Table 5.9 as characteristic component resistances in terms of EN 1993-1-9:2010 [156] along with the associated safety factors. The individual worst-case considerations are summarized by:

*Fracture Mechanics Consideration:*

- For surface defects, a fracture-mechanics evaluation independent of volume or crack-like defects is an appropriate assumption as long as the validity limits of fracture mechanics are met.
- In the case of internal defects, this consideration does not take the identified crack initiation phase into account (see Section 4.1). However, in general applications, internal crack-like defects can also occur.

*Unfavorable Crack Configuration:*

- The evaluation was performed on elliptical or semi-elliptical cracks with a ratio $a/c = 0.4$, causing the highest stress intensities in the initial state.

- This approach considers cracks at the stress hotspots of the individual component in the most unfavorable orientation.
- To account for interacting defects, all defect sizes given in Tables 5.10 and 5.11 contain an overall stress increase of 15 %. This increase was derived from a worst-case consideration of two semi-elliptical surface defects of the same size with minimum spacing and was increased by additional 5 % due to the strong simplification of the complex processes.
- Defects in the rim zone were assessed as surface defects.

*Material Resistance:*

- The material resistance was derived from the minimum requirements of the technical delivery conditions. To reflect the actual solidification conditions, these must be verified on the component itself [103]. The fracture toughness was conservatively estimated after [161] from the minimum requirements of the Charpy tests. However, the investigations in [103] have shown that actual resistances (KV or $J_{IC}$) are considerably higher in most applications for the material 1.6220. For the material 1.6760 the minimum requirements are higher, correlate well with real applications and are strongly dependent on the component thickness.
- In general, crack propagation behavior is not represented by any technological substitute quantity. Its application in the approach was discussed in Section 5.2.2. The applied Paris parameters were taken from BS 7910:2013 [138] and are characterized herein as mean values that represent a wide range of metallic materials. However, these are valid for R = 0.5, so its application for R = 0.1 is conservative. The final crack sizes in Tables 5.10 and 5.11, resulted from an evaluation of $\Delta K_{th}$ with the specific value according to BS 7910:2013 [138]. Its actual characteristic being a mean or design value is not evident. In BS 7910:2013 [138] it is mentioned, that these can be used for different metallic materials, even under negative conditions (e.g. maritime environment).
- As discussed in Section 5.2.3.1, setting the endurance limit to $N_{endu} = 1 \cdot 10^7$ load cycles causes smaller defect sizes than determined by crack growth calculation and the consideration of the actual, higher load cycles in this stage. Then again, the application of $N_{endu} = 5 \cdot 10^6$ according to EN 1993-1-9:2010 [156] would lead to significantly smaller permissible defect sizes.

*Additional Effects:*

- In addition to the resistances, geometric boundary conditions were decisive for the definition of the permissible defect sizes.
- Geometric tolerances were taken into account in an unfavorable way.
- The applied linearization of stress gradients causes lower calculative service lives compared to actual stress distributions and other simplifications.

- The investigations on the interaction of static loads (brittle fracture) and fatigue loads, considered the occurrence of the static design load (ULS) at the end of the calculated lifetime.
- The defect sizes summarized in Tables 5.10 and 5.11 result from a conservative linearization of the calculated defect sizes. Here, the focus was on a simple applicability. Additionaly in cases wehen the influence of the wall thickness was small, a constant size corresponding to the minimum value was specified.

Despite of all conservative approaches, the derived defect sizes represent economically reasonable quality requirements for practical construction applications. In some cases not the calculated resistances but geometrical boundary conditions became decisive. Increasing the resistance would not bring any additional benefit in these cases. For an optimization to other fields of application, in which higher resistances at the same defect sizes or larger permissible defect sizes are required, the summarized aspects need to be re-evaluated.

### 5.5.2 Comparison with Experimental Results

In Figure 5.29a, the design curves for RC 1 or FAT 100 and RC 5 or FAT 45 are plotted as boundary values. These are compared with the results of the experimental investigations on tensile specimens with real casting defects from Section 3.2. The results are shown separately for internal and surface defects as nominal stress ranges. The 95 % probability of survival line is plotted as well. These are evaluated with a free gradient in the case of internal defects and with k = 3 in the case of surface defects. It should also be noted that all specimens and thus all defect sizes are evaluated in combination. The stress intensity factor, which depends on the defect size, led to the scatter discussed in Section 3.2.2 and thus to low $\Delta\sigma_{95\%}$ values. The existing cross-sectional weakening $A_{fail,mic}/A_{sec}$ of the examined specimens is between 1 % and 20 % and thus in the order of magnitude of the sizes $A_{def}/A_{ref}$ = 1 % to 13 % assumed in the design procedure. A direct transfer between test results and design curves is not possible. However, it is evident that both the internal defects of the test results and their 95 % survival probability are conservatively covered by the design method. Especially in the area of low stress ranges, the design values are strongly on the safe side due to the crack initiation phase not being considered. In the case of surface defects, the test results are above the applicable maximum loads, and their 95 % survival probability is within the order of magnitude of the design curves. This again illustrates the special caution that must be applied when evaluating surface defects.

Figure 5.29b provides an analogous comparison with the literature values discussed in Section 2.3.1. Here, data labeled with "Ma 1997" [58] represent tests without casting defects. The remaining data points indicate specimens with defects. In the case of the values marked with "Vishnevsky 1966" [117], the regression curves are given for the different material states and defect locations discussed in Section 2.3.1. The statements correspond to the illustrations in Figure 5.29a. The test results for the flawless base

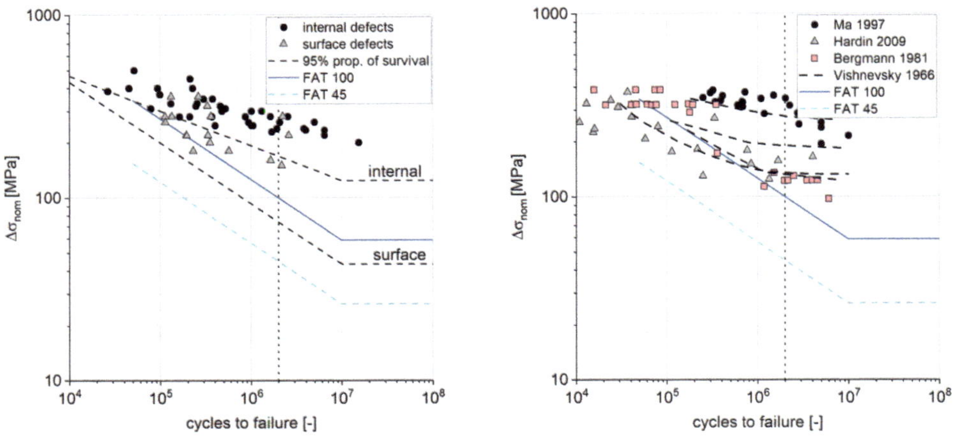


(a) (b)

**Figure 5.29:** Comparison of the resistance curves with a) test results on tensile specimens and b) literature data

materials show significantly higher fatigue resistances than covered by the present design concept. This underlines the conservative nature of this method at the absence of defects. Specimen with larger defects, as considered in the remaining studies, caused fatigue resistances that are within the range of this concept. The experiments marked with “Hardin 2009” [37] were performed at specimens of a high strength material with nominal tensile strength $R_m > 1100$ MPa. It is questionable if the present assumptions and the resulting concept could be transferred to materials of higher strength than considered in these present investigations. As discussed in Section 3.2.2.8 there was not identified any benefit of using high strength material with respect to fatigue resistance. Rather the less ductile residual fracture behavior of the high strength material as well as the resistance against brittle fracture [103] make their use in practical construction applications, especially for parts exposed to low temperature and large wall thicknesses, questionable.

The results of the experimental investigations on the components from Section 3.3 are shown for individual component sizes in Figure 5.30a. In addition, the 95 % probability of survival is given for all experiments (Foundries B and C cf. Section 3.3.2). The artificially introduced defects fall between the maximum allowable defect sizes for RC 2 and RC 3. To be on the safe side, the resistances for RC 2 resp. FAT 90 are determined and specified in this analysis as a function of the component thickness.

The following additional assumptions are made. Since the evaluation of the test results is based on the actual dimensions, a correction using $R_{tol,1}$ according to Eq. (5.14) to correct any deviations in the component thickness was not necessary. Additional stresses from eccentricity and deviations from straightness were taken into account on the resistance side using

$$\Delta\sigma = \Delta\sigma_c/(1 + R_{tol,2}). \tag{5.20}$$

For a better assessment, a curve without the reduction by $R_{tol,2}$ is also shown. The comparison was based on the maximum stress ranges, taking into account the stress increase caused by the geometric notch. This adaptation was carried out using

$$\Delta\sigma_{max} = R_{SCF} \cdot SCF \cdot \Delta\sigma_{nom} \qquad \text{with SCF} = 2.0. \tag{5.21}$$

In this case, all test results and the derived 95 % survival probability were represented by the design values.

### 5.5.3 Comparison with Existing Design Concepts

Figure 5.30b compares the resistances with the existing design methods compiled in Section 2.3.2. It is important to note that only Seeger's method [96] allows the direct consideration of different defect sizes. This comparison shows the curve for the highest quality requirements. All other methods assume that the material is almost free of defects. As expected, these methods allow higher design resistances compared to larger permissible initial defects. The requirements of RC 1 are in the orders of magnitude of the other methods and provide slightly conservative results.

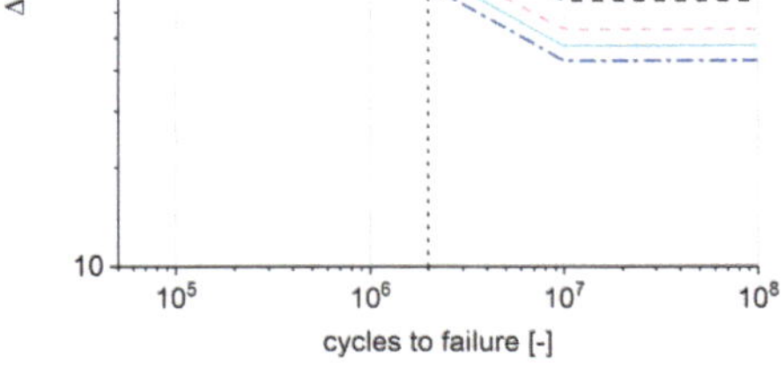


(a)

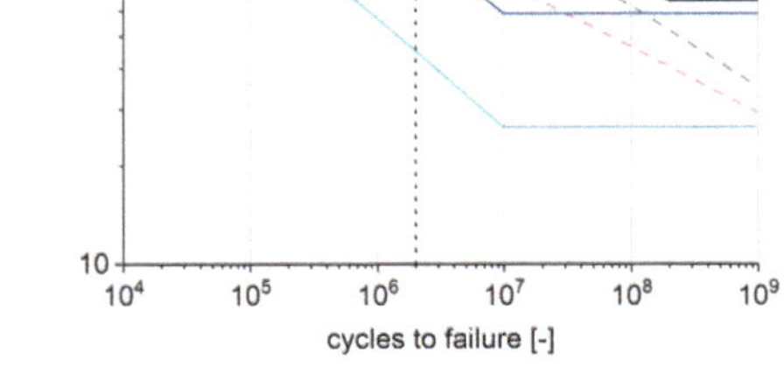


(b)

**Figure 5.30:** Comparison of the resistance curves with a) component tests and b) existing design concepts for defect-free material

Finally, a simplified fracture mechanics based procedure from BS 7910:2013 [138] was applied for comparison. The basic idea of this method is the classification of a component with a specific, known defect size into quality categories respectively into an equivalent design class according to BS 7608:2014 [136]. The actual component and defect must be transferred to reference cases analogous to Section 5.2.3. In addition to the geometries examined in Section 5.2.3, extensions for welded joints are included. The resulting SN curves have a gradient of k = 3.0, an endurance limit is not specified for the considered applications. The evaluation is carried out by means of nomograms. In the first step, an elliptical defect is related to a continuous crack. The assignment to a resistance class is part of a second step. These nomograms are based on crack propagation calculations with the Paris parameters marked in Table 3.2 as no. 16. As discussed in Section 5.2.2 this is a very conservative but universal crack propagation approach.

Table 5.12 assesses the allowable defect sizes according to Table 5.10 using the simplified procedure from BS 7910:2013 [138] and lists the resulting fatigue resistances for $\Delta\sigma_c$ at $2 \cdot 10^6$ cycles. The actual considered defect sizes are supplemented. In all cases the worst case was evaluated with a wall thickness of t = 30 mm.

**Table 5.12:** Evaluation of the maximum permissible defect sizes according to the simplified approach from BS 7910:2013 [138] and comparison of the resulting fatigue resistances with the present approach

| present approach | | simplified approach from BS 7910:2013 | | | | | |
|---|---|---|---|---|---|---|---|
| | | internal defect | | | surface defect | | |
| RC | FAT | 2a | 2c | $\Delta\sigma_c$ | a | 2c | $\Delta\sigma_c$ |
| [-] | [MPa] | [mm] | [mm] | [MPa] | [mm] | [mm] | [MPa] |
| 1 | 100 | 4 | 10 | 68 | 1.6 | 8 | 68 |
| 2 | 90 | 5 | 13 | 60 | 2.0 | 10 | 68 |
| 3 | 71 | 7 | 18 | 60 | 3.2 | 16 | 60 |
| 4 | 56 | 11 | 28 | 43 | 5.0 | 25 | 50 |
| 5 | 45 | 11 | 28 | 43 | 5.0 | 25 | 50 |

Identical defects were evaluated significantly lower by BS 7910:2013 [138] then by the present approach. This effect was more pronounced for internal defects than for surface defects. Regardless of the defect location, the evaluation of small initial defects led to greater differences than that of large initial defects.

The poorer rating of the defects primarily resulted from the selected Paris parameters. In addition, the clustering in BS 7910:2013 [138] caused significantly different initial defects to be assigned to the same category. The decreasing deviation, with increasing initial crack sizes despite identical gradients of the SN curves, resulted from the additional consideration of $\Delta K_{th}$ in this present approach. As discussed in Section 5.2.3.1, this consideration led to increasingly conservative values compared to a crack propagation calculation with increasing initial crack size.

## 5.6 Example of Application

In the following, this simplified design approach was applied to two typical cast steel components and is compared with the results of a crack propagation simulation and the literature data. The results illustrate both its conservative character as well as the simple applicability, resulting in economically reasonable quality requirements.

### 5.6.1 Fork Head

In this first example, a generic fork head from a tension rod system of material 1.6220 was considered. The system was assumed to have a screwed, threaded rod of diameter D = 120 mm. The aim of this comparative calculation was to determine local RCs in a way that the thread of the tension rod represents the decisive failure location. The geometric situation resulting from numerical stress analyses and the derived local RC is summarized in Figure 5.31.

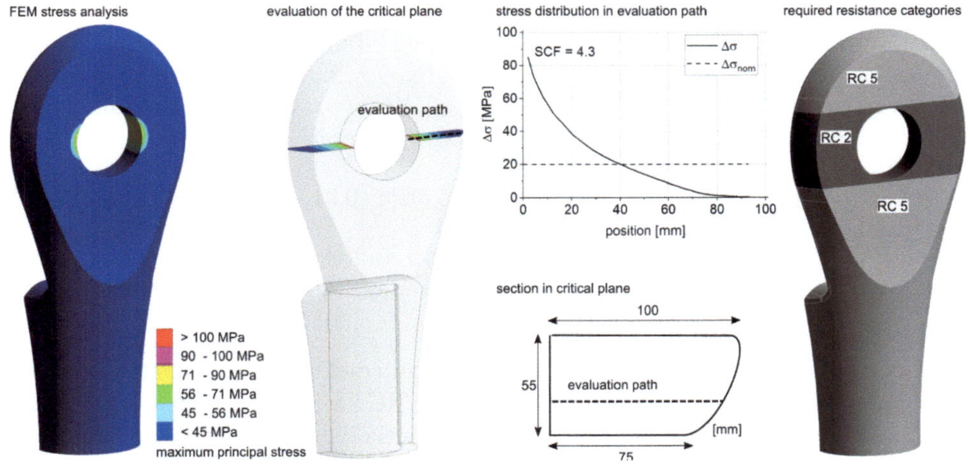


**Figure 5.31:** Evaluation of a generic fork head

*Modeling:*
The semi-model shown in Figure 5.31 was examined and meshed with volumetric elements of quadratic shape functions. The boundary and symmetry conditions were applied according to the installation situation. A linear elastic material model was used, and geometrically non-linear effects were avoided.

*Loading:*
The calculation was carried out at a loading that corresponds to the nominal resistance of the thread at $2 \cdot 10^6$ cycles, FAT 50. The result could be transmitted to any other load range due to the identical gradient k = 3 for both the SN curve of the thread and

the SN curve of the cast component. The resistance curve of the thread according to EN 1993-1-9:2010 [156] was reduced by a diameter-dependent factor (Table 1 Constr. Detail 14) from FAT 50 to $\Delta\sigma_c = 36$ MPa. The stress ratio was assumed to be $R = 0.1$.

*Other influencing factors:*
As material 1.6220 was used, the ULS was not considered in the fatigue assessment according to Section 5.2.4; it needed to be limited to 0.75 $\sigma_y$. Deviations from straightness had a negligible effect on the relevant stresses and were not considered in this example.

*Results:*
The legend of the stress plot in Figure 5.31 was set to represent the minimum values of the FAT classes. In the evaluation, only maximum principal stresses were considered. Although the strengthening and reducing effects according to Section 5.3 were not yet considered, this allowed for a quick identification of the decisive areas and a preliminary assessment on the basis of the maximum stress amplitudes.

To consider the load increasing influence in the specific application case, the factor $R_{tol,1} = 1.1$ according to Table 5.7 was applied in post-processing and considered in the stress plot. Since the stresses in the entire component except for the hole area were lower than FAT 45, only the hole area was considered in detail. With an SCF = 4.3, (5.4) and $L = (100 + 75)/2 = 88$ mm, a factor of $R_{SCF} = 0.91$ considering stress support effects was calculated. According to Eq. (5.22), FAT 90 covered the determining stress range, and thus RC 2 was selected in the high-stress region.

$$\Delta\sigma_{Ek} = \Delta\sigma_{max} \cdot R_{SCF} \cdot R_{tol,1} = 85\,\text{MPa} \cdot 0.91 \cdot 1.1 = 85\,\text{MPa} \tag{5.22}$$

As the FAT resistances of different areas (thread and clevis) were compared and no rated loads were considered, no safety factors were taken into account.

*Comparative calculations:*
In comparison to the simple approach, a numerical crack propagation simulation was carried out. The initial crack was set according to RC 2 with a = 2 mm and c = 5 mm. The simulation was carried out according to the method described in Section 4.1.2 and approach 3 in Figure 2.4 with a variable step size of 1 mm to 10 mm and 30 calculation steps. The crack propagation parameters were selected according to Table 5.2. This resulted in a service life of 6,050,000 load cycles.

Since only values corresponding to the FAT classes were considered in the simplified approach, the cycles were transferred to the actual stress range using

$$N = 2 \cdot 10^6 \cdot \left(\frac{\Delta\sigma_c}{\Delta\sigma_{Ek}}\right)^3 = 2 \cdot 10^6 \cdot \left(\frac{90\,\text{MPa}}{77\,\text{MPa}}\right)^3 = 3,200,000. \tag{5.23}$$

The growth calculation was based on the nominal dimensions; for this reason it applies $R_{tol,1} = 1.0$ resulting in $\Delta\sigma_{Ek} = 77$ MPa analogous to Eq. (5.22).

Finally, the service life calculated by the simplified procedure was

$$3,200,000/6,050,000 = 53\,\% \tag{5.24}$$

of the numerically determined value. These differences were caused by an underestimation of the support effects from stress gradients, the approach of $A_{Kth}$ as well as load transfer effects that were not taken into account. Nevertheless, an economical component design was achieved.

### 5.6.2 K Joints

The aim of this example was to compare the present design approach with the specific approach in [35] for K joints in truss bridges. The K joint was taken from the exemplary truss bridge investigated in [35] and labeled as joint 416. All relevant dimensions are given in Figure 5.32. The material was 1.6220, the design temperature was $T_{Ed}$ = -30 °C and the utilization factor at ULS was less than 75 %; thus, brittle fracture did not need to be considered. Dimensional derivations were not considered, as they are not part of [35].

Originally, nine defect locations were investigated using a simplified approach and with the boundary element method. Within this comparative calculation, only four critical positions were investigated, as illustrated in Figure 5.32.

Note, that position 4 in [35] did not correspond to the location of the local maximum stress range but to a point that would be easily accessible for NDT. The loading situation for the main girder and the trusses is summarized in Table 5.14 according to the indices given in Figure 5.32. The crack sizes were determined to achieve $2 \cdot 10^6$ load cycles, so a direct comparison of the FAT classes was possible.

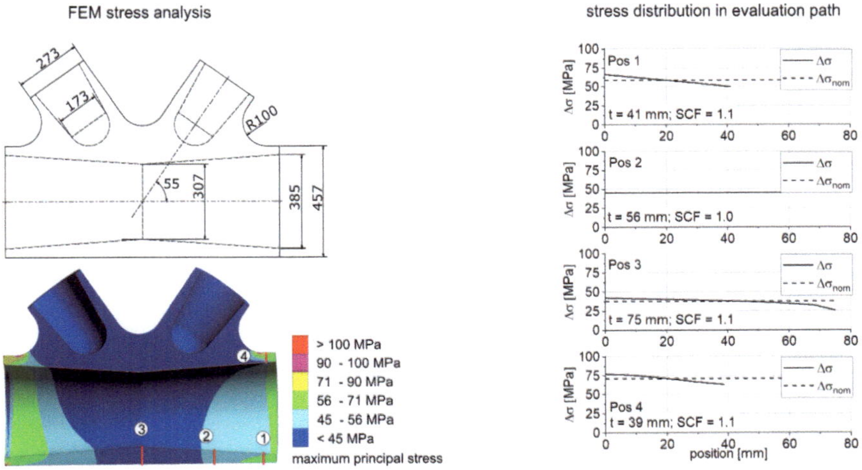


**Figure 5.32:** K joint geometry, stress analysis and resulting requirement categories

**Table 5.13:** Loading situation according to [35]

| | $N_{ULS}$ [kN] | $\Delta N_{FAT}$ [kN] |
|---|---|---|
| chord 1 | 7602 | 3132 |
| chord 2 | 7510 | 3165 |
| truss 3 | -67 | 396 |
| truss 4 | 91 | -400 |

For each defect position, the stress distribution was evaluated and presented in Figure 5.32. Resulting stresses, reduction factors, necessary FAT classes and associated RCs are summarized in Table 5.14.

A concluding comparison of the maximal tolerable defect sizes showed differences up to a factor of ten that were caused by the following:

- For the stresses at the nodal boundaries specified in [35], higher stress and stress intensity factors occurred in these calculations despite almost identical calculation models.
- Each model is based on different crack geometries. In [35], a half-circular crack was chosen, whereas in the present approach, the more critical semi-elliptical surface crack with a ratio $a/c = 0.4$ was selected. Along with the Paris exponent, even small differences in $\Delta K$ cause larger differences in the calculated service life.
- In Section 5.2.3 the increasing conservatism of the $\Delta K_{th}$ based approach compared to a crack growth-based approach is discussed. These effects are particularly significant for the large permissible initial defects occurring in this example
- Other than in[35], the present approach did not determine a separate defect quantity for each point. In this case, they were available in clusters so that further conservative assumptions were made for stresses that did not correspond to the cluster boundary.
- The additionally defined requirements (e.g., that initial defects may only extend over a certain percentage of the wall thickness) led to smaller permissible defect sizes at low loads.
- In [35], interaction of multiple cracks was not considered.
- The defect sizes determined in [35] illustrate that very large defects could be partly tolerated despite fatigue loading. In contrast to established quality requirements, considerably lower qualities were permitted. Considering production quality according to EN 1090-2:2018 [143], special attention and high quality standards are required for fatigue-stressed components independent of the manufacturing method. The quality requirements derived according to this concept correspond much better with this philosophy.

**Table 5.14:** Result summary of the fatigue analysis on an exemplary K joint

| Pos. | | 1 | 2 | 3 | 4 |
|---|---|---|---|---|---|
| t | [mm] | 51 | 56 | 75 | 39 |
| $\Delta\sigma_{\mathrm{max}}$ | [MPa] | 67 | 45 | 43 | 78 |
| $R_{SCF}$ | [-] | 0.97 | 1.00 | 0.97 | 0.97 |
| $\Delta\sigma_{\mathrm{max}}$ | [MPa] | 65 | 45 | 42 | 75 |
| FAT | [MPa] | 71 | 45 | 45 | 90 |
| RC | [-] | 3 | 5 | 5 | 2 |
| $A_{\mathrm{max,RC}}$ | [$mm^2$] | 41 | 98 | 98 | 17 |
| $A_{\mathrm{max,HS}}$ from [35] | [$mm^2$] | 208 | 774 | 2731 | 312 |
| fact | [-] | 5 | 8 | 28 | 18 |
| N(RC,$\Delta\sigma$) | [-] | $2.6 \cdot 10^6$ | $2.0 \cdot 10^6$ | $2.5 \cdot 10^6$ | $3.4 \cdot 10^6$ |
| fact | [-] | 1.3 | 1.0 | 1.3 | 1.7 |

# 6 Conclusion

The inner and outer structure of cast steel components define their resistance to fatigue. To address this issue, designers must define manufacturing quality requirements as early as possible in the design process. A generally applicable and easy-to-use method that links fatigue resistance to quality grades is not yet available. The definition is mainly based on experience, transferred to the entire component and verified afterward by extensive experimental investigations. In exceptional cases, fatigue strength is estimated using complex fracture mechanics calculations during the design process.

The aim of this thesis was to develop an easy-to-use design approach that meets the special requirements of the construction industry. For this purpose, fatigue resistance was directly related to maximum permissible defect sizes, which were determined by fracture mechanics. In this last chapter, the key aspects for creating and validating this concept are summarized. Further research needs that have shown up in the course of this work are presented. Finally, potential enhancements and the transfer of the basic concept to other manufacturing processes are discussed.

## 6.1 Summary

To quantify the influence of casting defects on fatigue behavior, tensile specimens with real casting defects of different geometric properties were manufactured and tested under cyclic loading. By considering two materials, G20Mn5 and G22NiMoCr5-6, a wide range of applications could be covered, and different failure mechanisms could be revealed. The exact nature of the casting defects was recorded by CT and confirmed by fractographic examinations of the fracture surfaces. The objective of a subsequent numerical analysis was the evaluation of crack propagation behavior. The result of these investigations were SN curves determined by the size and position of the defect. For surface defects, crack growth was identified as the decisive damage mechanism. The considered internal shrinkage cavities showed a pronounced crack initiation phase, depending on the stress level. These fundamentally different behaviors were reflected in the SN curve gradients. Fatigue strength was almost independent of the material. In contrast, the residual fracture behavior of G20Mn5 was found to be ductile, whereas G22NiMoCr5-6 was found to be more brittle.

To transfer these findings to realistic geometric dimensions and verify the design concept, specimens with wall thicknesses of 35 mm, 55 mm and 100 mm were investigated. On the one hand, this served as an integral evaluation of technological thickness effects. On the other hand, geometric notches with a stress concentration factor of 2.0 covered geometric influences on the stress state relevant to usual civil engineering components subject to fatigue. The effect of casting defects was considered by artificially introducing semi-elliptical surface defects into the root of the geometric notch. Influences of geometric deviations were taken into account individually for all specimens by numerically mapping the measured strain states. The results of these investigations confirmed the independence of cyclic crack propagation behavior from the component thickness, the effect of additional stress concentrations and the negative influence of geometric deviations. Despite the unfavorable combination of major influencing factors, nominal stress ranges of approximately 90 MPa at $2 \cdot 10^6$ cycles to failure and an average gradient of the SN curve of 3 were achieved.

These findings resulted in a stress-based design concept and in permissible defect sizes to be met by foundries. The linking of both perspectives was achieved by five newly defined resistance categories (RC). In the future, designers will not have to think about the actual defects, their location or their mechanical effects and will be able to design components in the ways they are familiar with. The permissible defect sizes were derived from fracture mechanics investigations on reference components for internal and surface defects. Both crack growth and behavior near the threshold value were analytically considered. In addition to permissible defect sizes of single defects, requirements were defined for an overall defect size as well as geometric requirements for the interaction of adjacent defects. Influences such as the mean stress dependency, local stress peaks and geometric deviations were taken into account in the fatigue check by reduction factors. In addition, the derived defect sizes include the interaction of multiple defects as well as brittle fracture. The comparison of component resistance calculated in this way with the experimental investigations and literature data showed a consistently conservative approach. The derived permissible individual defect sizes are within the range of established quality grades and can thus be economically implemented.

## 6.2 Research Needs

The defect sizes resulting from the presented design approach exclusively represent minimal mechanical requirements. For specific applications, these must be converted into indication characteristics, registrations and acceptance limits depending on the NDT method used. The reliable detection of the casting defects, which are plane cracks in the worst case, must be ensured. The influence of typical casting component properties, such as curved surfaces, non-parallel component areas or complex geometries, plays a significant role in this context. An additional reduction factor as suggested by [96], which takes the complexity and testability of a component into account, can be a suitable tool. For practical implementations, clear criteria and assignments must be defined.

Along with the NDT requirements derived in this way, it is desirable to establish a general design approach for cast steel components. Based on the resistance categories defined herein and the underlying fracture mechanics evaluation strategies, the investigations into the plastic design and brittle fracture behavior of [103] can be transferred to this concept. Particular potential can be found in the evaluation of different stress concentrations in the brittle fracture analysis. For this purpose, the detail of the unloaded longitudinal stiffener applied in [103] can be replaced by the approach from Section 5.2.5. Although the unloaded longitudinal stiffener covers almost all geometries, it is too conservative for components made of cast steel that are designed properly with respect to castability and the flow of forces. In addition, the discussed approach for the interaction of multiple defects can simplify the method in [103].

The fracture mechanics calculations on which the present design approach is based were carried out for a uniaxial stress state. Usually, this does not represent a significant limitation for practical building applications. However, the effects of more complex multiaxial states can be unfavorable for special defect orientations. Their behavior and possible negative effects have not yet been considered and should be investigated.

Investigations such as [58, 117] show a negative influence of production welding (the fixing of casting defects that do not fulfill quality requirements) on fatigue strength compared to flawless base material. Analogous to the casting defects discussed in this thesis, designers are also faced with the challenge of assuming component resistances without a data-driven basis. This results in extended welding prohibition zones. For a calculative analysis, the resistance of production-welded components must be set in relation to the component resistance permissible for RC 1. Since the highest RC used in the present concept does not assume a flawless base material, but rather allows certain imperfections, no major restrictions are assumed if the welding is carried out correctly according to professional workmanship. Features such as over- and under-matching, the scope of subsequent heat treatment methods and imposed residual stresses need to be investigated. This quantification could reduce the scope of heat treatment after the welding process or invalidate welding prohibition zones.

## 6.3 Perspectives

The performance of this design method is fundamentally determined by the crack propagation properties of the defect-free base material. For the purpose of general applicability, these have been chosen very conservatively and not specifically for cast materials. If technical applications not considered in this context require component resistances higher than FAT 100 or greater than the defined permissible defect sizes, this offers the potential for optimization. Extensive investigations on cast steel materials can provide a more accurate basis of argumentation for a general increase of the fatigue classification from a statistical point of view and would correspond to the experimental investigations carried out herein.

In addition, this approach can be the starting point for the transfer to other manufacturing processes with similar conditions. Examples are forging or additive manufacturing. These structures can be adapted to the flow of forces, and manufacturing defects can be revealed and subjected to NDT. Material properties, typical defect sizes, practice-relevant FAT classes and NDT detection limits would have to be adapted to the specific conditions.

# References

## Scientific & Technical Publications


[1] A. Al Tamimi and M. Modarres: *Coalescence and Growth of Two Semi-Elliptical Coplanar Cracks in API-5L Grade B Steel. Fracture, fatigue, failure, and damage evolution.* Ed. by J. D. Carroll and S. H. Daly. Vol. 60. Conference Proceedings of the Society for Experimental Mechanics Series. Cham, US: Springer, 2015, pp. 57–66.

[2] T. L. Anderson: *Fracture Mechanics: Fundamentals and Applications.* 3rd ed. Boca Raton, Florida, US: CRC Press, 2005.

[3] ANSYS inc.: *Ansys Workbench 2019 R2: Software Manual.* Canonsburg, US, 2019.

[4] ANSYS inc.: *Fracture Analysis Guide: Software Manual.* Canonsburg, US, 2019.

[5] J. Baier, C. Grefhorst, W. Kleinmann, M. Köppen, and O. Podobed: *Handbuch der Gußfehler.* 3rd ed. Marl: S&B Industrial Minerals GmbH, 2005.

[6] A. C. Bannister: *SINTAP: Structural Integrity Assessment Procedures for European Industry: Sub-Task 3.3 Determination of Fracture Toughness from Charpy Impact Energy: Procedure and Validation.* Rotherham, UK: British Steel, 1998.

[7] Y. Bao and T. Wierzbicki: *On fracture locus in the equivalent strain and stress triaxiality space.* International Journal of Mechanical Sciences 46 (2004), pp. 81–98.

[8] J. M. Barsom: *Fatigue-crack propagation in steels of various yield strengths.* Journal of Engineering for Industry 93.4 (1971), pp. 1190–1196.

[9] J. Baumgartner: *Schwingfestigkeit von Schweißverbindungen unter Berücksichtigung von Schweißeigenspannungen und Größeneinflüssen: Doctoral Thesis.* Vol. 238. LBF-Berichte, Darmstadt. Stuttgart: Fraunhofer-Verlag, 2014.

[10] J. Bergmann, K. Biswas, K. H. Hoff, and T. Seeger: *Fatigue Behaviour of HOESCH GS Ark 10. - Part 1: Experimental. Report FI-13/1981.* Fachgebiet Werkstoffmechanik, TU Darmstadt, 1981.

[11] M. Blair, R. Monroe, C. Beckermann, R. Hardin, K. Carlson, and C. Monroe: *Predicting the Occurrence and Effects of Defects in Castings.* Journal of the Minerals 57.5 (2005), pp. 29–34.

[12] M. Blair and T. L. Stevens, eds.: *Steel Castings Handbook.* 6th ed. Materials Park, Ohio, US: Steel Founders' Society of America and ASM International, 1995.

[13] H. Blumenauer and G. Pusch: *Technische Bruchmechanik.* 3rd ed. Leipzig: Deutscher Verlag für Grundstoffindustrie, 1993.

[14] R. Branco and F. V. Antunes: *Finite Element Modelling and Analysis of Crack Shape Evolution in Mode-I Fatigue Middle Cracked Tension Specimens.* Engineering Fracture Mechanics 75.10 (2008), pp. 3020–3037.

[15] E. S. Breznak, C. Vishnevsky, and J. F. Wallace: *The Effect of Internal Shrinkage Discontinuities on the Fatigue and Impact Properties of Cast Steel Sections: Final Report.* Rocky River, Ohio, US: Steel Foundry Research Foundation, 1969.

[16] K. B. Broberg: *Cracks and Fracture.* San Diego, US: Academic Press, 1999.

[17] W. Brocks: *Plasticity and Fracture.* Solid Mechanics and Its Applications, Vol. 244. Cham, US: Springer, 2018.

[18] M. W. Brown and K. J. Miller: *A Theory for Fatigue under Multiaxial Stress-Strain Conditions.* Proceedings of the Institute of Mechanical Engineers 187.1 (1973), pp. 745–756.

[19] C. Bueno: *Bewertung, Optimierung und Anwendung etablierter Ermüdungskonzepte auf ungänzenbehaftete Stahlgussbauteile: Master Thesis.* KIT Steel & Lightweight Structures. Karlsruhe, 2018.

[20] A. de Castro e Sousa and A. Nussbaumer: *Multiaxial Ultra Low Cycle Fatigue in Welded High Strength Steel Structural Components.* Journal of Constructional Steel Research 153 (2019), pp. 473–482.

[21] L. F. J. Coffin: *A Study of the Effects of Cyclic Thermal Stresses on a Ductile Metal.* Transactions of A.S.M.E. 76 (1954), pp. 931–950.

[22] K. Drebenstedt and M. Euler: *Statistical Analysis of Fatigue Test Data according to Eurocode 3.* Proceedings of International Conference on Bridge Maintenance, Safety and Management IABMAS, Melbourne, 9–13 July 2018 (2018).

[23] K.-O. Edel: *Einführung in die bruchmechanische Schadensbeurteilung.* Berlin: Springer Vieweg, 2015.

[24] A. Fatemi and D. F. Socie: *A Critical Plane Approach to Multiaxial Fatigue Damage Including Out-of-Phase Loading.* Fatigue Fracture of Engineering Materials and Structures 11.3 (1988), pp. 149–165.

[25] M. Feldmann, H. Bartsch, T. Ummenhofer, B. Seyfried, U. Kuhlmann, and K. Drebenstedt: *Final Report on Research Project: IGF 19178 N Neubewertung und Erweiterung des Kerbfallkatalogs nach Eurocode 3 für eine zukunftsfähige Auslegung hochbeanspruchter Stahlkonstruktionen.* Forschungsvereinigung Stahlanwendung e.V. et al., 2019.

[26] R. G. Forman, V. E. Kearney, and R. M. Engle: *Numerical Analysis of Crack Propagation in Cyclic-Loaded Structures.* Journal of Basic Engineering 89.3 (1967), pp. 459–463.

[27] Foundry Technologies & Engineering (FT&E) GmbH: *Picture of cold crack: https://www.giessereilexikon.com/giesserei-lexikon/Encyclopedia.* Downloaded at: 23.06.2020.

[28] D. Groß and T. Seelig: *Fracture Mechanics: With an Introduction to Micromechanics.* 3rd ed. Cham, US: Springer, 2018.

[29] T. R. Gurney: *An Analysis of some Fatigue Crack Propagation Data for Steels Subjected to Pulsating Tension Loading.* Welding Research International 6.4 (1976), pp. 45–59.

[30] A. L. Gurson: *Continuum Theory of Ductile Rupture by Void Nucleation and Growth: Part I - Yield Criteria and Flow Rules for Porous Ductile Media.* Journal of Engineering Materials and Technology 99.1 (1977), pp. 2–15.

[31] Gussstahl: *Picture of Einschluss in metallischer Grundmasse: https://de.wikipedia.org/wiki/Datei:Einschluss1.jpg.* Downloaded at: 23.06.2020, (2009).

[32] P. Haasen: *Physikalische Metallkunde.* 2nd ed. Berlin: Springer, 1984.

[33] M. Häbig: *Versuchstechnische Bestimmung und numerische Evaluierung von Werkstoffparametern für die Anwendung in plastischen Schädigungsmodellen: Master Thesis.* KIT Steel & Lightweight Structures. Karlsruhe, 2016.

[34] E. Haibach: *Betriebsfestigkeit: Verfahren und Daten zur Bauteilberechnung.* 3rd ed. Berlin: Springer, 2006.

[35] S. C. Haldimann-Sturm: *Ermüdungsverhalten von Stahlgussknoten in Brücken aus Stahlhohlprofilen: Doctoral Thesis.* École Polytechnique Fédérale de Lausanne. Lausanne, CH, 2005.

[36] R. Hardin and C. Beckermann: *Effect of Shrinkage on Service Performance of Steel Castings.* Proceedings of the 56th SFSA Technical and Operating Conference, Steel Founders' Society of America, Chicago (2002), Paper No. 4.5.

[37] R. Hardin and C. Beckermann: *Prediction of the Fatigue Life of Cast Steel Containing Shrinkage Porosity.* Metallurgical and Materials Transactions 40 (2009), pp. 581–597.

[38] R. Hardin and C. Beckermann: *Validation and Gage R+R Studies of a New Radiography Standard for Steel Castings.* Proceedings of the 64th SFSA Technical and Operating Conference, Steel Founders' Society of America, Chicago (2010), Paper No. 2.6.

[39] R. Hardin and C. Beckermann: *Effect of Porosity on Deformation, Damage, and Fracture of Cast Steel.* Supplemental Proceedings: Volume 2: Materials Properties, Characterization, and Modeling (2012), pp. 217–224.

[40] R. Hardin and C. Beckermann: *Integrated Design of Castings: Effect of Porosity on Mechanical Performance.* IOP Conference Series: Materials Science and Engineering 33 (2012).

[41] R. Hardin and C. Beckermann: *Effect of Porosity on the Stiffness of Cast Steel.* Metallurgical and Materials Transactions A 38.12 (2007), pp. 2992–3006.

[42] P. Heuler, C. Berger, and J. M. Motz: *Oberflächennahe Fehlstellen im Stahlguß. Bewertung mit Hilfe von Kerb- und Bruchmechanikkonzepten. Teil 1: Fehlstellen und ihre Wirkung, Versuchskonzept.* Materialprüfung 33.9 (1991), pp. 246–252.

[43] P. Heuler, C. Berger, and J. M. Motz: *Oberflächennahe Fehlstellen im Stahlguß. Bewertung mit Hilfe von Kerb- und Bruchmechanikkonzepten. Teil 2: Rechnerische Bewertung von Fehlstellen.* Materialprüfung 33.10 (1991), pp. 292–296.

[44] P. Heuler, C. Berger, and J. M. Motz: *Oberflächennahe Fehlstellen im Stahlguß. Bewertung mit Hilfe von Kerb- und Bruchmechanikkonzepten. Teil 3: Bruchmechanisches Verhalten von Warmrissen, Zusammenfassung.* Materialprüfung 33.11/12 (1991), pp. 329–331.

[45] J. von Hirsch: *Die Gießerei und die Gütestufe.* Gießerei-Erfahrungsaustausch 6 (1995), pp. 213–214.

[46] A. Hobbacher, ed.: *Recommendations for Fatigue Design of Welded Joints and Components: IIW document IIW-2249-15, ex XIII-2460-13/XV-1440-13.* 2nd ed. Vol. 2259-15. International Institute of Welding. Cham, US: Springer, 2016.

[47] J. C. Hrabowski: *Ermüdungsverhalten von Schweißverbindungen aus höchstfestem Stahl im Kurzzeitfestigkeitsbereich: Doctoral Thesis.* Research Center for Steel, Timber & Masonry - Berichte zum Stahl- und Leichtbau, Vol. 9. Karlsruhe: KIT Scientific Publishing, 2019.

[48] P. Hübner and G. Pusch: *Bruchsicherheitsanalyse eines rissbehafteten Kugelschieberzapfens.* konstruieren + gießen 25.3 (2000), pp. 4–9.

[49] B. Jonsson, J. Samuelsson, and G. B. Marquis: *Development of Weld Quality Criteria based on Fatique Performance.* Welding in the Word 55.11 (2011), pp. 79–88.

[50] M. Kachanov: *Elastic Solids with Many Cracks and Related Problems.* Advances in Applied Mechanics 30 (1993), pp. 259–445.

[51] M. Kamaya, E. Miyokawa, and M. Kikuchi: *Growth Prediction of Two Interacting Surface Cracks of Dissimilar Sizes.* Engineering Fracture Mechanics 77.16 (2010), pp. 3120–3131.

[52] M. Klesnil and P. Lukáš: *Fatigue of Metallic Materials.* Vol. 7. Materials science monographs. Amsterdam, NL: Elsevier, 1980.

[53] R. Kuguel: *A Relation between Theoretical Stress Concentration Factor and Fatigue Notch Factor Deduced From the Concept of Highly Stressed Volume.* ASTM Proc. 61 (1961), pp. 732–744.

[54] B. Kühn: *Beitrag zur Vereinheitlichung der europäischen Regelungen zur Vermeidung von Sprödbruch: Doctoral Thesis.* RWTH Aachen, Schriftenreihe Stahlbau, Vol. 54. Aachen: Shaker, 2005.

[55] M. Kuroda: *Extremely Low Cycle Fatigue Life Prediction Based on a New Cumulative Fatigue Damage Model.* International Journal of Fatigue 24 (2001), pp. 699–703.

[56] K.-E. Kurrer: *The History of the Theory of Structures: Searching for Equilibrium.* 2nd ed. Berlin: Wiley, 2018.

[57] P. Lukas and L. Kunz: *Specific Features of High-Cycle and Ultra-High-Cycle Fatigue.* Fracture of Engineering Materials and Structures 25.8-9 (2002), pp. 747–753.

[58] A. Ma and J. V. Sharp: *Fatigue Design of Cast Steel Nodes in Offshore Structures Based on Research Data.* Proceedings of the Institution of Civil Engineers - Water and Maritime Engineering 2 (1997), pp. 112–126.

[59] S. S. Manson: *Behavior of Materials under Conditions of Thermal Stress. Report 1170.* Cleveland, US: National Advisory Committee for Aeronautics, 1954.

[60] MathWorks: *Matlab R2017b: User Documentation.* Natick, US, 2017.

[61] K. J. Miller and E. R. De los Rios, eds.: *The Behaviour of Short Fatigue Cracks.* Vol. 1. EGF publication. London, UK: Mechanical Engineering Publications, 1986.

[62] I. Milne, R. A. Ainsworth, A. R. Dowling, and A. T. Stewart: *Assessment of the Integrity of Structures Containing Defects.* International Journal of Pressure Vessels and Piping 32.3 (1988), pp. 3–104.

[63] R. R. Moore: *Material Testing Machine: Patent Number: US 2154277.* 1939.

[64] Y. Murakami, T. Namoto, and T. Ueda: *Factors Influencing the Mechanism of Superlong Fatigue Failure in Steel.* Fatigue and Fracture of Engineering Materials and Structures 22 (1999), pp. 581–590.

[65] Y. Murakami: *Stress Intensity Factors Handbook.* Oxford, UK: Pergamon Press, 1987.

[66] Y. Murakami: *Metal Fatigue: Effects of Small Defects and Nonmetallic Inclusions.* 1st ed. Amsterdam, NL: Elsevier, 2002.

[67] Y. Murakami and S. Nemat-Nasser: *Interacting Dissimilar Semi-Elliptical Surface Flaws under Tension and Bending.* Engineering Fracture Mechanics 16.3 (1982), pp. 373–386.

[68] R. Murayama and K. Yamaguchi: *Nonlinear Ultrasonic Wave Detection by Electromagnetic Acoustic Transducer [EMAT] for Guided Wave.* Japanese Journal of Applied Physics 44 (2005), pp. 4385–4388.

[69] S. Nagel, C. Rauber, I. Veile, P. Knödel, and T. Ummenhofer: *Influence of Internal Imperfections on the Fatigue Resistance of Cast Steel – Testing Methodology.* Proceedings on ICMFF12, MATEC Web of Conferences (2019).

[70] S. Nagel, M. Spannaus, and T. Ummenhofer: *Stahlguss – ein unterschätzter Werkstoff. Stahlbau-Kalender 2019.* Ed. by U. Kuhlmann. Vol. 21. Berlin: Ernst & Sohn, 2019, pp. 433–479.

[71] H. Neuber: *Theory of Stress Concentration for Shear-Strained Prismatical Bodies with Arbitrary Nonlinear Stress-Strain Law.* Journal of Applied Mechanics 28 (1961), pp. 544–550.

[72] T. Nicholas: *Critical Issues in High Cycle Fatigue.* International Journal of Fatigue 21 (1999), pp. 221–231.

[73] A. Nussbaumer and H.-P. Günther: *Stahlbaunormen - Kommentar zu DIN EN 1993-1-9: Ermüdung Grundlagen und Erläuterungen. Stahlbau-Kalender 2012.* Ed. by U. Kuhlmann. Vol. 12. Berlin: Ernst & Sohn, 2012, pp. 255–351.

[74] P. C. Paris and F. Erdogan: *A Critical Analysis of Crack Propagation Laws.* Transactions of the ASME, Journal of Basic Engineering 85 (1963), pp. 528–534.

[75] R. E. Peterson: *Notch Sensitivity. Metal Fatigue.* Ed. by G. Sines and J. L. Waismann. New York: McGraw-Hill, 1959.

[76] R. E. Peterson: *Stress Concentration Factors.* New York: John Wiley & Son, 1974.

[77] L. P. Pook: *Some Implications of Corner Point Singularities.* Engineering Fracture Mechanics 48.3 (1994), pp. 367–378.

[78] R. Puthli, S. Herion, M. Veselcic, M. A. Hirt, A. Nussbaumer, S. Haldimann, J. Schlaich, M. Schlaich, and H. Schober: *Research Project: FOSTA P591: Economic use of Structural Hollow Sections for Highway and Railway Bridges.* Düsseldorf: Forschungsvereinigung für Stahlanwendungen e.V., 2010.

[79] B. Pyttel, D. Schwerdt, and C. Berger: *Very High Cycle Fatigue – Is there a Fatigue Limit?* International Journal of Fatigue 33.1 (2011), pp. 49–58.

[80] D. Radaj: *Ermüdungsfestigkeit: Grundlagen für Leichtbau, Maschinen- und Stahlbau.* Berlin: Springer, 1995.

[81] D. Radaj and W. Möhrmann: *Kerbwirkung an Schulterstäben unter Querschub.* Konstruktion 36.10 (1984), pp. 399–402.

[82] D. Radaj and G. Schilberth: *Kerbspannungen an Ausschnitten und Einschlüssen.* Düsseldorf: DVS-Verlag, 1977.

[83] D. Radaj and M. Vormwald: *Ermüdungsfestigkeit: Grundlagen für Ingenieure.* 3rd ed. Berlin: Springer, 2007.

[84] J. R. Rice and D. M. Tracey: *On the Ductile Enlargement of Voids in Triaxial Stress Fields.* Journal of the Mechanics and Physics of Solids 17.3 (1969), pp. 201–217.

[85] R. O. Ritchie: *Mechanisms of Fatigue-Crack Propagation in Ductile and Brittle Solids.* International Journal of Fracture 100.1 (1999), pp. 55–83.

[86] M. T. Rowley, ed.: *International Atlas of Casting Defects.* Des Plaines, Illinois, US: American Foundrymen's Society, 1993.

[87] H. Saal, G. Steidl, and M. Volz: *Guss im Konstruktiven Ingenieurbau - Gussstücke in Bauwerken.* konstruieren + giessen 31.2 (2006), pp. 2–13.

[88] R. H. Sailors and H. T. Corten: *Relationship Between Material Fracture Toughness using Fracture Mechanics and Transition Temperature Tests.* Proceedings of the 1971 fifth National symposium on fracture mechanics (1972), pp. 164–191.

[89] Sande Stahlguss GmbH: *Picture of a Steam Turbine.* (2020).

[90] M. Sander: *Sicherheit und Betriebsfestigkeit von Maschinen und Anlagen: Konzepte und Methoden zur Lebensdauervorhersage.* Berlin: Springer, 2008.

[91] G. N. Sawin: *Spannungserhöhung am Rande von Löchern.* Berlin: VEB Verlag Technik, 1956.

[92] G. S. Schajer: *H-Drill Version 3.2: Software Manual.* Vancouver, CAN, 2009.

[93] J. Schijve: *Fatigue of Structures and Materials.* Dordrecht, NL: Springer, 2009.

[94] E. Schmidtmann: *Bruchmechanikuntersuchungen an Stählen - Einfluss der Belastungsgeschwindigkeit.* Beiträge zur Bruchmechanik metallischer Werkstoffe, Fortschritt-Berichte der VDI-Zeitschriften (1975), pp. 77–88.

[95] G. Sedlacek, M. Feldmann, B. Kühn, D. Tschickardt, S. Höhler, and S. Müller: *Commentary and Worked Examples to EN 1993-1-10 "Material Toughness and Through Thickness Properties" and other Toughness Oriented rules in EN 1993.* 1st ed. Luxembourg: Official Publications of the European Communities, 2008.

[96] T. Seeger: *Ermüdungsverhalten von HOESCH GS Ark 10. – Teil 2: Konzept für den Ermüdungsnachweis. Report: FI-14/1981.* Fachgebiet Werkstoffmechanik, TU Darmstadt, 1981.

[97] K. M. Sigl, R. A. Hardin, R. I. Stephens, and C. Beckermann: *Fatigue of 8630 Cast Steel in the Presence of Porosity.* International Journal of Cast Metals Research 17.3 (2004), pp. 130–146.

[98] G. C. Sih: *Handbook of Stress-Intensity Factors.* Bethlehem, Pennsylvania, US: Lehigh University, Institute of Fracture and Solid Mechanics, 1973.

[99] C. M. Smith, G. Deierlein, and A. M. Kanvinde: *A Stress-Weighted Damage Model for Ductile Fracture Initiation in Structural Steel under Cyclic Loading and Generalized Stress States: Technical Report 187.* Stanford, California, US: John A. Blume Earthquake Engineering Center, 2014.

[100] W. O. Soboyejo, J. F. Knott, M. J. Walsh, and K. R. Cropper: *Fatigue Crack Propagation of Coplanar Semi-Elliptical Cracks in Pure Bending.* Engineering Fracture Mechanics 37.2 (1990), pp. 323–340.

[101] C. Sonsino: *Course of SN-Curves Especially in the High-Cycle Fatigue Regime with Regard to Component Design and Safety.* International Journal of Fatigue 29.12 (2007), pp. 2246–2258.

[102] W. Sörgel and W. Günther: *Zur Wechselwirkung bei Mehrfachrissen.* Technische Mechanik 4.3 (1983), pp. 5–11.

[103] M. J. Spannaus: *Bemessung von Erzeugnissen aus Stahlguss unter vorwiegend ruhender Beanspruchung: Doctoral Thesis.* Research Center for Steel, Timber & Masonry - Berichte zum Stahl- und Leichtbau, Vol. 5. Karlsruhe: KIT Scientific Publishing, 2016.

[104] M. O. Speidel and R. B. Scarlin: *Auswirkungen des Gefüges auf das Wachstum von Ermüdungsrissen.* proceedings of: Werkstoffprüftagung Leoben (1976).

[105] G. Steidl: *Guss im konstruktiven Ingenieurbau: Bauteile aus Eisen- und Aluminiumwerkstoffen in Tragwerken.* Die schweißtechnische Praxis, Vol. 33. Düsseldorf: DVS-Verlag, 2006.

[106] M. Stern, E. B. Becker, and R. S. Dunham: *A Contour Integral Computation of Mixed-Mode Stress Intensity Factors.* International Journal of Fracture 12.3 (1976), pp. 359–368.

[107] S. Suresh: *Fatigue of Materials.* 2nd ed. Cambridge, US: Cambridge Univ. Press, 2004.

[108] K. Szielasko, R. Tschuncky, I. Alpeter, G. Dobmann, and C. Boller: *Real-Time Monitoring of Crack Growth Behaviour During Impact and Compact Tension Test with Non-Destructive Testing. Electromagnetic Nondestructive Evaluation (XVI).* Ed. by F. Kojima, J. M. A. Rebello, and T. Chady. Studies in applied electromagnetics and mechanics. Amsterdam, NL: IOS Press, 2014, pp. 238–246.

[109] H. Tada, G. R. Irwin, and P. C. Paris: *The Stress Analysis of Cracks Handbook.* 3rd ed. New York, US: American Society of Mechanical Engineers, 2000.

[110] P. Tempel, L. Reißig, C. Eichheimer, M. Farajian, M. Luke, D. Siegele, I. Veile, and F. Weber: *Einfluss von herstellungsbedingten Ungänzen auf das Werkstoffverhalten von Stahlguss: Berichte aus der FVG-Gemeinschaftsforschung: Final Report on Research Project: IGF No. 469 ZN.* Düsseldorf: Forschungsvereinigung Gießereitechnik e.V., 2017.

[111] L. Trško, O. Bokůvka, F. Nový, and M. Guagliano: *Effect of Severe Shot Peening on Ultra-High-Cycle Fatigue of a Low-Alloy Steel.* Materials & Design 57 (2014), pp. 103–113.

[112] UEG Offshore Research: *Design of Tubular Joints for Offshore Structures.* UEG publication. London, UK, 1985.

[113] T. Ummenhofer, M. Albiez, Ö. Bucak, H. Ehard, B. Mayer, and H. Fricke: *Final Report on Research Project: FOSTA P884 / IGF-Nr. 17199 N: Adhesive Bonded Tubular Cast Steel – Steel Joints in Framework Structures.* Düsseldorf: Forschungsvereinigung für Stahlanwendungen e.V., 2015.

[114] T. Ummenhofer, S. Herion, M. Veselicic, R. J. Dietrich, A. Nussbaumer, and F. Zammiri: *Final Report on Research Project: FOSTA P816: Optimaler Einsatz von Hohlprofilen und Gussknoten im Brückenbau aus Stahls S355 bis S690.* Düsseldorf: Forschungsvereinigung für Stahlanwendungen e.V., 2014.

[115] T. Ummenhofer and D. Kohler: *"DeQaGuss" On the Definition of New Quality Criteria for the Assessment of the Load-Bearing Capacity of Cast Steel Components in Terms of Non-Destructive and Destructive Testing: Requested Research Project.* Düsseldorf, Karlsruhe, Offenburg: Forschungsvereinigung Gießereitechnik e.V., KIT Research Center for Steel, Timber & Masonry, Offenburg University of Applied Sciences, 2020-2023.

[116] T. Ummenhofer, S. Nagel, I. Veile, C. Rauber, M. Farajian, and M. Jung: *"ErStaGu" On the Influence of Casting Defects on the Fatigue Resistance of Cast Steel Components: Research Project: IGF-No. 19691N.* Düsseldorf, Karlsruhe, Freiburg, Saarbrücken: Forschungsvereinigung Gießereitechnik e.V., KIT Research Center for Steel, Timber & Masonry, Fraunhofer IWM, and Fraunhofer IZFP, 2018-2020.

[117] C. Vishnevsky, N. F. Bertolino, and J. F. Wallace: *The Effects of Surface Discontinuities on the Fatigue Properties of Cast Steel Sections.* Rocky River, Ohio, US: Steel Founders' Society of America, 1966.

[118] C. Vishnevsky, N. F. Bertolino, and J. F. Wallace: *The Evaluation of Discontinuities in Commercial Steel Castings by Dynamic Loading to Failure in Fatigue.* Rocky River, Ohio, US: Steel Founders' Society of America, 1967.

[119] M. Volz, H. Saal, and G. Steidl: *Schweissen von Guss im konstruktiven Ingenieurbau.* Schweißtechnik Soundure 98.2 (2009), pp. 6–15.

[120] G. Z. Voyiadjis, ed.: *Handbook of Damage Mechanics: Nano to Macro Scale for Materials and Structures.* New York, US: Springer, 2015.

[121] J. Watrin: *Numerische Bewertung von Ermüdungsversuchen an Stahlgussbauteilen unter Berücksichtigung künstlich eingebrachter Ersatzungänzen: Master Thesis.* KIT Steel & Lightweight Structures. Karlsruhe, 2020.

[122] J. Weertman: *Dislocation Crack Tip Shielding and the Paris Exponent.* Materials Science and Engineering (2007), pp. 59–63.

[123] P. Weidner: *Zum Ermüdungsverhalten einseitig geschweißter Stumpfstöße von Kreishohlprofilen: Doctoral Thesis.* KIT Steel & Lightweight Structures. Karlsruhe, 2019.

[124] R. Wlodawer: *Die gelenkte Erstarrung von Stahlguß.* 2nd ed. Düsseldorf: Giesserei-Verlag GmbH, 1967.

[125] U. Zerbst, M. Madia, B. Schork, J. Hensel, P. Kucharczyk, D. Ngoula, D. Tchuindjang, J. Bernhard, and C. Beckmann: *Fatigue and Fracture of Weldments: The IBESS Approach for the Determination of the Fatigue Life and Strength of Weldments by Fracture Mechanics Analysis.* Cham, US: Springer, 2019.

# Codes and Standards

[126] ASTM E125:1963-01, Standard Reference Photographs for Magnetic Particle Indications on Ferrous Castings.

[127] ASTM E1820:2018-02, Standard Test Method for Measurement of Fracture Toughness.

[128] ASTM E186:1962-01, Standard Reference Radiographs for Heavy-Walled (2 to 41/2 -in. (51 to 114-mm)) Steel Castings.

[129] ASTM E280:1965-01, Standard Reference Radiographs for Heavy-Walled (41/2 to 12-in. (114 to 305-mm)) Steel Castings.

[130] ASTM E399:2019-08, Test Method for Linear-Elastic Plane-Strain Fracture Toughness KIc of Metallic Materials.

[131] ASTM E446:1972-01, Standard Reference Radiographs for Steel Castings Up to 2 in. (50.8 mm) in Thickness.

[132] ASTM E647:2015-07, Standard Test Method for Measurement of Fatigue Crack Growth Rates.

[133] ASTM E71:1964-08, Reference Radiographs for Steel Castings up to 2 in. (50.8 mm) in Thickness.

[134] ASTM E837:2013-10, Standard Test Method for Determining Residual Stresses by the Hole-Drilling Strain-Gage Method.

[135] BS 6208:1990-04, Method for Ultrasonic Testing of Ferritic Steel Castings Including Quality Levels.

[136] BS 7608:2014-03, Guide to Fatigue Design and Assessment of Steel Products.

[137] BS 7910:2005-07, Guide to Methods for Assessing the Acceptability of Flaws in Metallic Structures.

[138] BS 7910:2013-12, Guide to Methods for Assessing the Acceptability of Flaws in Metallic Structures.

[139] DIN 1690-2:1985-06, Technical Delivery Conditions for Castings made from Metallic Materials; Steel Castings; Classification into Severity Levels on the Basis of Non-Destructive Testing.

[140] DIN 17182:1992-05, General-Purpose Steel Castings with Enhanced Weldability and Higher Toughness; Technical Delivery Conditions.

[141] DIN EN 10293:2015-04, Steel Castings for General Engineering Uses: German Version.

[142] DIN EN 10340:2008-01, Steel Castings for Structural Uses: German Version.

[143] DIN EN 1090-2:2018-09, Execution of Steel Structures and Aluminium Structures – Part 2: Technical Requirements for Steel Structures: German Version.

[144] DIN EN 12680-1:2003-06, Founding – Ultrasonic Examination – Part 1: Steel Castings for General Purposes: German Version.

[145] DIN EN 12680-2:2003-06, Founding – Ultrasonic Examination – Part 2: Steel Castings for Highly Stressed Components: German Version.

[146] DIN EN 12681-1:2018-02, Founding – Radiographic Testing – Part 1: Film Techniques: German Version.

[147] DIN EN 12681-2:2018-02, Founding – Radiographic Testing – Part 2: Techniques with Digital Detectors: German Version.

[148] DIN EN 1369:2013-01, Founding – Magnetic Particle Testing: German Version.

[149] DIN EN 1370:2012-03, Founding – Examination of Surface Condition: German Version.

[150] DIN EN 1371-1:2012-02, Founding – Liquid Penetrant Testing – Part 1: Sand, Gravity Die and Low Pressure Die Castings: German Version.

[151] DIN EN 1559-2:2000-04, Founding – Technical Conditions of Delivery – Part 2: Additional Requirements for Steel Castings: German Version.

[152] DIN EN 1559-2:2014-12, Founding – Technical Conditions of Delivery – Part 2: Additional Requirements for Steel Castings: German Version.

[153] DIN EN 1990:2010-12, Eurocode 0, Basis of Structural Design: German Version.

[154] DIN EN 1993-1-10:2010-12, Eurocode 3, Design of Steel Structures – Part 1-10: Material Toughness and Through-Thickness Properties: German Version.

[155] DIN EN 1993-1-8/NA:2010-12, Eurocode 3, Design of Steel Structures – Part 1-8: Design of Joints: German National Annex.

[156] DIN EN 1993-1-9:2010-12, Eurocode 3, Design of Steel Structures – Part 1-9: Fatigue: German Version.

[157] DIN EN ISO 8062-3:2008-09, Geometrical Product Specifications (GPS) – Dimensional and Geometrical Tolerances for Moulded Parts – Part 3: General Dimensional and Geometrical Tolerances and Machining Allowances for Castings: German Version.

[158] DIN EN ISO 9934-1:2017-03, Non-Destructive Testing – Magnetic Particle Testing – Part 1: General Principles: German Version.

[159] DNV RP-C203:2001-10, Det Norske Veritas, Fatigue Strength Analysis of Offshore Steel Structures.

[160] DVS 0905:2017-02, Deutscher Verband für Schweißen und verwandte Verfahren e.V., Merkblatt zur Industrielle Anwendung des Kerbspannungskonzeptes für den Ermüdungsfestigkeitsnachweis von Schweißverbindungen.

[161] FKM-Richtlinie-BM:2009, Forschungskuratorium Maschinenbau, Bruchmechanischer Festigkeitsnachweis für Maschinenbauteile.

[162] FKM-Richtlinie:2012, Forschungskuratorium Maschinenbau, Rechnerischer Festigkeitsnachweis für Maschinenbauteile aus Stahl, Eisenguss- und Aluminiumwerkstoffen.

[163] OTH 92 390:1999-02, Health and Safety Executive, Background to New Fatigue Guidance for Steel Joints and Connections in Offshore Structures.

[164] SEW 520:2017-12, Stahl-Eisen-Werkstoffblätter, High-Strength Cast Steel with Good Weldability – Technical Delivery Conditions.

[165] SIA 261:2003-01, Swiss Society of Engineers and Architects, Actions on Structures.

[166] UK Health and Safety Executive:1995, Offshore Installations: Guidance on Design, Construction and Certification 4th ed.

# A Material Properties

This annex summarizes the material properties and chemical compositions according to inspection certificate 3.1 and additional examinations.

**Table A.1:** Mechanical properties according to inspection certificate 3.1

| Mat. | Foundry | t [mm] | $R_{p0.2}$ [MPa] | $R_m$ [MPa] | A [%] | T [°C] | KV [J] | T [°C] | KV* [J] |
|---|---|---|---|---|---|---|---|---|---|
| min requirements [142] | | ≤ 100 | 360** | 500-650 | 22 | -40 | 27 | | |
| | A | 35 | 452 | 567 | 26 | -20 | 86/83/107 | -40 | 95 |
| | | 55 | 414 | 573 | 27 | -20 | 99/114/126 | | |
| | | 100 | 427 | 586 | 26 | -20 | 74/93/107 | | |
| 1.6220 | B | 35 | 371 | 555 | 26 | -40 | 47/56/21 | -40 | 48 |
| | | 55 | 362 | 548 | 27 | -40 | 42/23/35 | | |
| | | 100 | 362 | 546 | 28 | -40 | 30/49/42 | | |
| | C | 35 | 377 | 556 | 29 | +20 | 138/136/140 | -40 | 45 |
| | D | 32 | 401 | 583 | 24 | -20 | 71/68/84 | | |
| | E | 32 | 449 | 617 | 22 | -40 | 53/61/59 | | |
| min requirements [164] | | ≤ 50 | 825 | 930-1080 | 10 | -40 | 27 | | |
| 1.6760 | D | 32 | 849 | 979 | 15 | -40 | 28 | | |
| | E | 32 | 963 | 1065 | 14 | -40 | 44 | | |
| | | 100 | 991 | 1108 | 12 | -40 | 50 | | |

* additional tests carried out at KIT, mean value of 3 test.
** This corresponds to the additional requirement from [103]. Original value to [142]: 300 MPa

**Table A.2:** Chemical composition (melt) according to inspection certificate 3.1 in M%

| Mat. | Foundry | C | Si | Mn | P | S | Cr | Mo | Ni |
|---|---|---|---|---|---|---|---|---|---|
| | | min/max | max | min/max | max | max | min/max | min/max | min/max |
| requirements [142] | | 0.17/0.23 | 0.60 | 1.00/1.60 | 0.020 | 0.020 | -/0.30 | -/0.12 | -/0.80 |
| | A | 0.21 | 0.52 | 1.27 | 0.010 | 0.002 | ND | ND | 0.07 |
| | B | 0.22 | 0.51 | 1.16 | 0.011 | 0.003 | 0.13 | 0.67 | 0.02 |
| 1.6220 | C | 0.21 | 0.46 | 1.32 | 0.009 | 0.007 | 0.05 | 0.01 | 0.27 |
| | D | 0.21 | 0.51 | 1.29 | 0.227* | 0.004 | 0.16 | 0.02 | 0.06 |
| | E | 0.20 | 0.53 | 1.26 | 0.014 | 0.002 | 0.69* | 0.55 | 0.15 |
| requirements [164] | | 0.18/0.24 | 0.60 | 0.8/1.20 | 0.015 | 0.005 | 0.5/1.0 | 0.5/1.0 | 0.8/1.3 |
| 1.6760 | D | 0.25* | 0.54 | 1.04 | 0.015 | 0.003 | 1.0 | 0.5 | 1.1 |
| | E | 0.22 | 0.51 | 1.09 | 0.016* | 0.003 | 0.7 | 0.6 | 1.0 |

ND = not determined;
* exceeded maximum value. These alloy element embrittles the matieral so that a lower performance is predicted. With this argument the transgression is accepted for the purpose of these investigations.

# B Fracture Mechanics Calculations

This annex summarizes the formulas for the fracture mechanics assessments. In the first section, the calculation of stress intensity factors on representative components and loading situations is presented. These are taken from FKM-Rili-BM:2009 [161] and are based on different analytical and numerical considerations. The second section covers special approaches that are necessary to evaluate these model geometries using the FAD. These are extracted from BS 7910:2005 [137].

# B.1 Stress Intensity Factors

## B.1.1 Plate with Semi-Elliptical Surface Crack under Tension and Bending

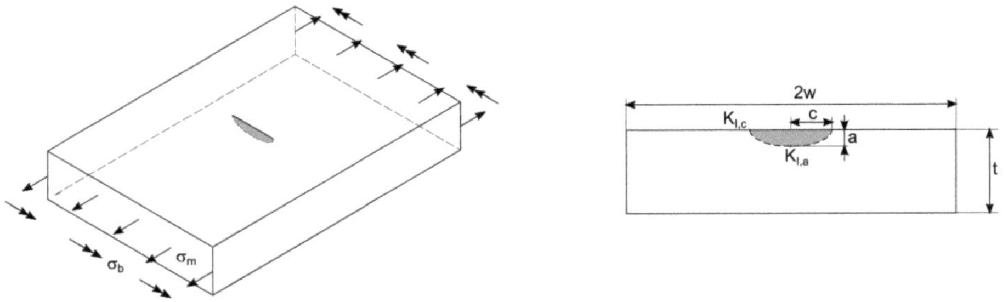


**Figure B.1:** Surface flaw [161]

$$K_a = K(\phi = \frac{\pi}{2}); \qquad K_c = K(\phi = 0); \qquad K = (\sigma_m F_m + \sigma_b F_b)\sqrt{\frac{\pi a}{Q}};$$

$$F_m = \left[M_1 + M_2\left(\frac{a}{t}\right)^2 + M_3\left(\frac{a}{t}\right)^4\right] g f_\phi f_w; \qquad F_b = F_m\left[H_1 + (H_2 - H_1)\sin^p\phi\right]$$

$$f_w = \sqrt{\sec\left(\frac{\pi c}{2w}\sqrt{\frac{a}{t}}\right)}$$

For a/c $\leq$ 1:

$$M_1 = 1.13 - 0.09\frac{a}{c}; \qquad M_2 = -0.54 + \frac{0.89}{0.2 + \frac{a}{c}}; \qquad M_3 = 0.5 - \frac{1}{0.65 + \frac{a}{c}} + 14\left(1 - \frac{a}{c}\right)^{24};$$

$$g = 1 + \left[0.1 + 0.35\left(\frac{a}{t}\right)^2\right](1 - \sin(\phi))^2; \qquad H_1 = 1 - 0.34\frac{a}{t} - 0.11\frac{a}{c}\frac{a}{t};$$

$$H_2 = 1 + \left(-1.22 - 0.12\frac{a}{c}\right)\frac{a}{t} + \left[0.55 - 1.05\left(\frac{a}{c}\right)^{0.75} + 0.47\left(\frac{a}{c}\right)^{1.5}\right]\left(\frac{a}{t}\right)^2;$$

$$Q = 1 + 1.464\left(\frac{a}{c}\right)^{1.65}; \quad f_\phi = \left(\sin^2(\phi) + \left(\frac{a}{c}\right)^2\cos^2(\phi)\right)^{0.25}; \quad p = 0.2 + \frac{a}{c} + 0.6\frac{a}{t}$$

For a/c $>$ 1:

$$M_1 = \sqrt{\frac{c}{a}}\left(1 + 0.04\frac{c}{a}\right); \qquad M_2 = 0.2\left(\frac{c}{a}\right)^4; \qquad M_3 = -0.11\left(\frac{c}{a}\right)^4;$$

$$g = 1 + \left[0.1 + 0.35\frac{c}{a}\left(\frac{a}{t}\right)^2\right](1 - \sin(\phi))^2;$$

$$H_1 = 1 + \left(-0.04 - 0.41\frac{c}{a}\right)\frac{a}{t} + \left[0.55 - 1.93\left(\frac{c}{a}\right)^{0.75} + 1.38\left(\frac{c}{a}\right)^{1.5}\right]\left(\frac{a}{t}\right)^2;$$

$$H_2 = 1 + \left(-2.11 - 0.77\frac{c}{a}\right)\frac{a}{t} + \left[0.55 - 0.72\left(\frac{c}{a}\right)^{0.75} + 0.14\left(\frac{c}{a}\right)^{1.5}\right]\left(\frac{a}{t}\right)^2;$$

$$Q = 1 + 1.464\left(\frac{c}{a}\right)^{1.65}; \quad f_\phi = \left(\cos^2(\phi) + \left(\frac{c}{a}\right)^2\sin^2(\phi)\right)^{0.25}; \quad p = 0.2 + \frac{c}{a} + 0.6\frac{a}{t}$$

Range of validity: $0 \leq \frac{a}{t} < 1$; $\qquad 0 \leq \frac{a}{c} < 2$; $\qquad \frac{c}{w} < 0.5$

## B.1.2 Plate with Elliptical Internal Crack under Tension and Bending

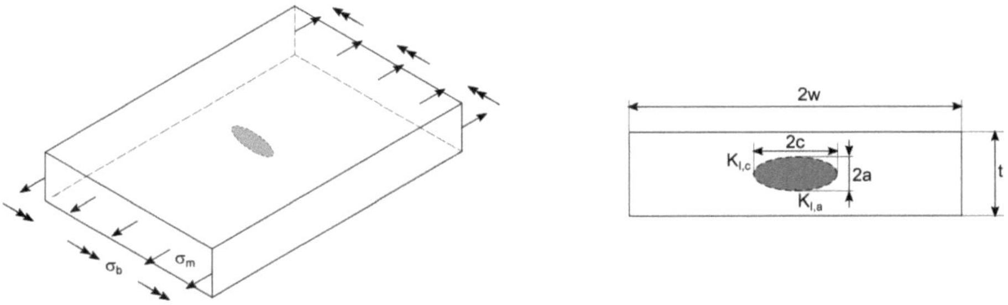


**Figure B.2:** Internal flaw [161]

$$K_a = K(\phi = \frac{\pi}{2}); \qquad K_c = K(\phi = 0); \qquad K = (\sigma_m F_m + \sigma_b \frac{2a}{t} F_b)\sqrt{\frac{\pi a}{Q}} f_\phi f_w;$$

$$F_m = \left[M_1 + M_2\left(\frac{2a}{t}\right)^2 + M_3\left(\frac{2a}{t}\right)^4\right] g f_\phi f_w;$$

$$F_b = -\left[0.5 + 0.2591\left(\frac{a}{c}\right)^{1.5} - 0.09189\left(\frac{a}{c}\right)^{2.5}\right] f_b \sin(\phi)$$

$$f_w = \sqrt{\sec\left(\frac{\pi c}{2w}\sqrt{\frac{2a}{t}}\right)}; \qquad f_b = 1 + \exp\left[-1.9249 - 3.9087\left(\frac{a}{c}\right)^{0.5} + 4.1067\left(\frac{2a}{t}\right)^3\right]$$

$$M_2 = 0.05\left[0.11 + \left(\frac{a}{c}\right)^{3/2}\right]^{-1}; \qquad M_3 = 0.29\left[0.23 + \left(\frac{a}{c}\right)^{3/2}\right]^{-1};$$

$$g = 1 - \left(1 + \frac{4a}{c}\right)^{-1}\left(\frac{2a}{t}\right)^4 \sqrt{2.6 - a\frac{a}{t}|\cos(\phi)|}$$

For $a/c \leq 1$:

$$M_1 = 1; \qquad f_\phi = \left[\sin^2(\phi) + \left(\frac{a}{c}\right)^2 \cos^2(\phi)\right]^{0.25}; \qquad Q = 1 + 1.464\left(\frac{a}{c}\right)^{1.65}$$

For $a/c > 1$:

$$M_1 = \sqrt{\frac{c}{a}}; \qquad f_\phi = \left[\left(\frac{c}{a}\right)^2 \sin^2(\phi) + \cos^2(\phi)\right]^{0.25}; \qquad Q = 1 + 1.464\left(\frac{c}{a}\right)^{1.65}$$

Range of validity at tension loading: $0 \leq \frac{a}{c} < \infty$; $\frac{c}{w} < 0.5$;
for $0 < \frac{a}{c} \leq 0.2$: $\frac{2a}{t} \leq 1.25\left(\frac{a}{c} + 0.6\right)$ else: $\frac{2a}{t} < 1$

Range of validity at bending loading: $0 \leq \frac{a}{c} < 1$; $\frac{2a}{t} < 0.8$; $\frac{c}{w} < 0.5$

### B.1.3 Cylinder with Semi-Circular Surface Crack

Shortened representation of the data from [161]

**Figure B.3:** Cylinder with semi-circular surface crack [161]

$$K = \sigma_m F_m \sqrt{\pi a}$$

**Table B.1:** Parameter $F_\mathrm{m}$ for determination of stress intensity factors for a cylinder with semi-circular curface crack

| a/c | 0.1 | 0.2 | 0.4 | 0.6 | 0.8 | 1.0 | 0.1 | 0.2 | 0.4 | 0.6 | 0.8 | 1.0 |
|---|---|---|---|---|---|---|---|---|---|---|---|---|
| 0.2 | .220 | .342 | .533 | .805 | 1.227 | 1.899 | 1.004 | 1.025 | 1.132 | 1.362 | 1.759 | 2.453 |
| 0.4 | .490 | .529 | .693 | .901 | 1.310 | 2.008 | .923 | .950 | 1.051 | 1.261 | 1.622 | 2.263 |
| 0.6 | .676 | .723 | .812 | 1.026 | 1.334 | 1.934 | .820 | .854 | .940 | 1.116 | 1.423 | 1.924 |
| 1.0 | .736 | .814 | .877 | 1.072 | 1.409 | 2.028 | .663 | .693 | .720 | .813 | .967 | 1.241 |

### B.1.4 Cylinder with Circular Central Internal Crack

Shortened representation of the data from [161]

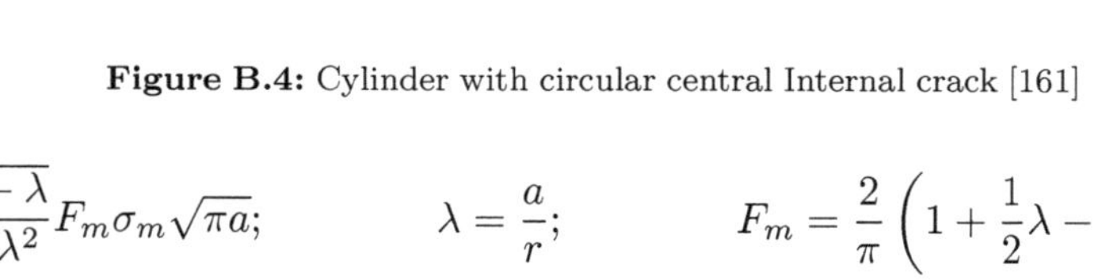


**Figure B.4:** Cylinder with circular central Internal crack [161]

$$K = \frac{\sqrt{1-\lambda}}{1-\lambda^2} F_m \sigma_m \sqrt{\pi a}; \qquad \lambda = \frac{a}{r}; \qquad F_m = \frac{2}{\pi}\left(1 + \frac{1}{2}\lambda - \frac{5}{8}\lambda^2\right) + 0.268\lambda^3$$

Range of validity: $0 \leq \frac{a}{r} < 1$;

# B.2 Failure Assessmet Diagramm

The calculations are taken from BS 7910:2005 [137].

## B.2.1 Semi elliptical surface crack

$$\sigma_{ref} = \frac{\sigma_b + \sqrt{\sigma_b^2 + 9\sigma_m^2(1-\alpha'')^2}}{3(1-\alpha'')^2} \tag{B.1}$$

$\alpha''$ consider geometric properties shown in Figure B.5

$$\alpha'' = \frac{\frac{a}{t}}{1+\frac{t}{c}} \qquad \text{for } W \geq (c+t)$$

$$\alpha'' = \frac{\frac{2a}{t}}{c/W} \qquad \text{for } W < (c+t)$$

## B.2.2 Embedded elliptical flaw

$$\sigma_{ref} = \frac{\sigma_b + 3\sigma_m\alpha'' + \sqrt{(\sigma_b + 3\sigma_m\alpha'')^2 + 9\sigma_m^2((1-\alpha'')^2 + 2(\frac{p\alpha''}{t})}}{3((1-\alpha'')^2 + 4(\frac{p\alpha''}{t}))} \tag{B.2}$$

$\alpha''$ and $t$ consider geometric properties shown in Figure B.5

$$\alpha'' = \frac{\frac{2a}{t}}{1+\frac{t}{c}} \qquad \text{for } W \geq (c+t)$$

$$\alpha'' = \frac{\frac{4a}{t}}{1+\frac{c}{2W}} \qquad \text{for } W < (c+t)$$

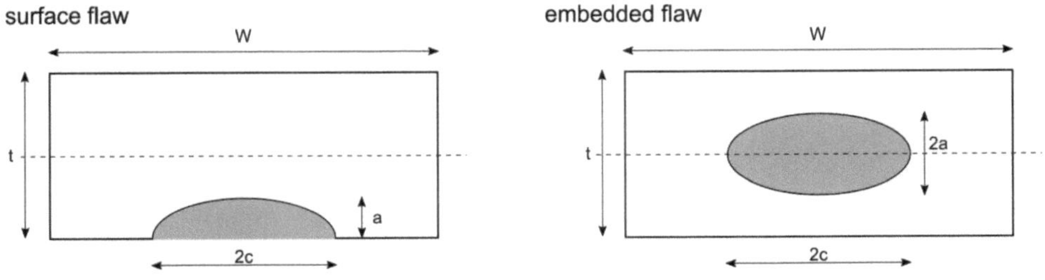


**Figure B.5:** Geometric properties for FAD evaluation [137]

# C Details of Tensile Specimen

This annex summarizes all tensile fatigue specimens discussed in Section 3.2. The structure of the data sheets is described in Table C.1 along with the notation used. The specimens are identified using the following concept.

U-(material)-(type)-(n)

| | |
|---|---|
| U | identifier for tensile specimen with casting defects |
| (material) | 20 = 1.6220; 22 = 1.6760 |
| (type) | O = without thickening; M = with thickening |
| (n) | consecutive number |

**Table C.1:** Structure of the data sheets – Tensile fatigue specimen

Table of Major Test Parameters

| | |
|---|---|
| $\Delta\sigma_{nom}$ | = Nominal stress range. Beach mark tests are identified in an additional index B. |
| $\Delta\sigma_{mic}$ | = Stress range related to the net section determined by microscopy according to Eq. (3.1). |
| foundry | = Anonymized index of the manufacturing foundry. Material properties according to Annex A. |
| $\sigma_y$, $\sigma_u$ | = Yield strength ($R_{eh}$ resp. $R_{p0.2}$ depending on the material) and tensile strength according to inspection certificate 3.1. |
| defect pos. | considering the defect location in the failure plane – internal and surface. |
| failure case | = according to Figure 3.6. |

| Computed Tomography Recording | Table with Defect Parameters |
|---|---|
| Display of the CT scans in all projections including the failure plane and the plane of maximum cross-sectional weakening. | $A_{proj,CT}$, $A_{max,ct}$, $z_{max,CT}$, $V_{max}/V_{tot}$ and $A_{fail,mic}$ as defined in Section 3.2.1<br>$z_{fail}$ = position of the failure plane<br>$\sigma_{max}/\sigma_y$, $\sigma_{max}/\sigma_u$ = percentage of the upper load related to the material strength in the residual fracture according to Eq. (3.4) |
| **Upper Fracture Surface** | **Lower Fracture Surface** |
| Visualization of the fracture surface topology and microscope image. The position and size of the defect, the region of stable crack growth and the residual fracture surface can be identified. | Visualization of the fracture surface topology and microscope image. The position and size of the defect, the region of stable crack growth and the residual fracture surface can be identified. |

## U20O1

| $\Delta\sigma_{nom}$ [MPa] | $\Delta\sigma_{mic}$ [MPa] | cycles to failure [-] | foundry [-] | $\sigma_y$ [MPa] | $\sigma_u$ [MPa] | defect pos. | failure case |
|---|---|---|---|---|---|---|---|
| 200 | 210 | 15,269,398 | D | 409 | 597 | int. | III |

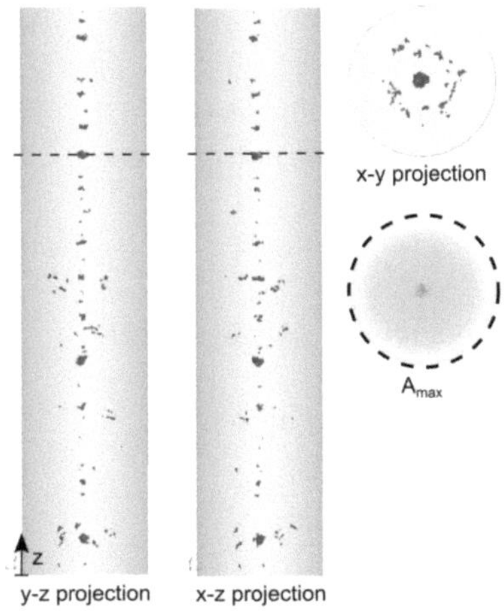


| $A_{proj,CT}$ [mm²]; [%] | $A_{max,CT}$ [mm²]; [%] | $z_{max,CT}$ [mm] |
|---|---|---|
| 1 | 1.5; 1 | -16.0 |

| $V_{max}/V_{tot}$ [%] | $A_{fail,mic}$ [mm]; [%] | $z_{fail}$ [mm] |
|---|---|---|
| 14 | 12.6; 5 | 59.2 |

| $\sigma_{max}/\sigma_y$ [%] | $\sigma_{max}/\sigma_u$ [%] |
|---|---|
| 138 | 95 |

Upper Fracture Surface

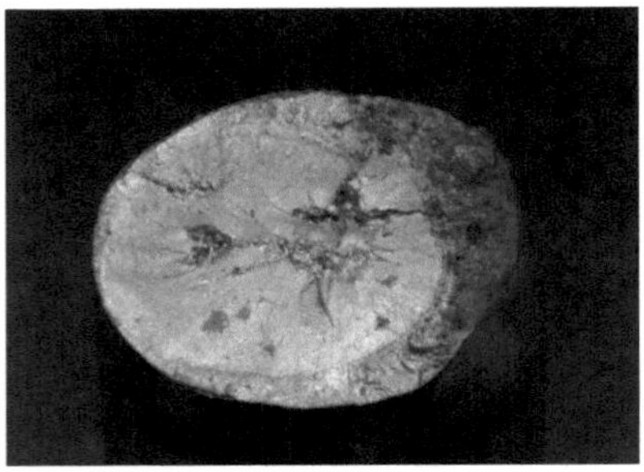

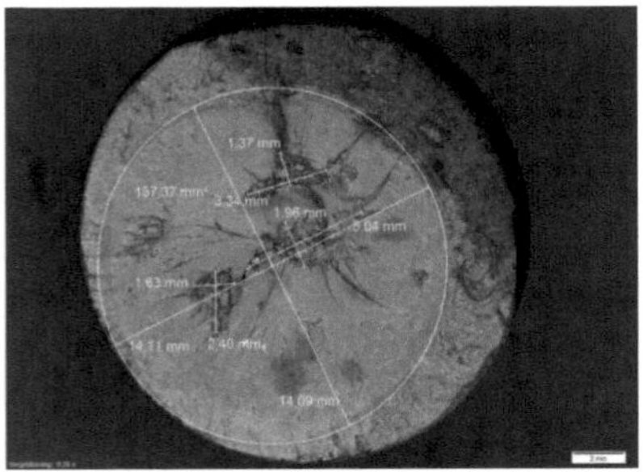

Lower Fracture Surface

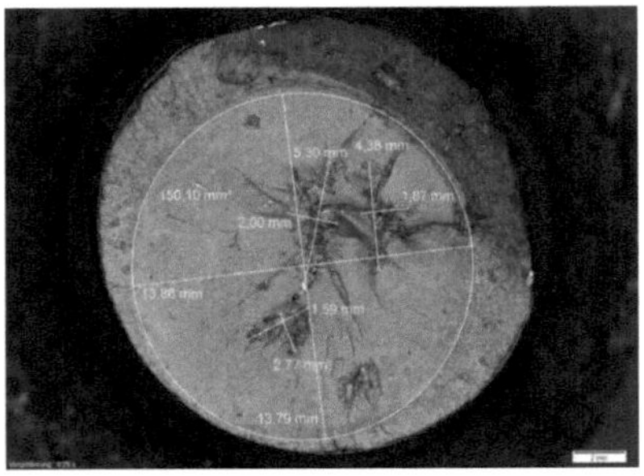

## U20O2

| $\Delta\sigma_{nom}$ [MPa] | $\Delta\sigma_{mic}$ [MPa] | cycles to failure [-] | foundry [-] | $\sigma_y$ [MPa] | $\sigma_u$ [MPa] | defect pos. | failure case |
|---|---|---|---|---|---|---|---|
| 280 | 286 | 2,522,936 | D | 409 | 597 | int. | III |

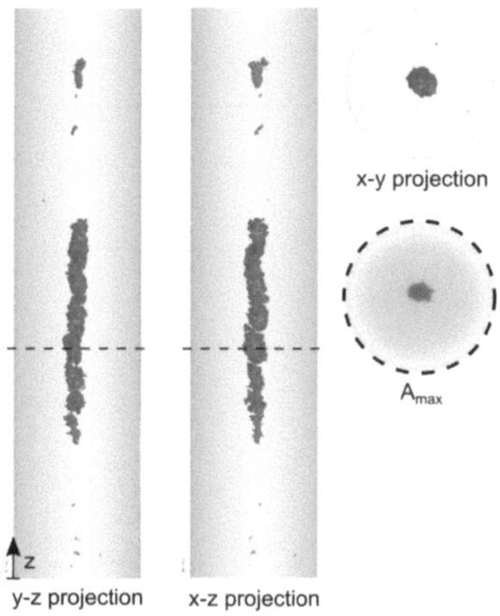


| $A_{proj,CT}$ [mm²]; [%] | $A_{max,CT}$ [mm²]; [%] | $z_{max,CT}$ [mm] |
|---|---|---|
| 1 | 5.0; 2 | -9.6 |

| $V_{max}/V_{tot}$ [%] | $A_{fail,mic}$ [mm]; [%] | $z_{fail}$ [mm] |
|---|---|---|
| 91 | 5.4; 2 | 32.3 |

| $\sigma_{max}/\sigma_y$ [%] | $\sigma_{max}/\sigma_u$ [%] |
|---|---|
| 149 | 102 |

Upper Fracture Surface

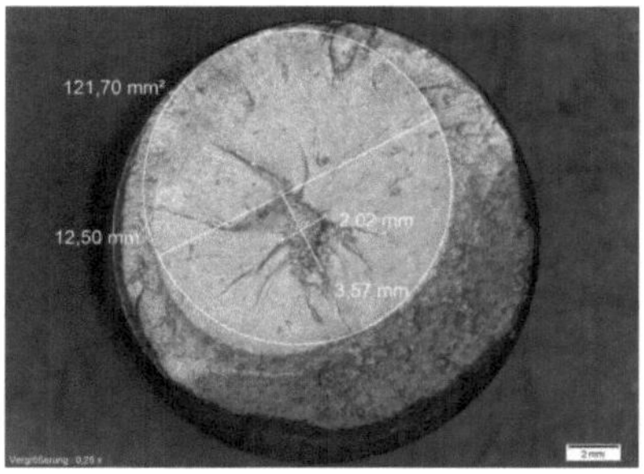


Lower Fracture Surface

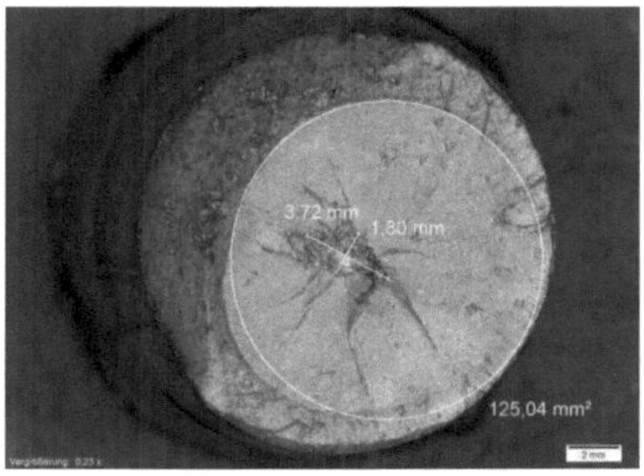

## U20O3

| $\Delta\sigma_{nom}$ [MPa] | $\Delta\sigma_{mic}$ [MPa] | cycles to failure [-] | foundry [-] | $\sigma_y$ [MPa] | $\sigma_u$ [MPa] | defect pos. | failure case |
|---|---|---|---|---|---|---|---|
| 250 | 281 | 1,049,028 | D | 409 | 597 | int. | III |

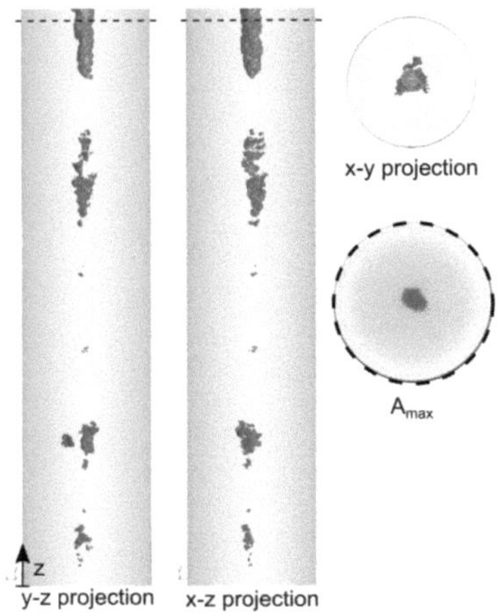


| $A_{proj,CT}$ [mm²]; [%] | $A_{max,CT}$ [mm²]; [%] | $z_{max,CT}$ [mm] |
|---|---|---|
| 1 | 5.8; 2 | 82.0 |

| $V_{max}/V_{tot}$ [%] | $A_{fail,mic}$ [mm]; [%] | $z_{fail}$ [mm] |
|---|---|---|
| 58 | 28.5; 11 | 78.4 |

| $\sigma_{max}/\sigma_y$ [%] | $\sigma_{max}/\sigma_u$ [%] |
|---|---|
| 150 | 103 |

Upper Fracture Surface

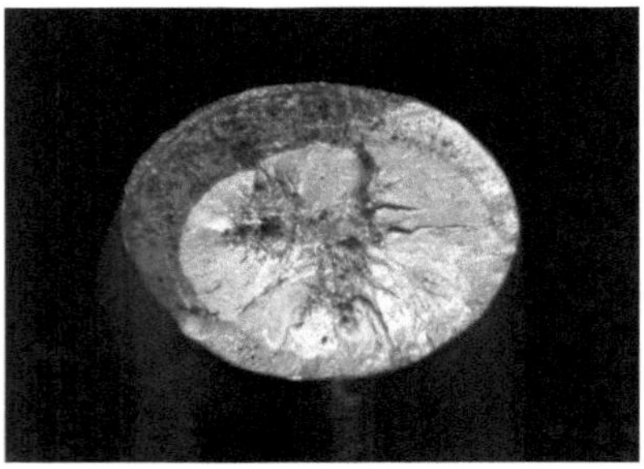

Lower Fracture Surface

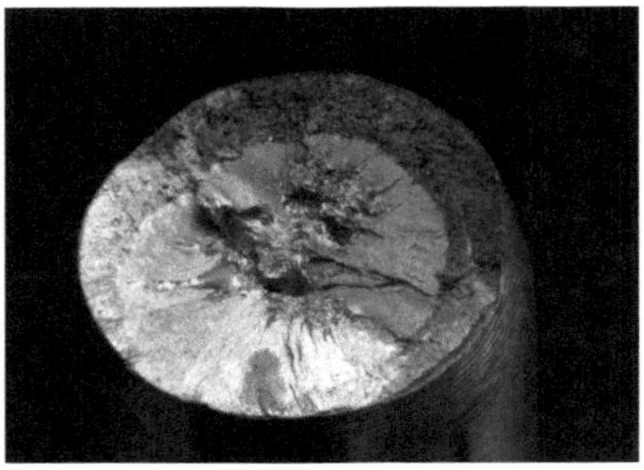

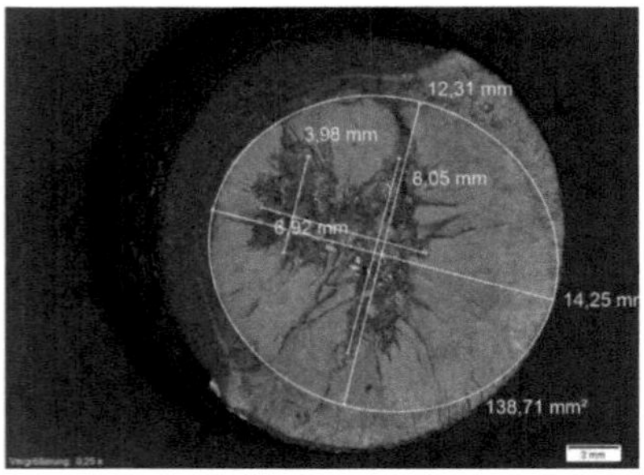


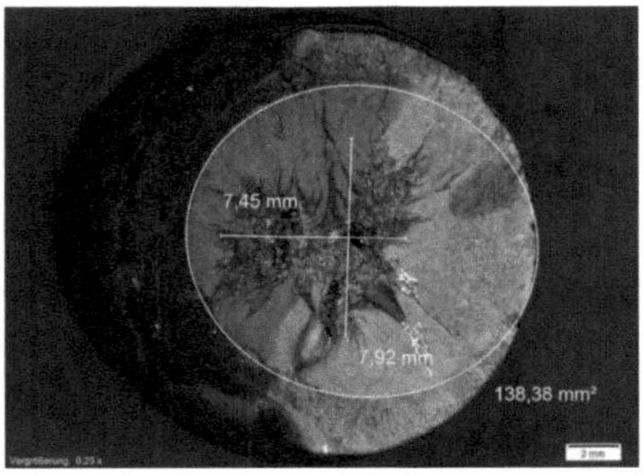

## U20O4

| $\Delta\sigma_{nom}$ [MPa] | $\Delta\sigma_{mic}$ [MPa] | cycles to failure [-] | foundry [-] | $\sigma_y$ [MPa] | $\sigma_u$ [MPa] | defect pos. | failure case |
|---|---|---|---|---|---|---|---|
| 300 | 325 | 479,484 | D | 409 | 597 | int. | III |

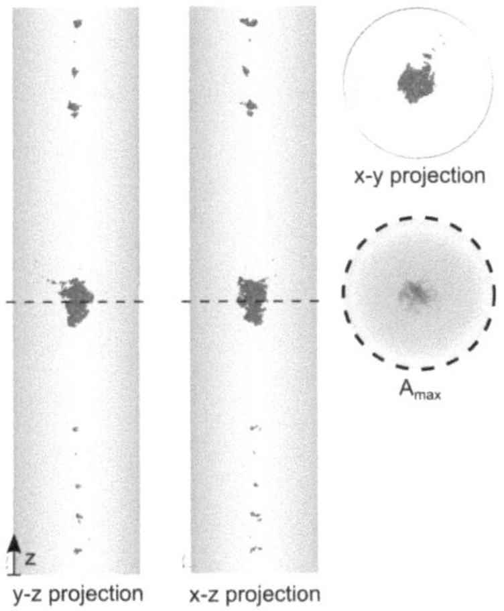


| $A_{proj,CT}$ [mm²]; [%] | $A_{max,CT}$ [mm²]; [%] | $z_{max,CT}$ [mm] |
|---|---|---|
| 1 | 9.2; 4 | 88.5 |

| $V_{max}/V_{tot}$ [%] | $A_{fail,mic}$ [mm]; [%] | $z_{fail}$ [mm] |
|---|---|---|
| 84 | 20.0; 8 | 38.6 |

| $\sigma_{max}/\sigma_y$ [%] | $\sigma_{max}/\sigma_u$ [%] |
|---|---|
| 182 | 125 |

Upper Fracture Surface

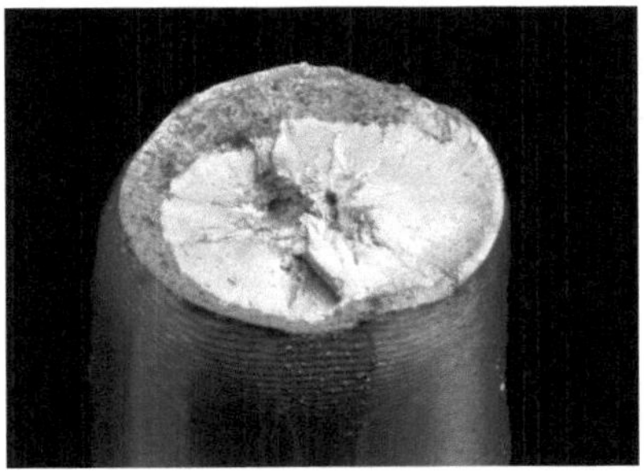

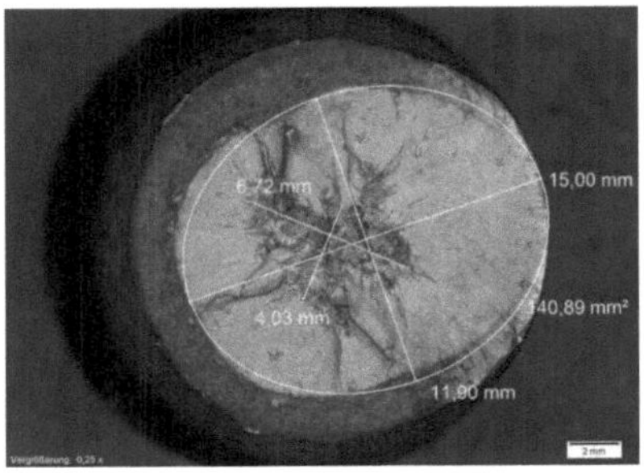


Lower Fracture Surface

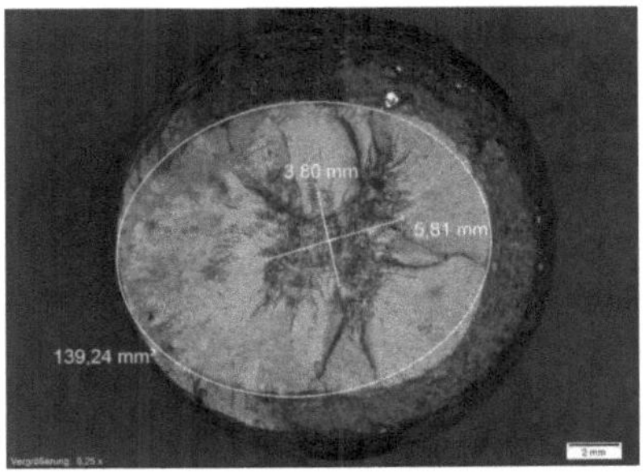

## U20O5

| $\Delta\sigma_{nom}$ [MPa] | $\Delta\sigma_{mic}$ [MPa] | cycles to failure [-] | foundry [-] | $\sigma_y$ [MPa] | $\sigma_u$ [MPa] | defect pos. | failure case |
|---|---|---|---|---|---|---|---|
| 350 | 369 | 376,179 | D | 409 | 597 | int. | II |

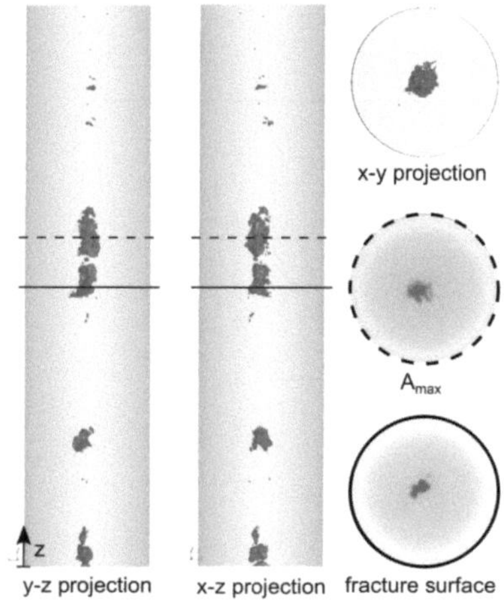


| $A_{proj,CT}$ [mm²]; [%] | $A_{max,CT}$ [mm²]; [%] | $z_{max,CT}$ [mm] |
|---|---|---|
| 1 | 5.2; 2 | 39.8 |

| $V_{max}/V_{tot}$ [%] | $A_{fail,mic}$ [mm]; [%] | $z_{fail}$ [mm] |
|---|---|---|
| 39 | 13.4; 5 | 46.9 |

| $\sigma_{max}/\sigma_y$ [%] | $\sigma_{max}/\sigma_u$ [%] |
|---|---|
| 160 | 110 |

Upper Fracture Surface

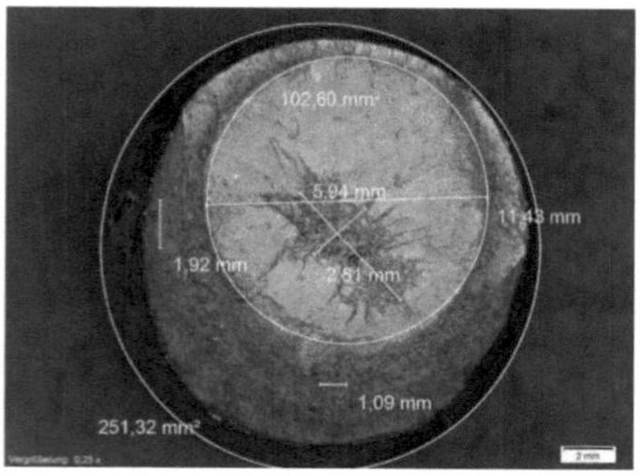


Lower Fracture Surface

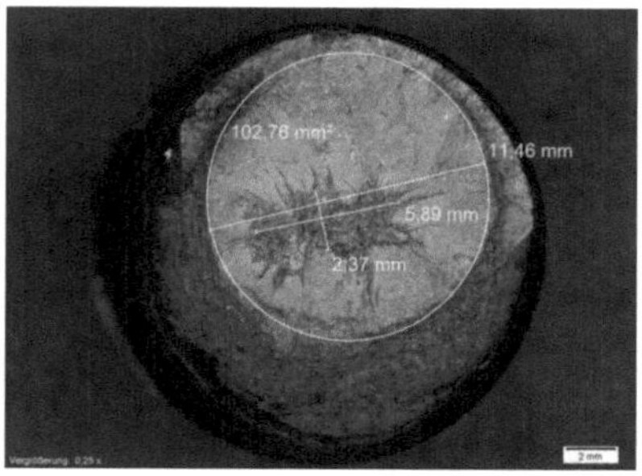

## U20O6

| $\Delta\sigma_{nom}$ [MPa] | $\Delta\sigma_{mic}$ [MPa] | cycles to failure [-] | foundry [-] | $\sigma_y$ [MPa] | $\sigma_u$ [MPa] | defect pos. | failure case |
|---|---|---|---|---|---|---|---|
| 385 | 459 | 26,531 | D | 409 | 597 | int. | I |

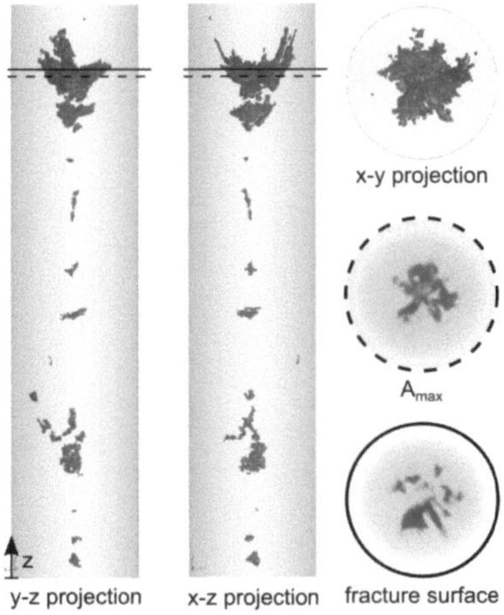


| $A_{proj,CT}$ [mm²]; [%] | $A_{max,CT}$ [mm²]; [%] | $z_{max,CT}$ [mm] |
|---|---|---|
| 1 | 24.0; 9 | 71.0 |

| $V_{max}/V_{tot}$ [%] | $A_{fail,mic}$ [mm]; [%] | $z_{fail}$ [mm] |
|---|---|---|
| 70 | 41.5; 16 | 70.1 |

| $\sigma_{max}/\sigma_y$ [%] | $\sigma_{max}/\sigma_u$ [%] |
|---|---|
| 174 | 119 |

Upper Fracture Surface

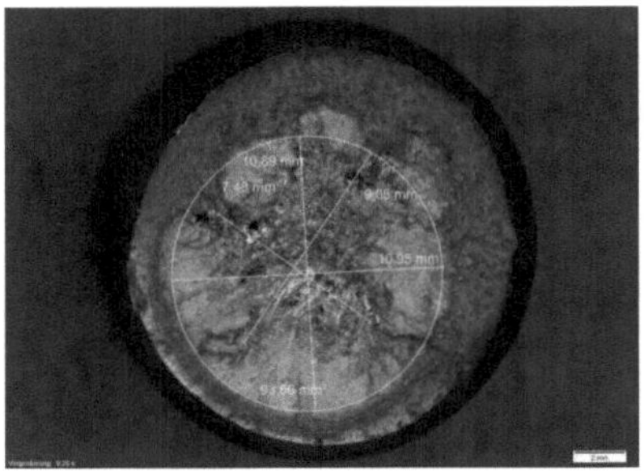

Lower Fracture Surface

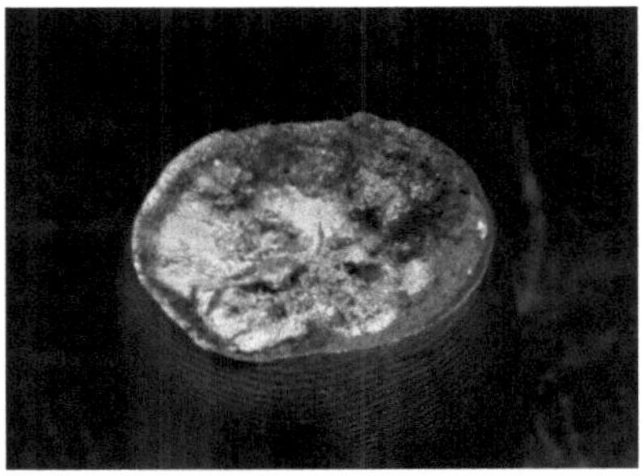

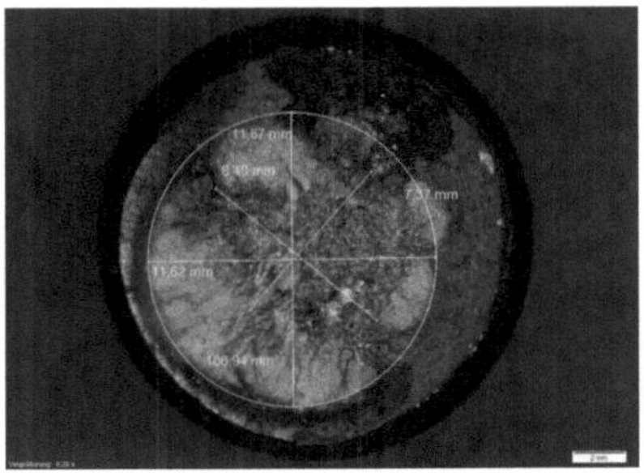

## U20O8

| $\Delta\sigma_{nom}$ [MPa] | $\Delta\sigma_{mic}$ [MPa] | cycles to failure [-] | foundry [-] | $\sigma_y$ [MPa] | $\sigma_u$ [MPa] | defect pos. | failure case |
|---|---|---|---|---|---|---|---|
| 370 | 410 | 98,990 | D | 409 | 597 | int. | I |

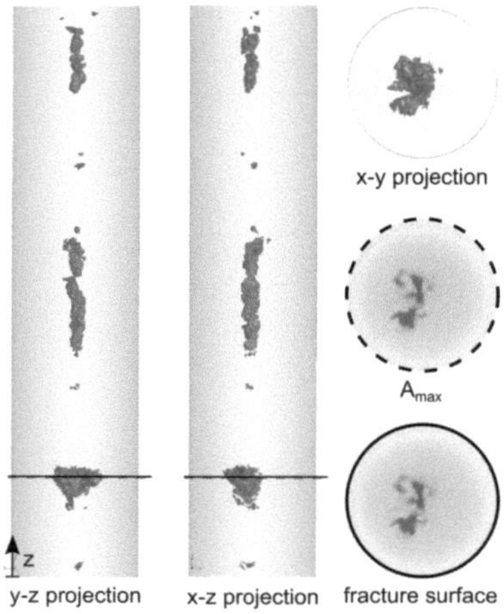


| $A_{proj,CT}$ [mm²]; [%] | $A_{max,CT}$ [mm²]; [%] | $z_{max,CT}$ [mm] |
|---|---|---|
| 1 | 11.8; 5 | 14.7 |

| $V_{max}/V_{tot}$ [%] | $A_{fail,mic}$ [mm]; [%] | $z_{fail}$ [mm] |
|---|---|---|
| 41 | 25.1; 10 | 14.0 |

| $\sigma_{max}/\sigma_y$ [%] | $\sigma_{max}/\sigma_u$ [%] |
|---|---|
| 166 | 114 |

Upper Fracture Surface

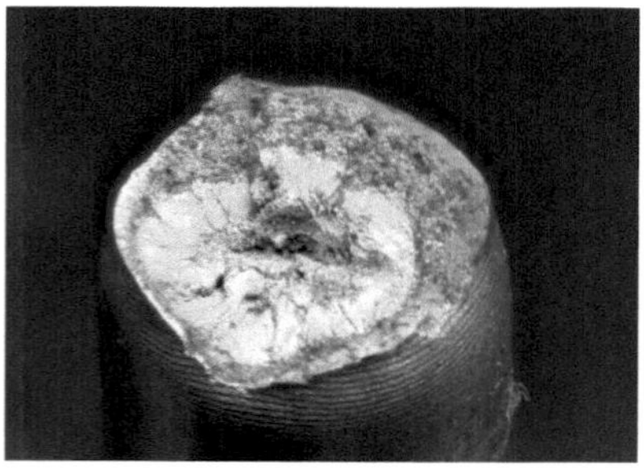

Lower Fracture Surface

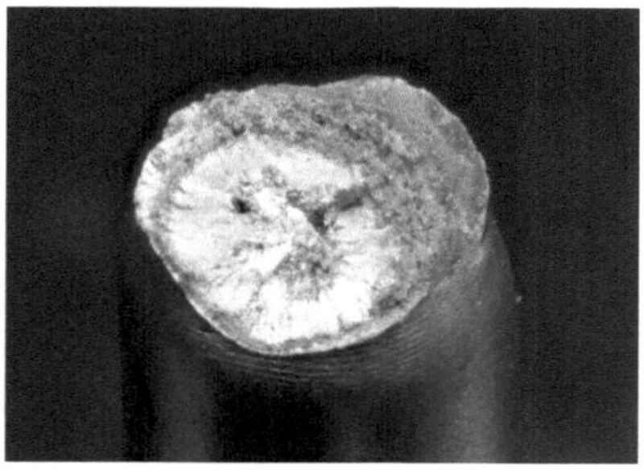

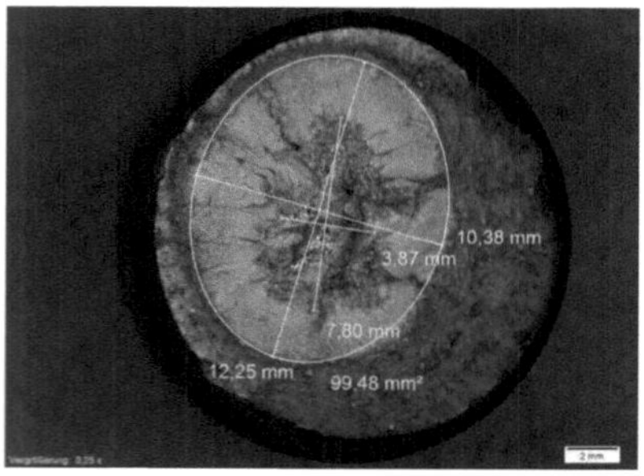


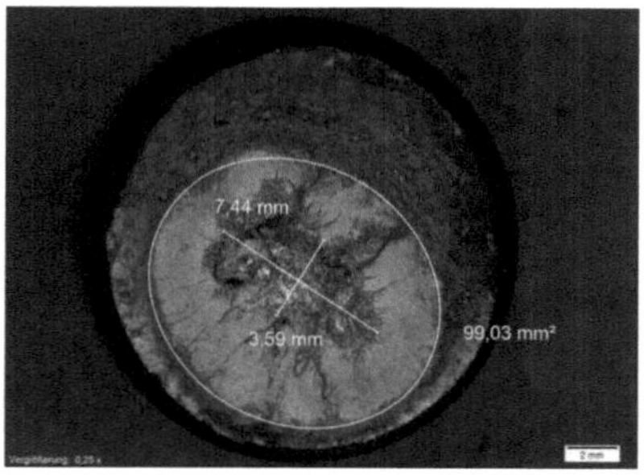

## U20O9

| $\Delta\sigma_{nom}$ [MPa] | $\Delta\sigma_{mic}$ [MPa] | cycles to failure [-] | foundry [-] | $\sigma_y$ [MPa] | $\sigma_u$ [MPa] | defect pos. | failure case |
|---|---|---|---|---|---|---|---|
| B280 | 344 | 212,437 | D | 409 | 597 | int. | II |

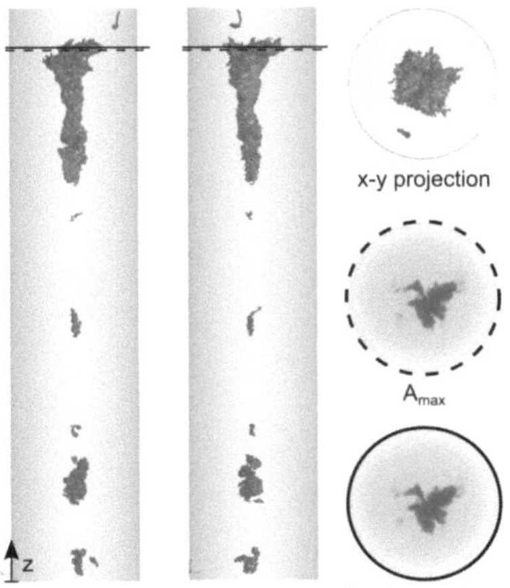


| $A_{proj,CT}$ [mm²]; [%] | $A_{max,CT}$ [mm²]; [%] | $z_{max,CT}$ [mm] |
|---|---|---|
| 1 | 19.8; 8 | 76.2 |

| $V_{max}/V_{tot}$ [%] | $A_{fail,mic}$ [mm]; [%] | $z_{fail}$ [mm] |
|---|---|---|
| 84 | 47.4; 18 | 74.1 |

| $\sigma_{max}/\sigma_y$ [%] | $\sigma_{max}/\sigma_u$ [%] |
|---|---|
| 150 | 103 |

Upper Fracture Surface

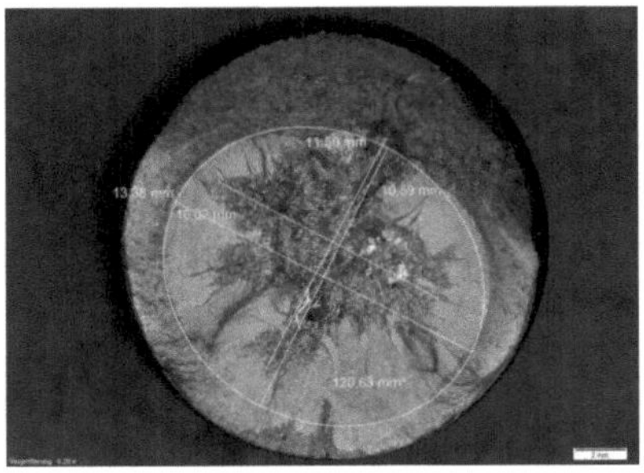

Lower Fracture Surface

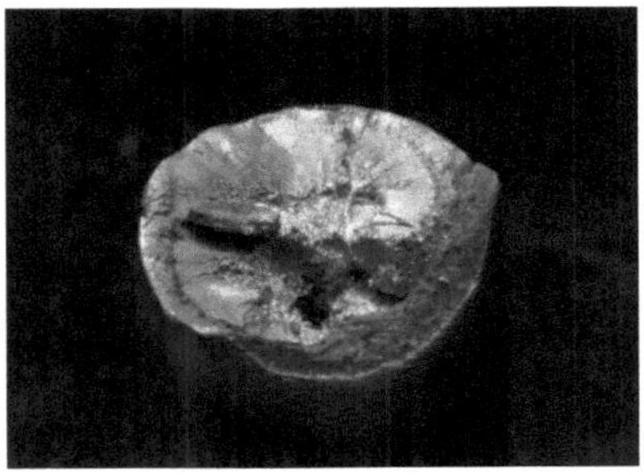

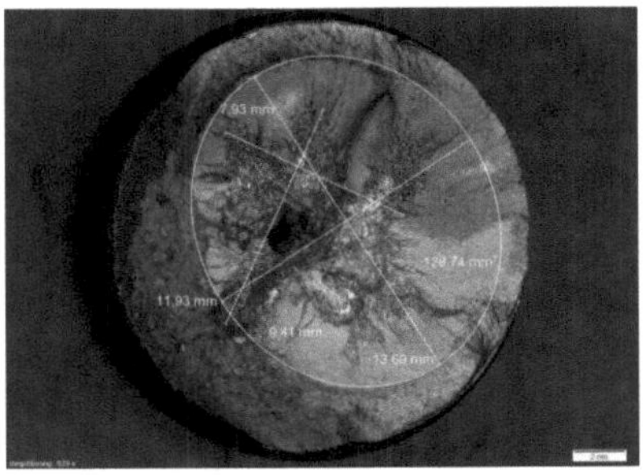

## U20O10

| $\Delta\sigma_{nom}$ [MPa] | $\Delta\sigma_{mic}$ [MPa] | cycles to failure [-] | foundry [-] | $\sigma_y$ [MPa] | $\sigma_u$ [MPa] | defect pos. | failure case |
|---|---|---|---|---|---|---|---|
| 220 | 241 | 335,470 | D | 409 | 597 | surf. | II |

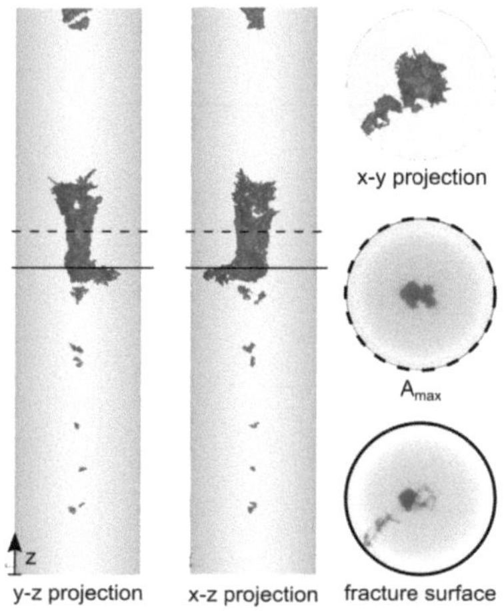


| $A_{proj,CT}$ [mm²]; [%] | $A_{max,CT}$ [mm²]; [%] | $z_{max,CT}$ [mm] |
|---|---|---|
| 1 | 10.1; 4 | 43.3 |

| $V_{max}/V_{tot}$ [%] | $A_{fail,mic}$ [mm]; [%] | $z_{fail}$ [mm] |
|---|---|---|
| 84 | 22.0; 9 | 48.4 |

| $\sigma_{max}/\sigma_y$ [%] | $\sigma_{max}/\sigma_u$ [%] |
|---|---|
| - | - |

Upper Fracture Surface

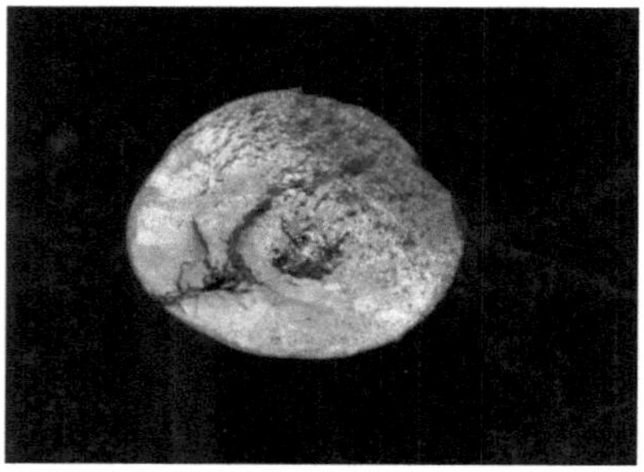

Lower Fracture Surface

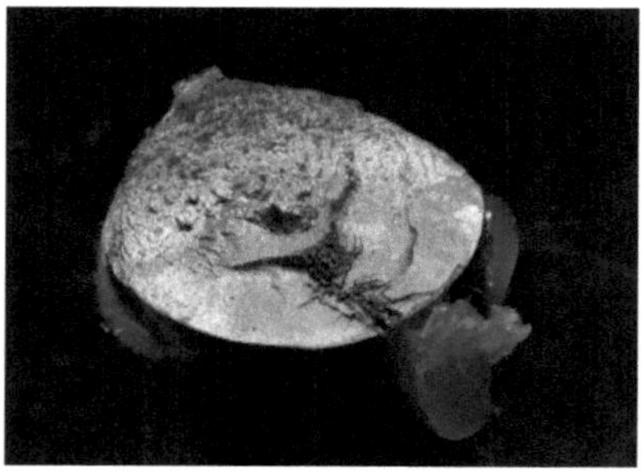

## U20O11

| $\Delta\sigma_{nom}$ [MPa] | $\Delta\sigma_{mic}$ [MPa] | cycles to failure [-] | foundry [-] | $\sigma_y$ [MPa] | $\sigma_u$ [MPa] | defect pos. | failure case |
|---|---|---|---|---|---|---|---|
| 280 | 314 | 112,335 | E | 409 | 597 | surf. | III |

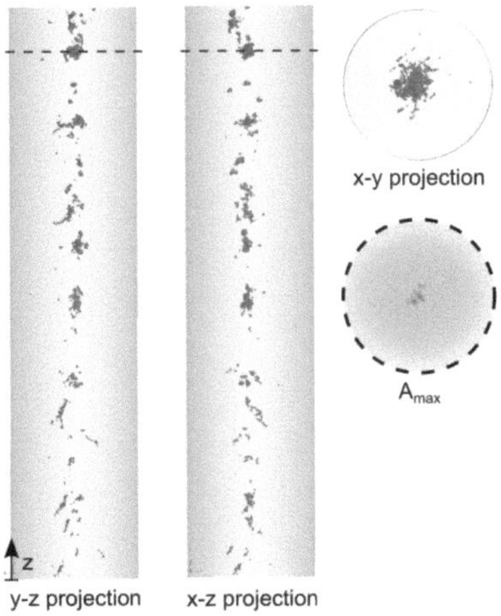


| $A_{proj,CT}$ [mm²]; [%] | $A_{max,CT}$ [mm²]; [%] | $z_{max,CT}$ [mm] |
|---|---|---|
| 1 | 1.6; 1 | -22.0 |

| $V_{max}/V_{tot}$ [%] | $A_{fail,mic}$ [mm]; [%] | $z_{fail}$ [mm] |
|---|---|---|
| 9 | 27.5; 11 | 73.6 |

| $\sigma_{max}/\sigma_y$ [%] | $\sigma_{max}/\sigma_u$ [%] |
|---|---|
| - | - |

Upper Fracture Surface

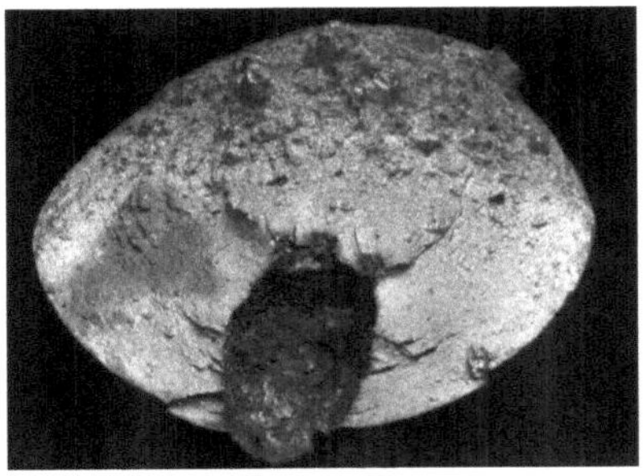

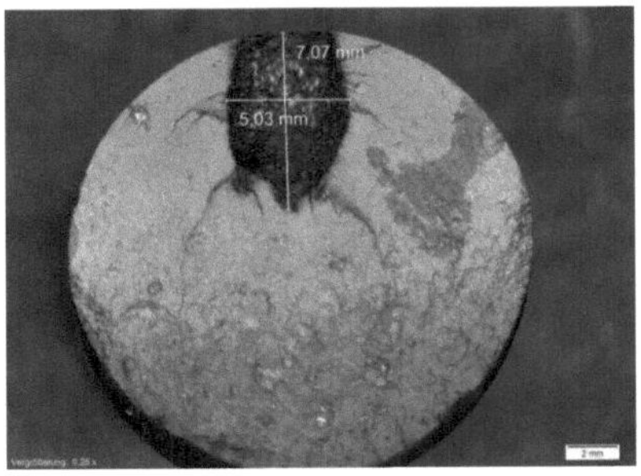


Lower Fracture Surface

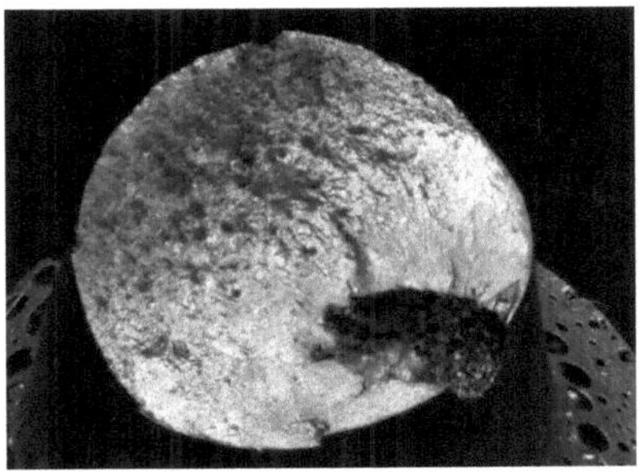

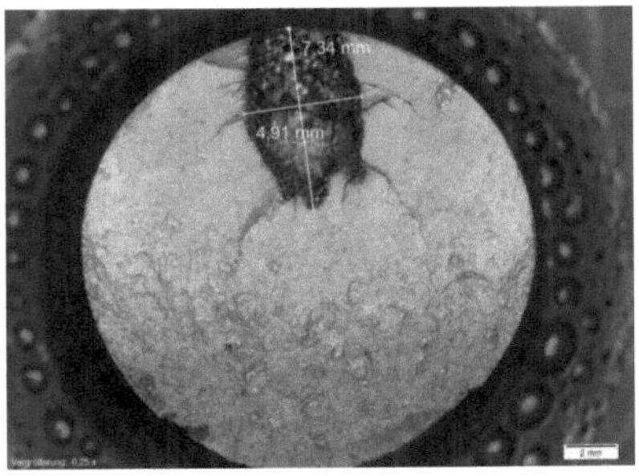

## U20O12

| $\Delta\sigma_{nom}$ [MPa] | $\Delta\sigma_{mic}$ [MPa] | cycles to failure [-] | foundry [-] | $\sigma_y$ [MPa] | $\sigma_u$ [MPa] | defect pos. | failure case |
|---|---|---|---|---|---|---|---|
| 220 | 225 | 6,407,335 | E | 409 | 597 | int. | I |

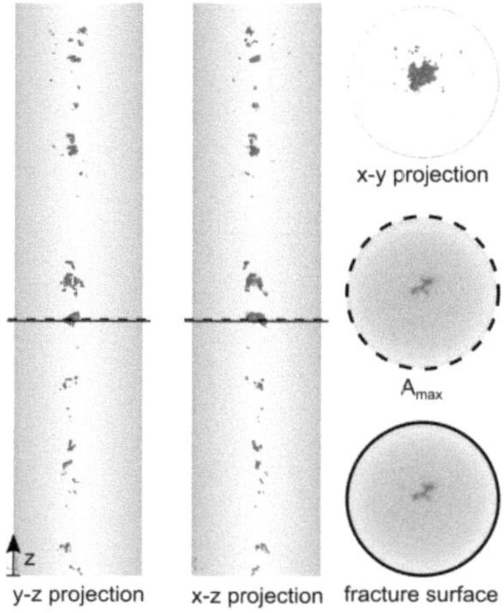


| $A_{proj,CT}$ [mm²]; [%] | $A_{max,CT}$ [mm²]; [%] | $z_{max,CT}$ [mm] |
|---|---|---|
| 1 | 2.2; 1 | 35.7 |

| $V_{max}/V_{tot}$ [%] | $A_{fail,mic}$ [mm]; [%] | $z_{fail}$ [mm] |
|---|---|---|
| 29 | 6.1; 2 | 35.8 |

| $\sigma_{max}/\sigma_y$ [%] | $\sigma_{max}/\sigma_u$ [%] |
|---|---|
| 120 | 82 |

Upper Fracture Surface

Lower Fracture Surface

## U20O13

| $\Delta\sigma_{nom}$ [MPa] | $\Delta\sigma_{mic}$ [MPa] | cycles to failure [-] | foundry [-] | $\sigma_y$ [MPa] | $\sigma_u$ [MPa] | defect pos. | failure case |
|---|---|---|---|---|---|---|---|
| 250 | 285 | 1,006,686 | E | 409 | 597 | int. | III |

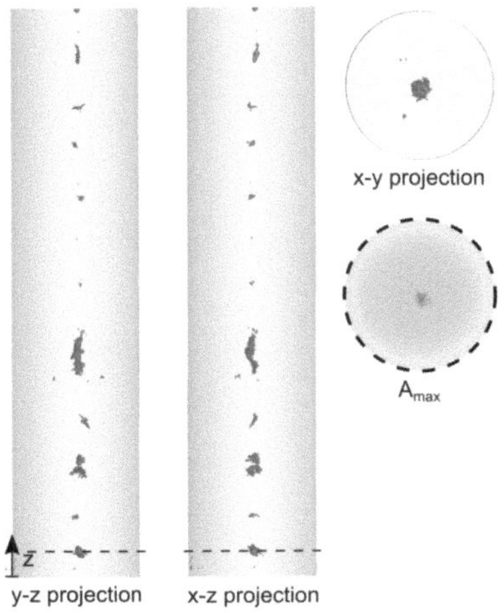


| $A_{proj,CT}$ [mm²]; [%] | $A_{max,CT}$ [mm²]; [%] | $z_{max,CT}$ [mm] |
|---|---|---|
| 1 | 1.4; 1 | 92.0 |

| $V_{max}/V_{tot}$ [%] | $A_{fail,mic}$ [mm]; [%] | $z_{fail}$ [mm] |
|---|---|---|
| 37 | 31.7; 12 | 3.4 |

| $\sigma_{max}/\sigma_y$ [%] | $\sigma_{max}/\sigma_u$ [%] |
|---|---|
| 140 | 96 |

Upper Fracture Surface

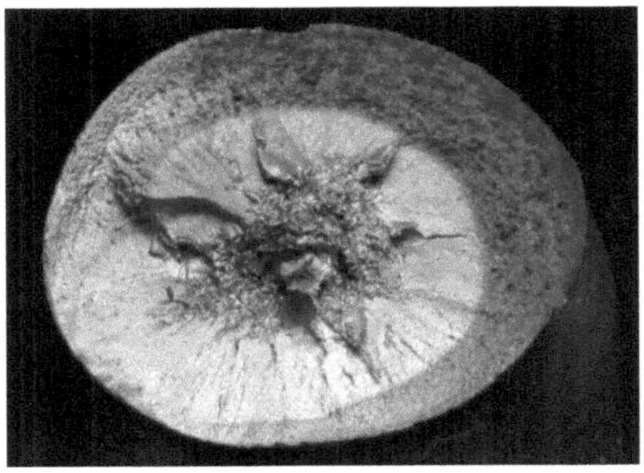

Lower Fracture Surface

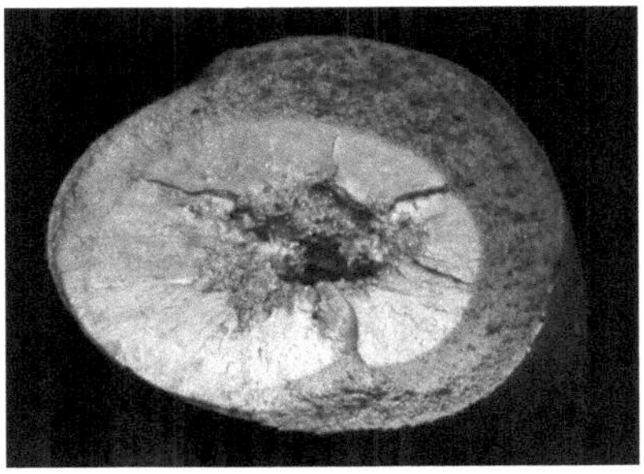

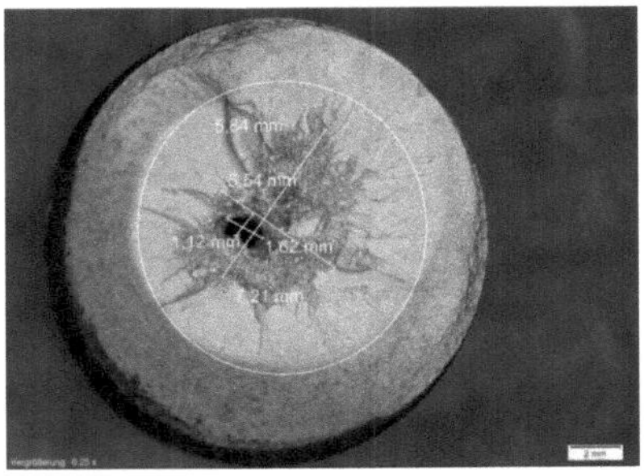

## U20O14

| $\Delta\sigma_{nom}$ [MPa] | $\Delta\sigma_{mic}$ [MPa] | cycles to failure [-] | foundry [-] | $\sigma_y$ [MPa] | $\sigma_u$ [MPa] | defect pos. | failure case |
|---|---|---|---|---|---|---|---|
| 280 | 293 | 362,554 | E | 409 | 597 | surf. | II |

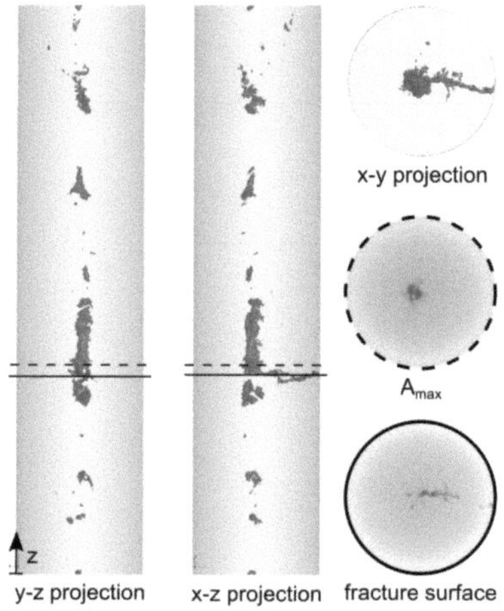


| $A_{proj,CT}$ [mm²]; [%] | $A_{max,CT}$ [mm²]; [%] | $z_{max,CT}$ [mm] |
|---|---|---|
| 1 | 2.7; 1 | 28.1 |

| $V_{max}/V_{tot}$ [%] | $A_{fail,mic}$ [mm]; [%] | $z_{fail}$ [mm] |
|---|---|---|
| 55 | 11.0; 4 | 30.1 |

| $\sigma_{max}/\sigma_y$ [%] | $\sigma_{max}/\sigma_u$ [%] |
|---|---|
| - | - |

Upper Fracture Surface

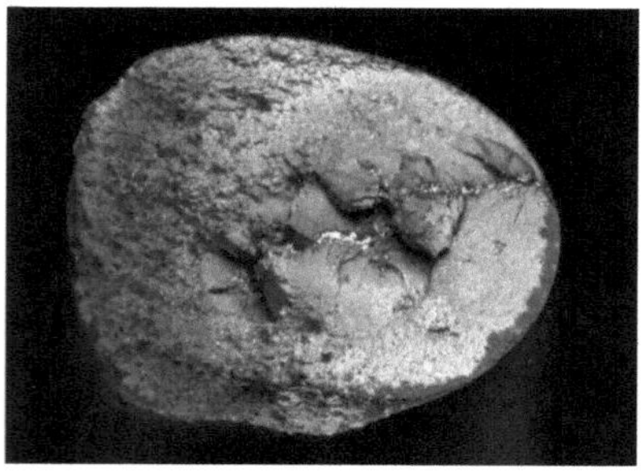

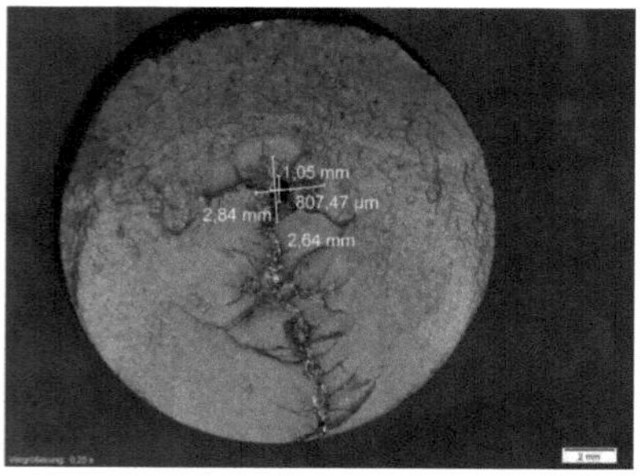


Lower Fracture Surface

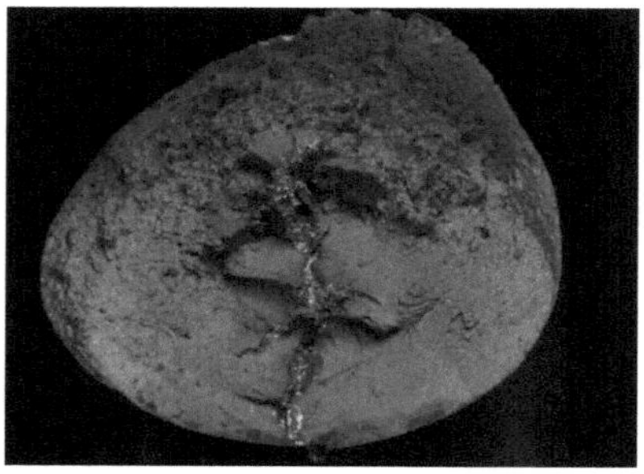

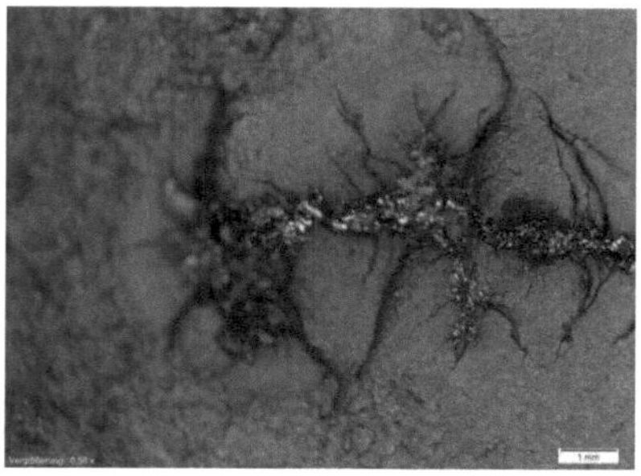

## U20O15

| $\Delta\sigma_{nom}$ [MPa] | $\Delta\sigma_{mic}$ [MPa] | cycles to failure [-] | foundry [-] | $\sigma_y$ [MPa] | $\sigma_u$ [MPa] | defect pos. | failure case |
|---|---|---|---|---|---|---|---|
| 330 | 361 | 255,594 | E | 409 | 597 | int. | I |

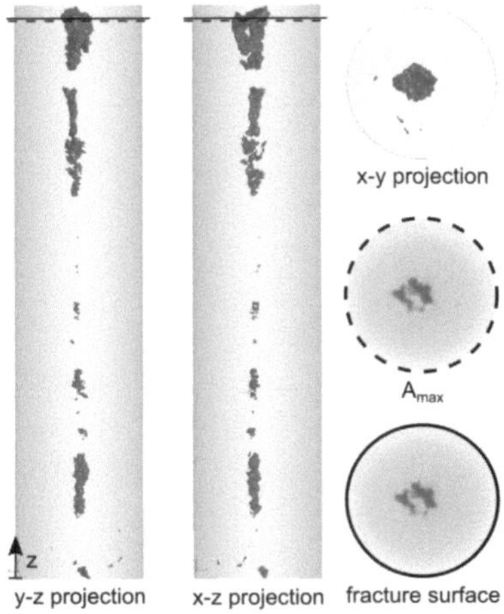


| $A_{proj,CT}$ [mm²]; [%] | $A_{max,CT}$ [mm²]; [%] | $z_{max,CT}$ [mm] |
|---|---|---|
| 1 | 8.2; 3 | 78.2 |

| $V_{max}/V_{tot}$ [%] | $A_{fail,mic}$ [mm]; [%] | $z_{fail}$ [mm] |
|---|---|---|
| 56 | 22.0; 9 | 78.2 |

| $\sigma_{max}/\sigma_y$ [%] | $\sigma_{max}/\sigma_u$ [%] |
|---|---|
| 159 | 109 |

Upper Fracture Surface

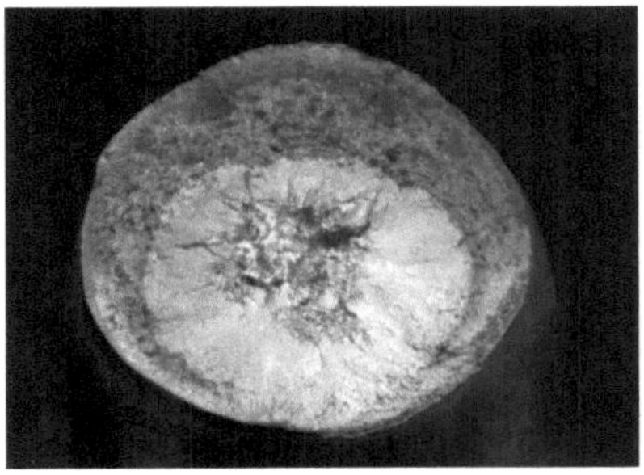

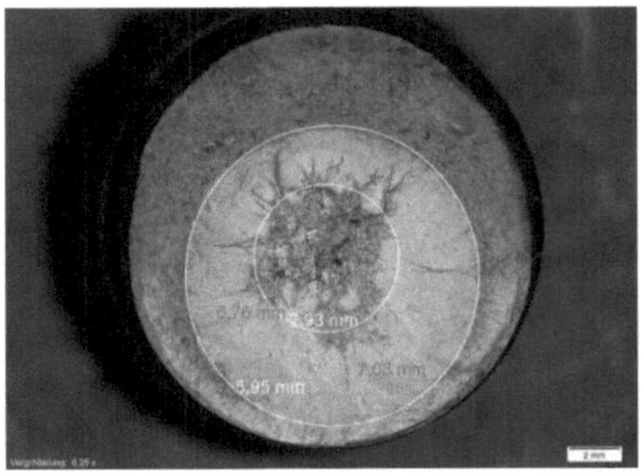


Lower Fracture Surface

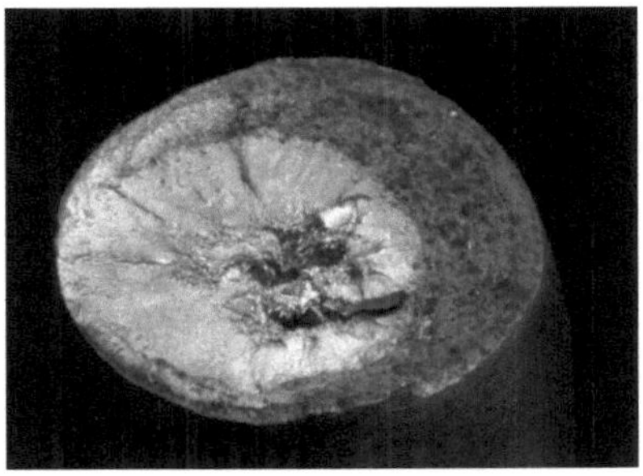

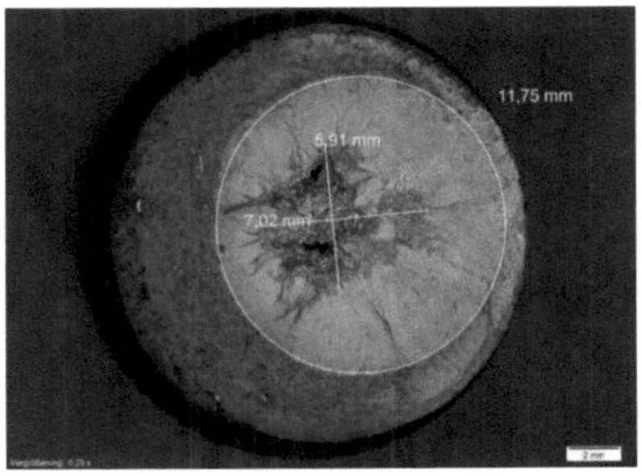

## U20O16

| $\Delta\sigma_{nom}$ [MPa] | $\Delta\sigma_{mic}$ [MPa] | cycles to failure [-] | foundry [-] | $\sigma_y$ [MPa] | $\sigma_u$ [MPa] | defect pos. | failure case |
|---|---|---|---|---|---|---|---|
| B350 | 375 | 300,000 | E | 409 | 597 | int. | III |

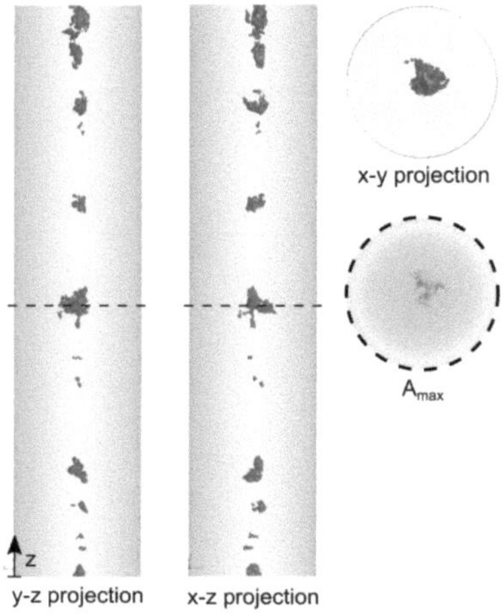


| $A_{proj,CT}$ [mm²]; [%] | $A_{max,CT}$ [mm²]; [%] | $z_{max,CT}$ [mm] |
|---|---|---|
| 1 | 4.0; 2 | -5.0 |

| $V_{max}/V_{tot}$ [%] | $A_{fail,mic}$ [mm]; [%] | $z_{fail}$ [mm] |
|---|---|---|
| 23 | 17.4; 7 | 37.9 |

| $\sigma_{max}/\sigma_y$ [%] | $\sigma_{max}/\sigma_u$ [%] |
|---|---|
| 173 | 118 |

Upper Fracture Surface

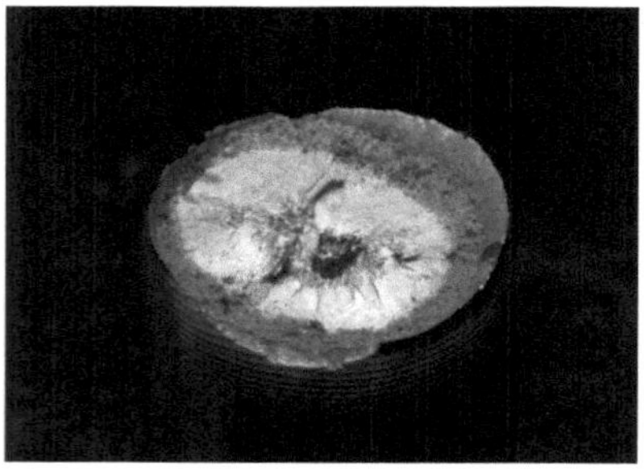

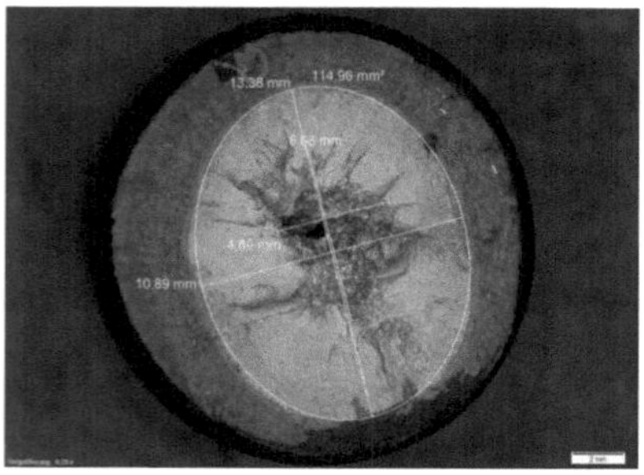

Lower Fracture Surface

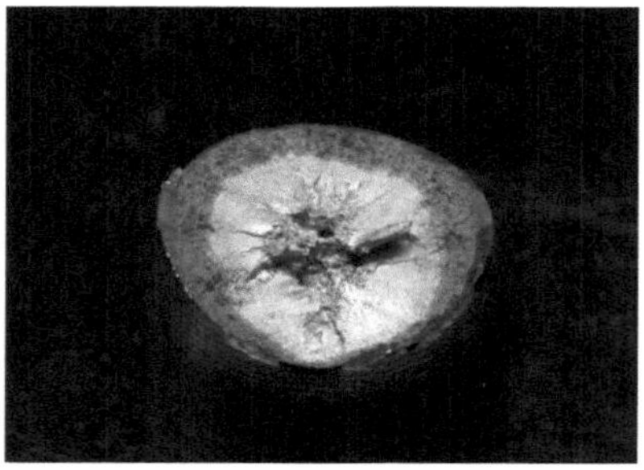

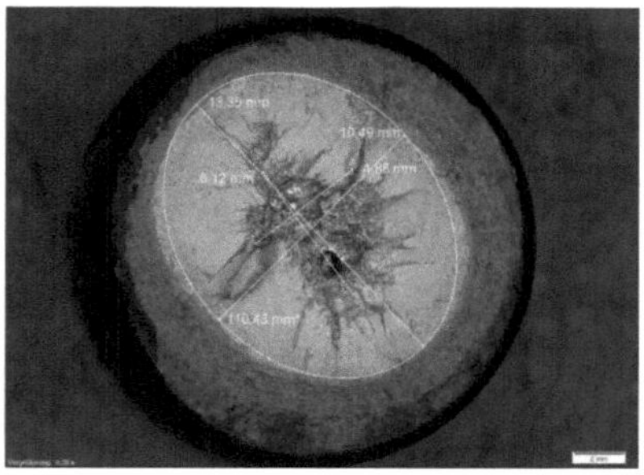

## U20O17

| $\Delta\sigma_{nom}$ [MPa] | $\Delta\sigma_{mic}$ [MPa] | cycles to failure [-] | foundry [-] | $\sigma_y$ [MPa] | $\sigma_u$ [MPa] | defect pos. | failure case |
|---|---|---|---|---|---|---|---|
| 235 | 248 | 3,972,402 | E | 409 | 597 | int. | III |

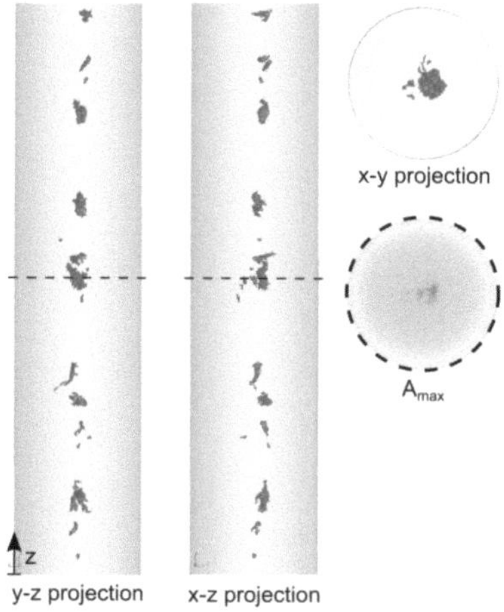


| $A_{proj,CT}$ [mm²]; [%] | $A_{max,CT}$ [mm²]; [%] | $z_{max,CT}$ [mm] |
|---|---|---|
| 1 | 2.4; 1 | 100.0 |

| $V_{max}/V_{tot}$ [%] | $A_{fail,mic}$ [mm]; [%] | $z_{fail}$ [mm] |
|---|---|---|
| 5 | 13.1; 5 | 41.6 |

| $\sigma_{max}/\sigma_y$ [%] | $\sigma_{max}/\sigma_u$ [%] |
|---|---|
| 144 | 98 |

Upper Fracture Surface

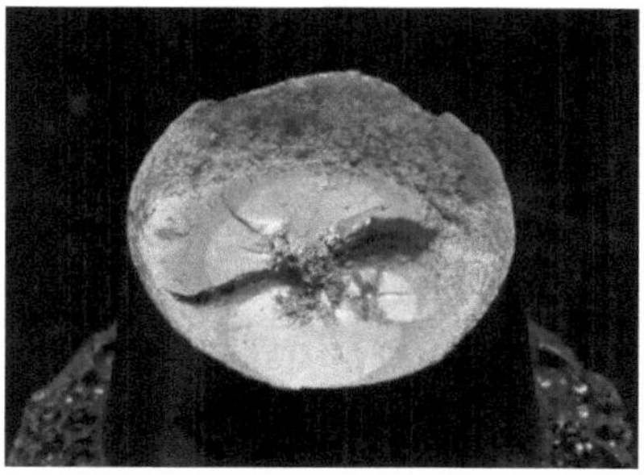

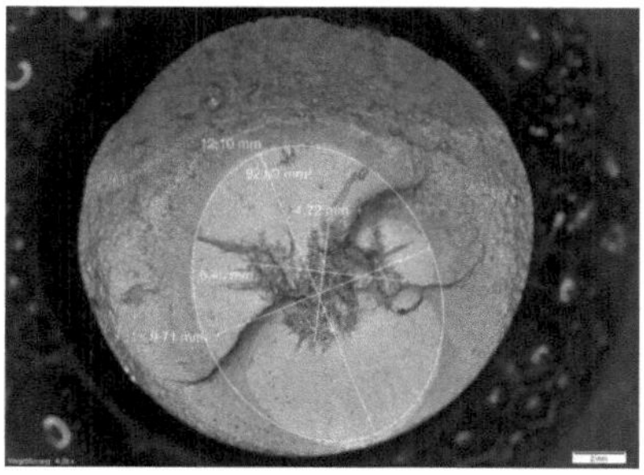

Lower Fracture Surface

## U20O19

| $\Delta\sigma_{nom}$ [MPa] | $\Delta\sigma_{mic}$ [MPa] | cycles to failure [-] | foundry [-] | $\sigma_y$ [MPa] | $\sigma_u$ [MPa] | defect pos. | failure case |
|---|---|---|---|---|---|---|---|
| B280 | 289 | 1,623,505 | E | 409 | 597 | int. | - |

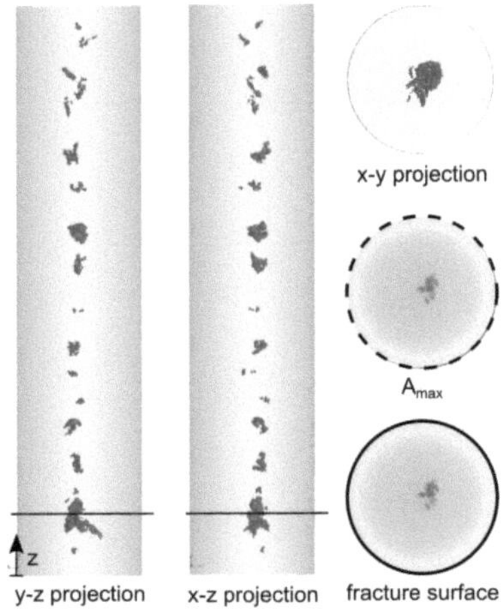


| $A_{proj,CT}$ [mm²]; [%] | $A_{max,CT}$ [mm²]; [%] | $z_{max,CT}$ [mm] |
|---|---|---|
| 1 | 3.0; 1 | 8.6 |

| $V_{max}/V_{tot}$ [%] | $A_{fail,mic}$ [mm]; [%] | $z_{fail}$ [mm] |
|---|---|---|
| 34 | 7.8; 3 | - |

| $\sigma_{max}/\sigma_y$ [%] | $\sigma_{max}/\sigma_u$ [%] |
|---|---|
| 144 | 99 |

Upper Fracture Surface

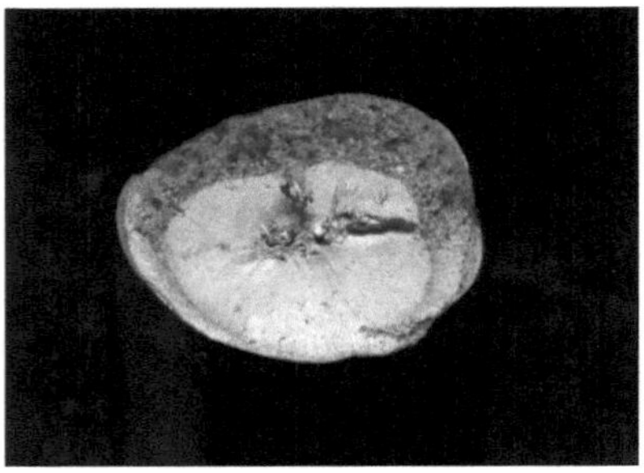

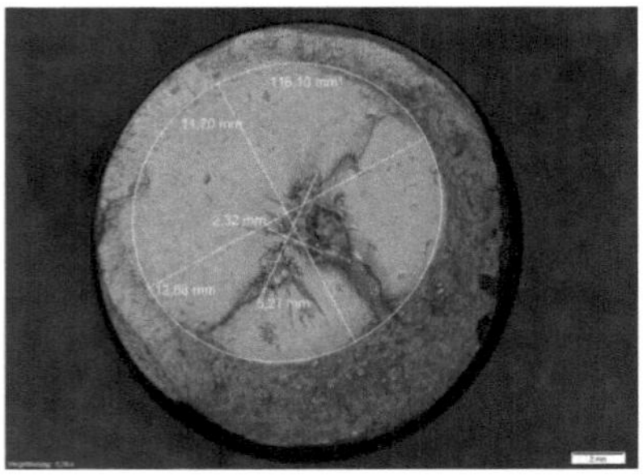

Lower Fracture Surface

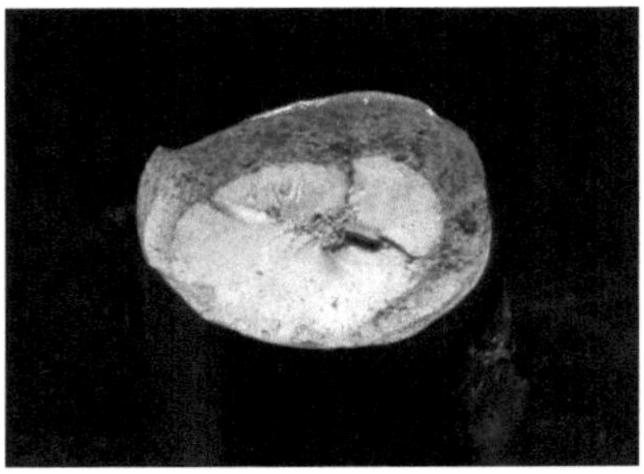

## U20O20

| $\Delta\sigma_{nom}$ [MPa] | $\Delta\sigma_{mic}$ [MPa] | cycles to failure [-] | foundry [-] | $\sigma_y$ [MPa] | $\sigma_u$ [MPa] | defect pos. | failure case |
|---|---|---|---|---|---|---|---|
| 235 | 241 | 6,427,123 | E | 409 | 597 | int. | II |

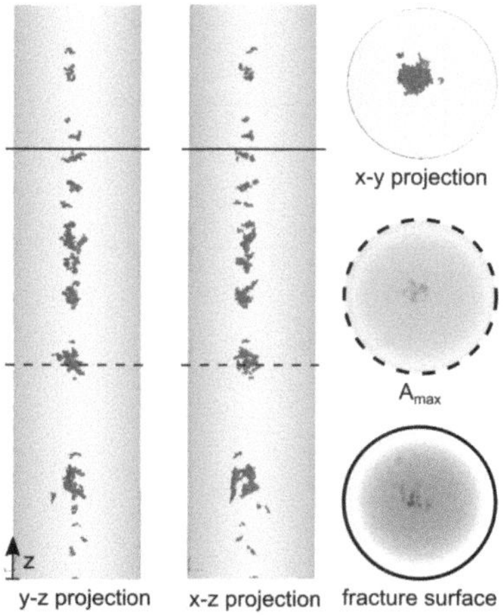


| $A_{proj,CT}$ [mm²]; [%] | $A_{max,CT}$ [mm²]; [%] | $z_{max,CT}$ [mm] |
|---|---|---|
| 1 | 2.5; 1 | 60.6 |

| $V_{max}/V_{tot}$ [%] | $A_{fail,mic}$ [mm]; [%] | $z_{fail}$ [mm] |
|---|---|---|
| 17 | 6.4; 2 | 30.4 |

| $\sigma_{max}/\sigma_y$ [%] | $\sigma_{max}/\sigma_u$ [%] |
|---|---|
| 192 | 131 |

Upper Fracture Surface

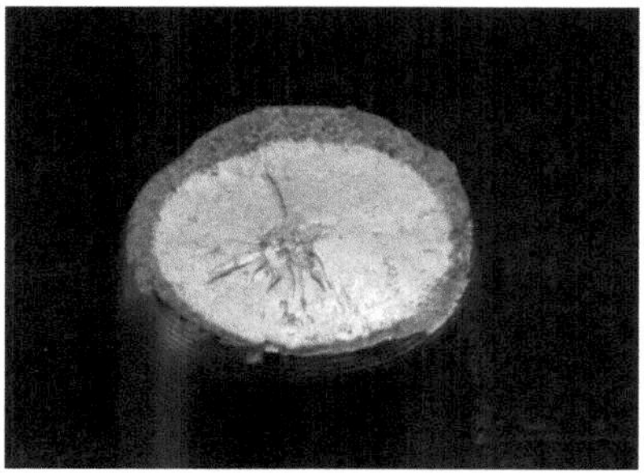

Lower Fracture Surface

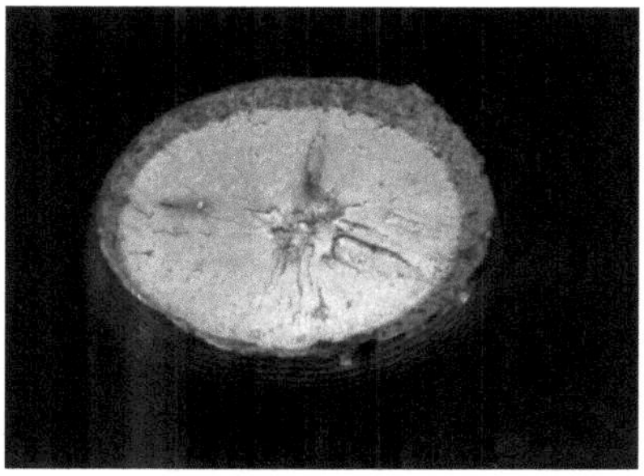

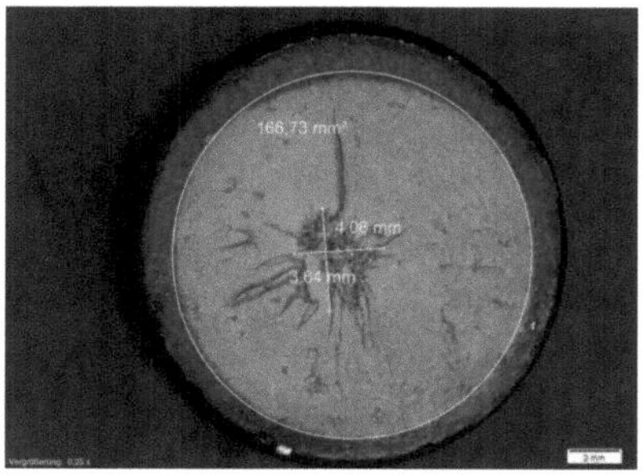


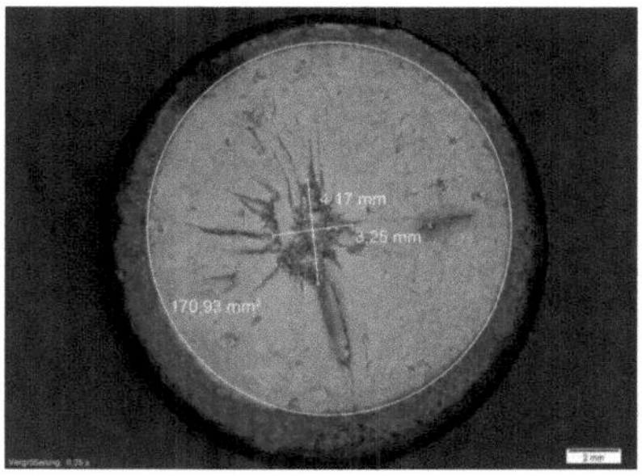

## U20M1

| $\Delta\sigma_{nom}$ [MPa] | $\Delta\sigma_{mic}$ [MPa] | cycles to failure [-] | foundry [-] | $\sigma_y$ [MPa] | $\sigma_u$ [MPa] | defect pos. | failure case |
|---|---|---|---|---|---|---|---|
| 200 | 219 | 353,178 | D | 409 | 597 | surf. | II |

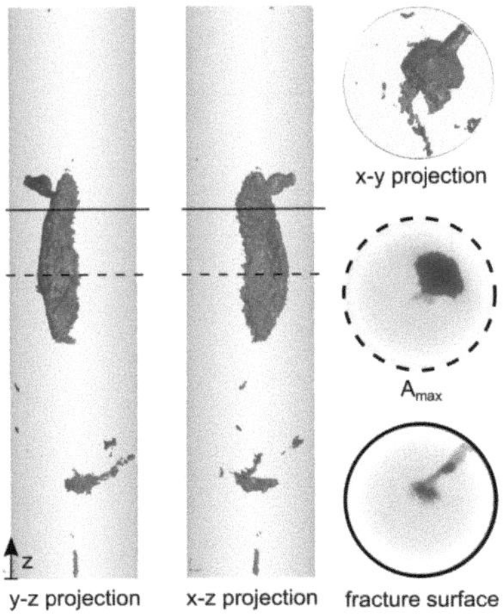


| $A_{proj,CT}$ [mm²]; [%] | $A_{max,CT}$ [mm²]; [%] | $z_{max,CT}$ [mm] |
|---|---|---|
| 1 | 30.4; 12 | 51.7 |

| $V_{max}/V_{tot}$ [%] | $A_{fail,mic}$ [mm]; [%] | $z_{fail}$ [mm] |
|---|---|---|
| 95 | 22.5; 9 | 42.5 |

| $\sigma_{max}/\sigma_y$ [%] | $\sigma_{max}/\sigma_u$ [%] |
|---|---|
| - | - |

Upper Fracture Surface

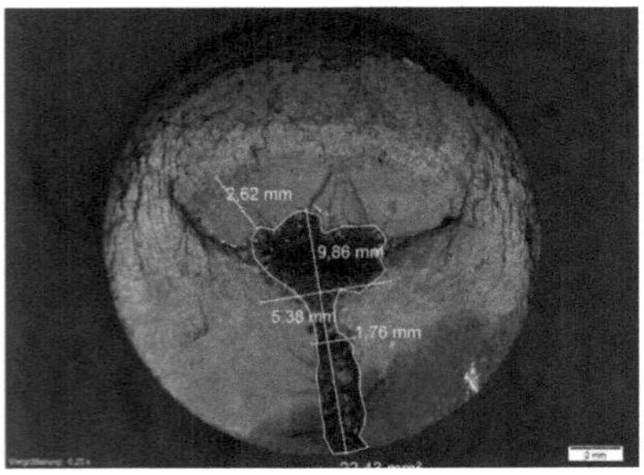


Lower Fracture Surface

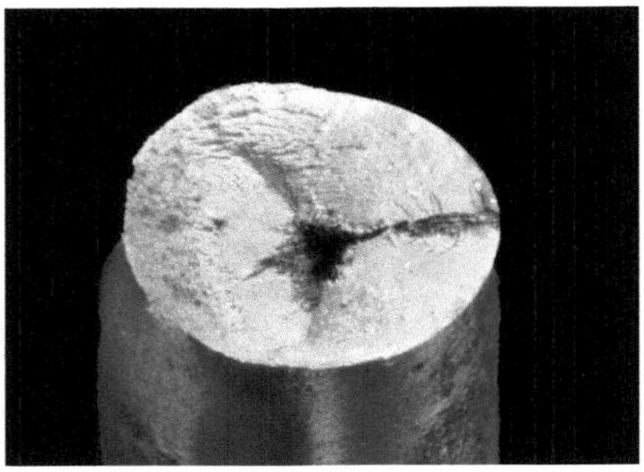

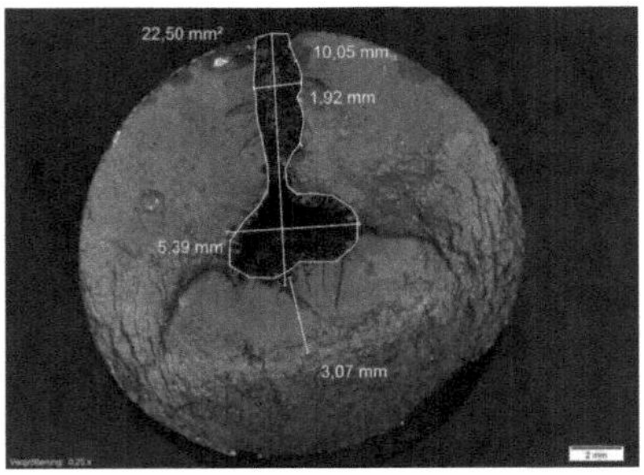

## U20M2

| $\Delta\sigma_{nom}$ [MPa] | $\Delta\sigma_{mic}$ [MPa] | cycles to failure [-] | foundry [-] | $\sigma_y$ [MPa] | $\sigma_u$ [MPa] | defect pos. | failure case |
|---|---|---|---|---|---|---|---|
| 180 | 197 | 572,882 | D | 409 | 597 | surf. | II |

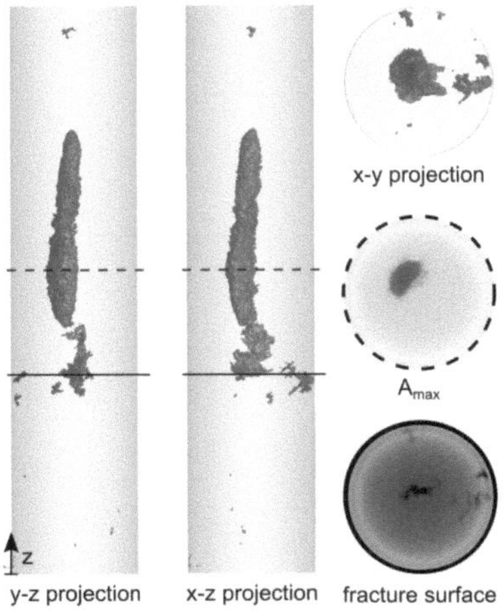


| $A_{proj,CT}$ [mm²]; [%] | $A_{max,CT}$ [mm²]; [%] | $z_{max,CT}$ [mm] |
|---|---|---|
| 1 | 12.9; 5 | 28.2 |

| $V_{max}/V_{tot}$ [%] | $A_{fail,mic}$ [mm]; [%] | $z_{fail}$ [mm] |
|---|---|---|
| 89 | 22.5; 9 | 42.9 |

| $\sigma_{max}/\sigma_y$ [%] | $\sigma_{max}/\sigma_u$ [%] |
|---|---|
| - | - |

Upper Fracture Surface

Lower Fracture Surface

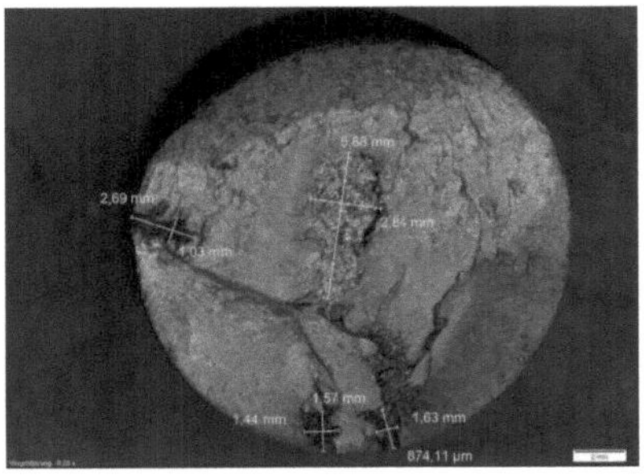

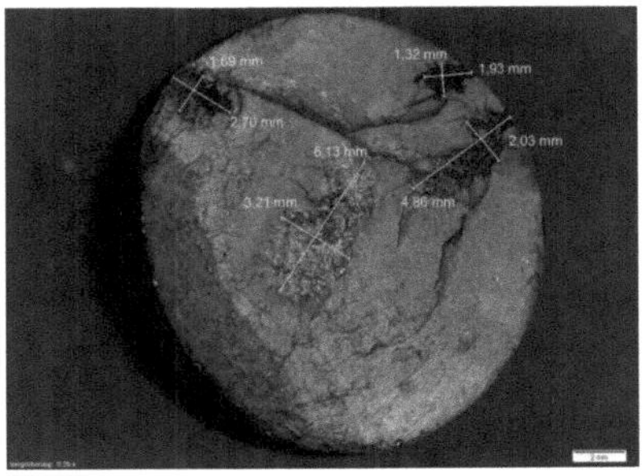

## U20M3

| $\Delta\sigma_{nom}$ [MPa] | $\Delta\sigma_{mic}$ [MPa] | cycles to failure [-] | foundry [-] | $\sigma_y$ [MPa] | $\sigma_u$ [MPa] | defect pos. | failure case |
|---|---|---|---|---|---|---|---|
| B280 | 303 | 133,224 | D | 409 | 597 | surf. | II |

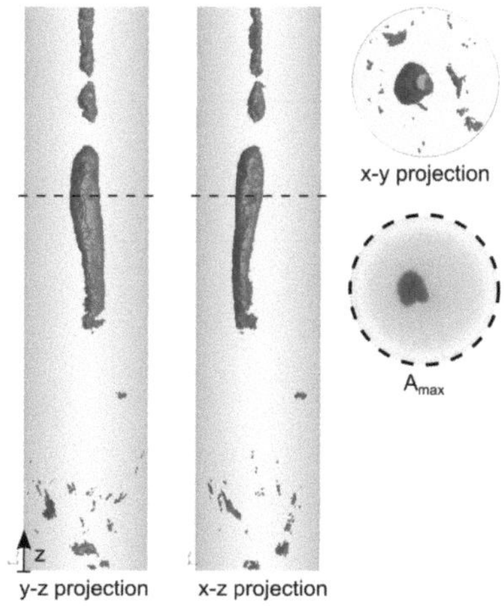


| $A_{proj,CT}$ [mm²]; [%] | $A_{max,CT}$ [mm²]; [%] | $z_{max,CT}$ [mm] |
|---|---|---|
| 1 | 17.5; 7 | 57.7 |

| $V_{max}/V_{tot}$ [%] | $A_{fail,mic}$ [mm]; [%] | $z_{fail}$ [mm] |
|---|---|---|
| 92 | 19.8; 8 | 45.1 |

| $\sigma_{max}/\sigma_y$ [%] | $\sigma_{max}/\sigma_u$ [%] |
|---|---|
| - | - |

Upper Fracture Surface

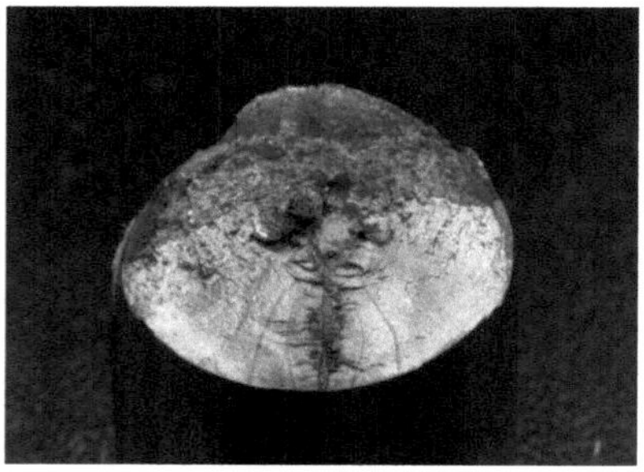

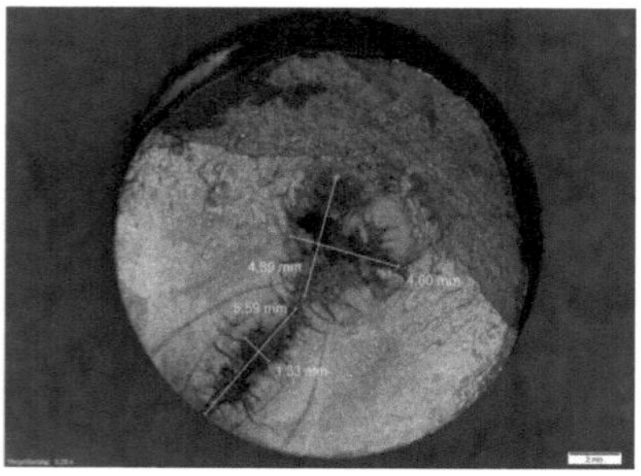

Lower Fracture Surface

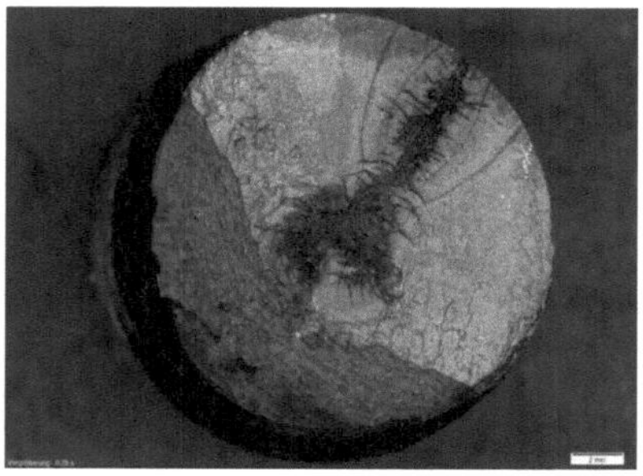

## U20M4

| $\Delta\sigma_{nom}$ [MPa] | $\Delta\sigma_{mic}$ [MPa] | cycles to failure [-] | foundry [-] | $\sigma_y$ [MPa] | $\sigma_u$ [MPa] | defect pos. | failure case |
|---|---|---|---|---|---|---|---|
| 150 | 152 | 2,156,913 | D | 409 | 597 | surf. | II |

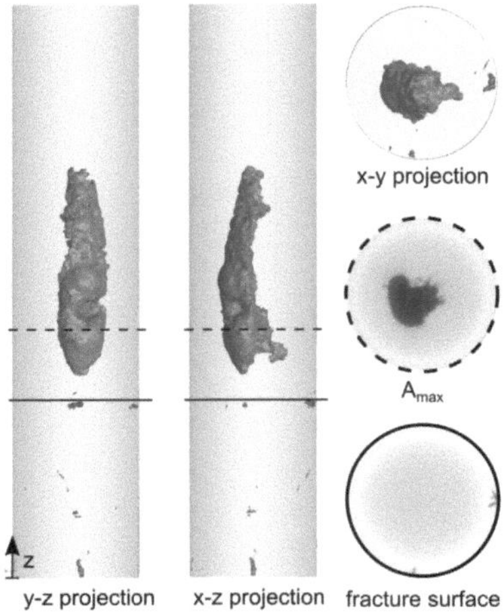


| $A_{proj,CT}$ [mm²]; [%] | $A_{max,CT}$ [mm²]; [%] | $z_{max,CT}$ [mm] |
|---|---|---|
| 1 | 28.2; 11 | 25.0 |

| $V_{max}/V_{tot}$ [%] | $A_{fail,mic}$ [mm]; [%] | $z_{fail}$ [mm] |
|---|---|---|
| 99 | 3.2; 1 | 34.8 |

| $\sigma_{max}/\sigma_y$ [%] | $\sigma_{max}/\sigma_u$ [%] |
|---|---|
| - | - |

Upper Fracture Surface

Lower Fracture Surface

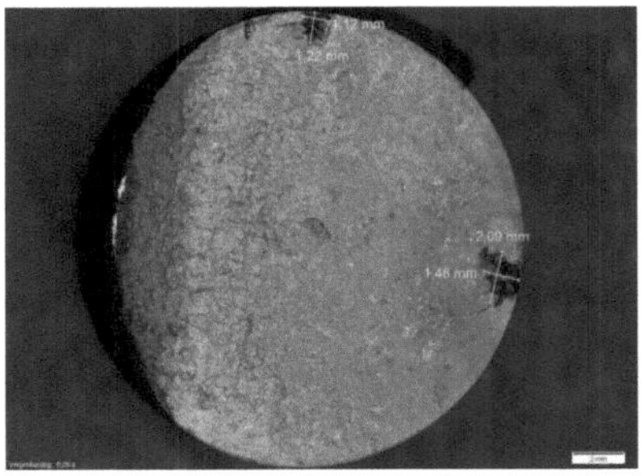

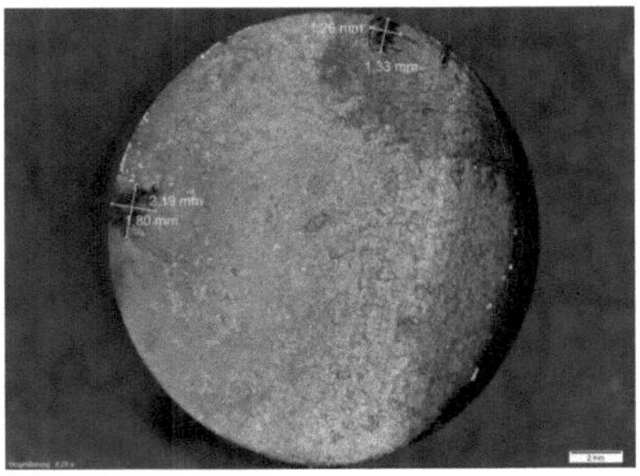

## U20M7

| $\Delta\sigma_{nom}$ [MPa] | $\Delta\sigma_{mic}$ [MPa] | cycles to failure [-] | foundry [-] | $\sigma_y$ [MPa] | $\sigma_u$ [MPa] | defect pos. | failure case |
|---|---|---|---|---|---|---|---|
| 330 | 384 | 130,405 | E | 409 | 597 | int. | II |

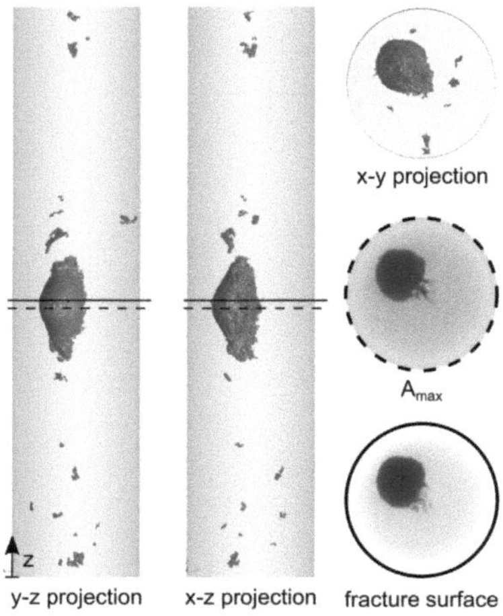


| $A_{proj,CT}$ [mm²]; [%] | $A_{max,CT}$ [mm²]; [%] | $z_{max,CT}$ [mm] |
|---|---|---|
| 1 | 31.8; 12 | 38.8 |

| $V_{max}/V_{tot}$ [%] | $A_{fail,mic}$ [mm]; [%] | $z_{fail}$ [mm] |
|---|---|---|
| 97 | 36.1; 14 | 37.7 |

| $\sigma_{max}/\sigma_y$ [%] | $\sigma_{max}/\sigma_u$ [%] |
|---|---|
| 132 | 91 |

Upper Fracture Surface

Lower Fracture Surface

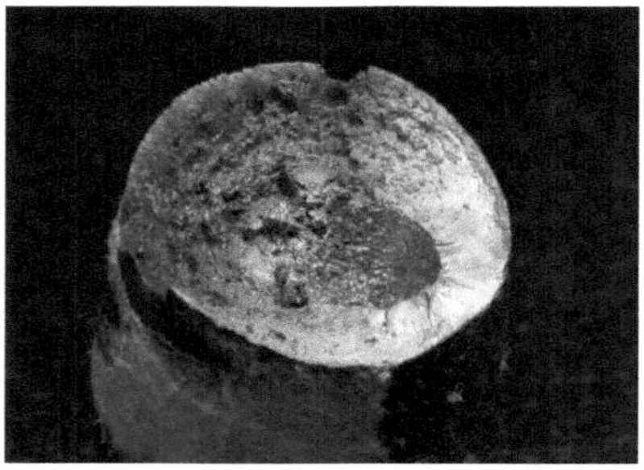

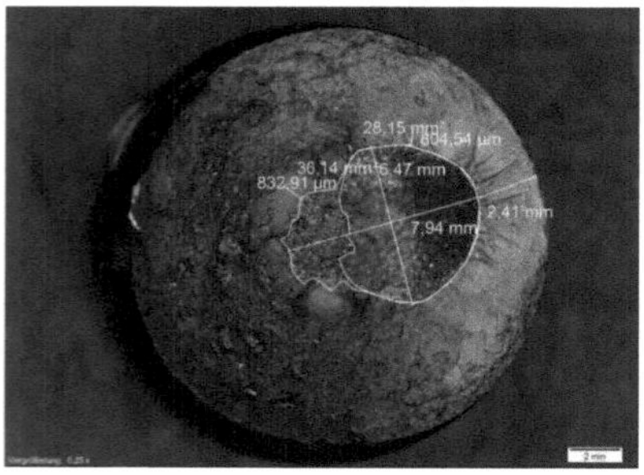

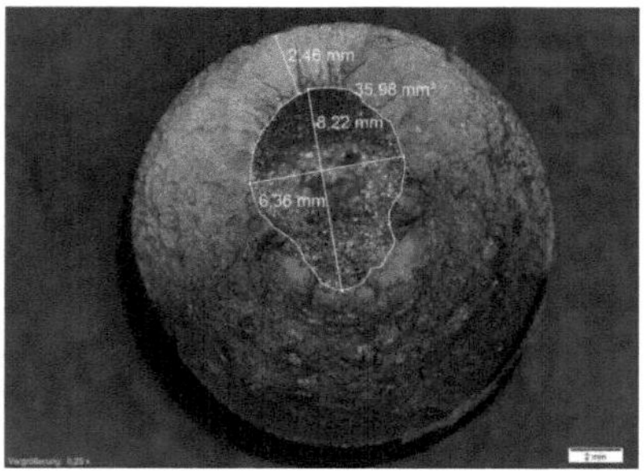

## U20M8

| $\Delta\sigma_{nom}$ [MPa] | $\Delta\sigma_{mic}$ [MPa] | cycles to failure [-] | foundry [-] | $\sigma_y$ [MPa] | $\sigma_u$ [MPa] | defect pos. | failure case |
|---|---|---|---|---|---|---|---|
| 250 | 294 | 403,379 | E | 409 | 597 | int. | I |

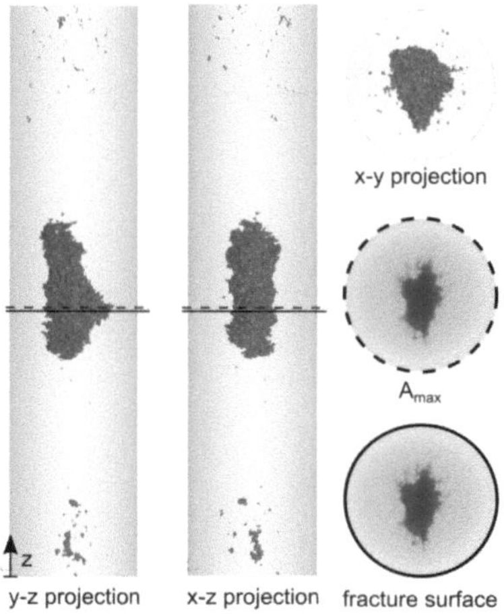


| $A_{proj,CT}$ [mm²]; [%] | $A_{max,CT}$ [mm²]; [%] | $z_{max,CT}$ [mm] |
|---|---|---|
| 1 | 30.9; 12 | 37.3 |

| $V_{max}/V_{tot}$ [%] | $A_{fail,mic}$ [mm]; [%] | $z_{fail}$ [mm] |
|---|---|---|
| 99 | 38.6; 15 | 37.8 |

| $\sigma_{max}/\sigma_y$ [%] | $\sigma_{max}/\sigma_u$ [%] |
|---|---|
| 113 | 78 |

Upper Fracture Surface

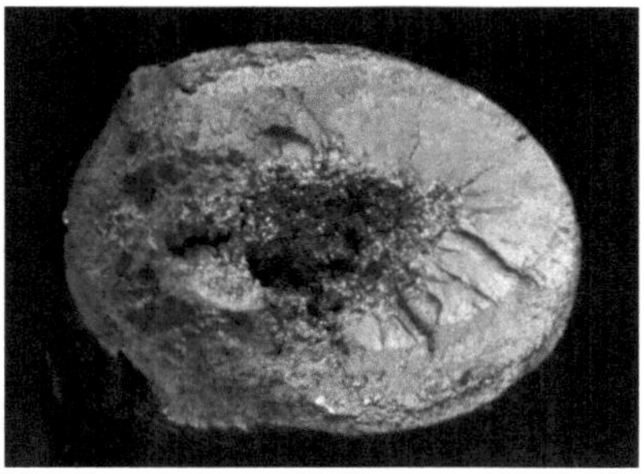

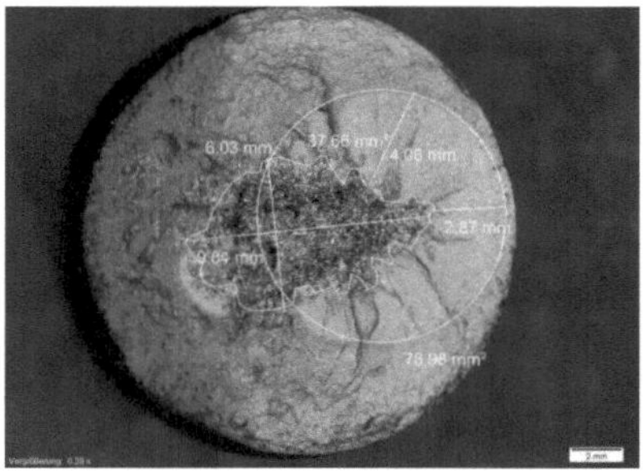

Lower Fracture Surface

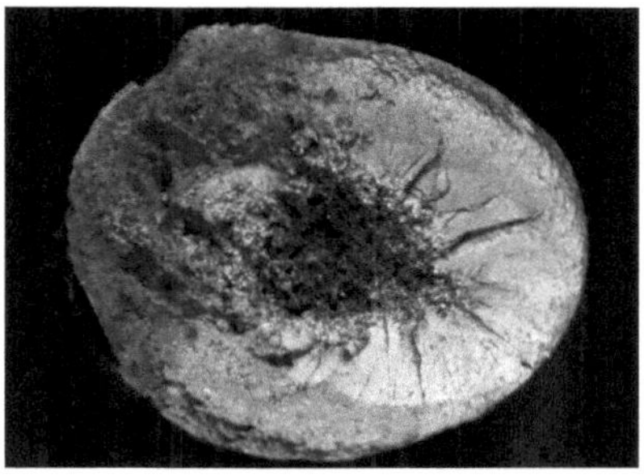

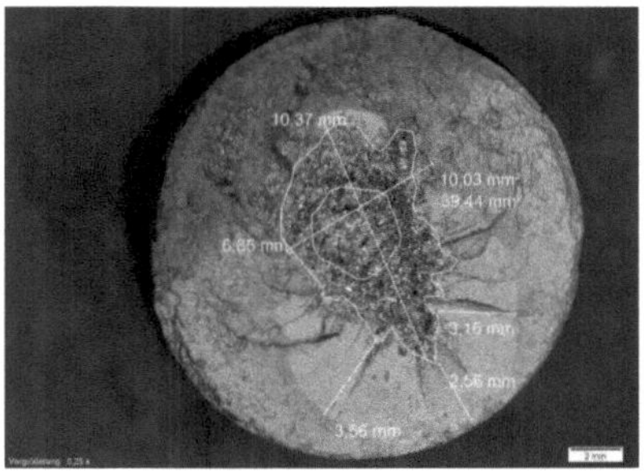

## U20M9

| $\Delta\sigma_{nom}$ [MPa] | $\Delta\sigma_{mic}$ [MPa] | cycles to failure [-] | foundry [-] | $\sigma_y$ [MPa] | $\sigma_u$ [MPa] | defect pos. | failure case |
|---|---|---|---|---|---|---|---|
| 160 | 169 | 1,644,239 | E | 409 | 597 | int. | I |

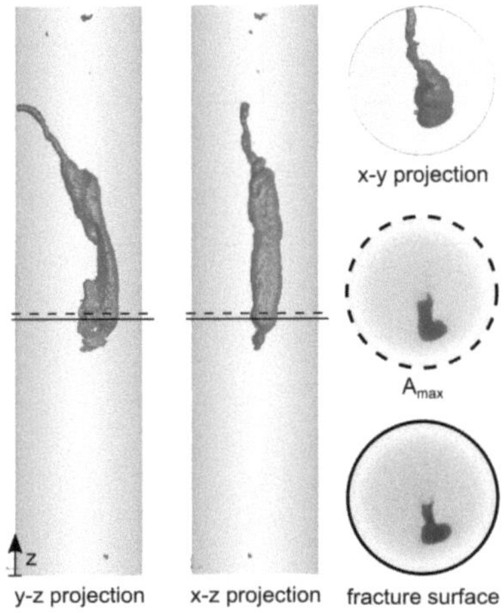


| $A_{proj,CT}$ [mm²]; [%] | $A_{max,CT}$ [mm²]; [%] | $z_{max,CT}$ [mm] |
|---|---|---|
| 1 | 12.2; 5 | 36.0 |

| $V_{max}/V_{tot}$ [%] | $A_{fail,mic}$ [mm]; [%] | $z_{fail}$ [mm] |
|---|---|---|
| 97 | 14.1; 6 | 36.8 |

| $\sigma_{max}/\sigma_y$ [%] | $\sigma_{max}/\sigma_u$ [%] |
|---|---|
| - | - |

Upper Fracture Surface

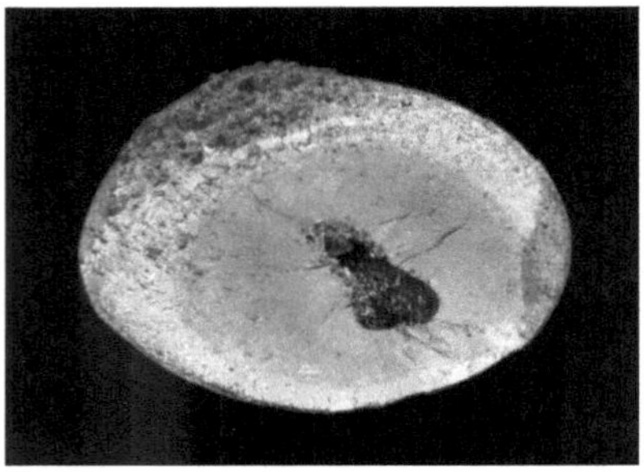

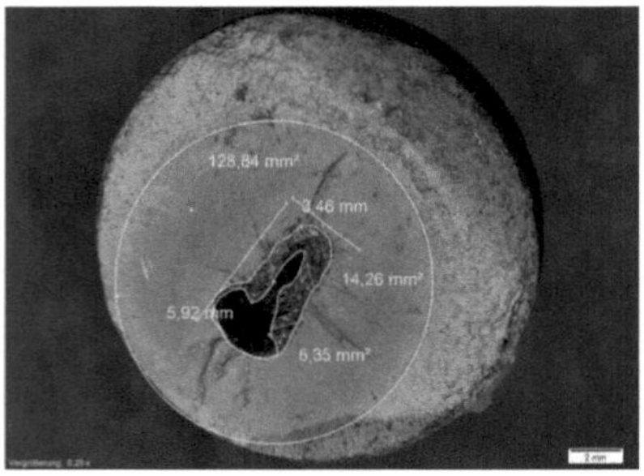


Lower Fracture Surface

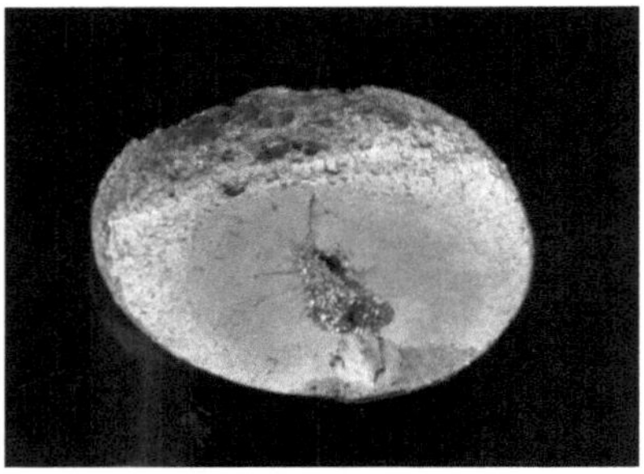

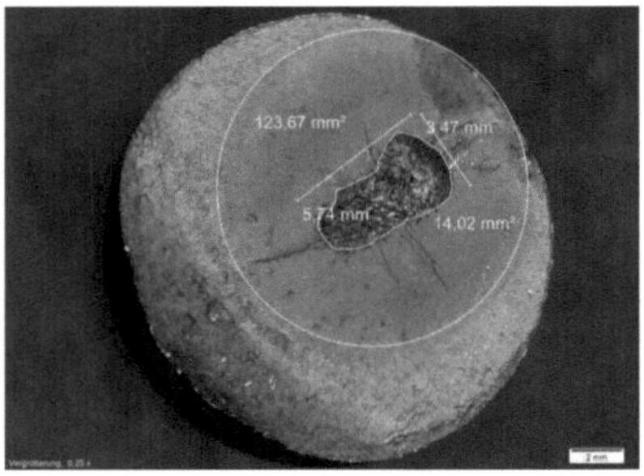

## U20M10

| $\Delta\sigma_{nom}$ [MPa] | $\Delta\sigma_{mic}$ [MPa] | cycles to failure [-] | foundry [-] | $\sigma_y$ [MPa] | $\sigma_u$ [MPa] | defect pos. | failure case |
|---|---|---|---|---|---|---|---|
| 180 | 223 | 232,230 | E | 409 | 597 | surf. | I |

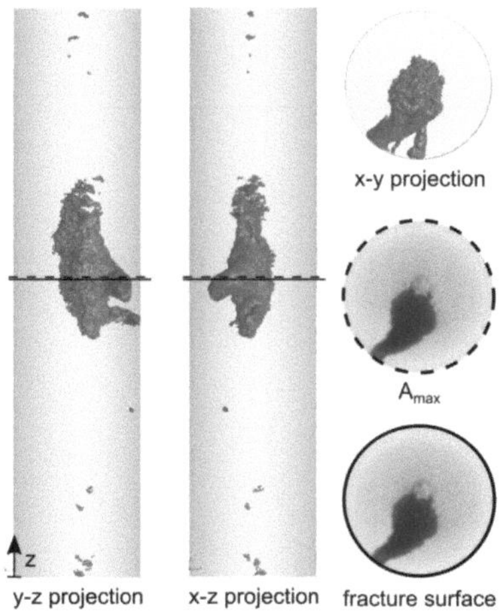


| $A_{proj,CT}$ [mm²]; [%] | $A_{max,CT}$ [mm²]; [%] | $z_{max,CT}$ [mm] |
|---|---|---|
| 1 | 35.0; 14 | 42.0 |

| $V_{max}/V_{tot}$ [%] | $A_{fail,mic}$ [mm]; [%] | $z_{fail}$ [mm] |
|---|---|---|
| 99 | 49; 19 | 42.1 |

| $\sigma_{max}/\sigma_y$ [%] | $\sigma_{max}/\sigma_u$ [%] |
|---|---|
| - | - |

Upper Fracture Surface

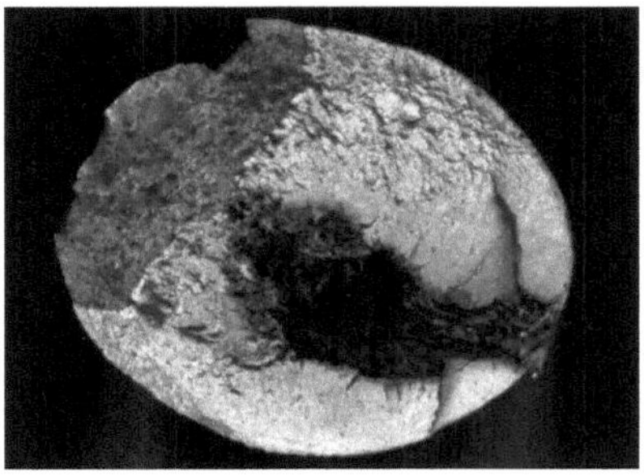

Lower Fracture Surface

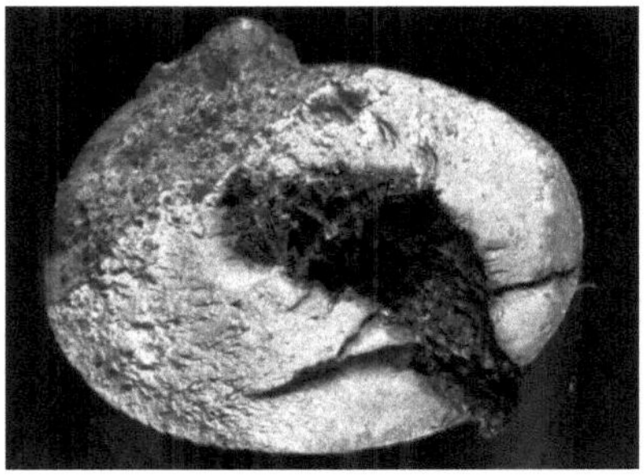

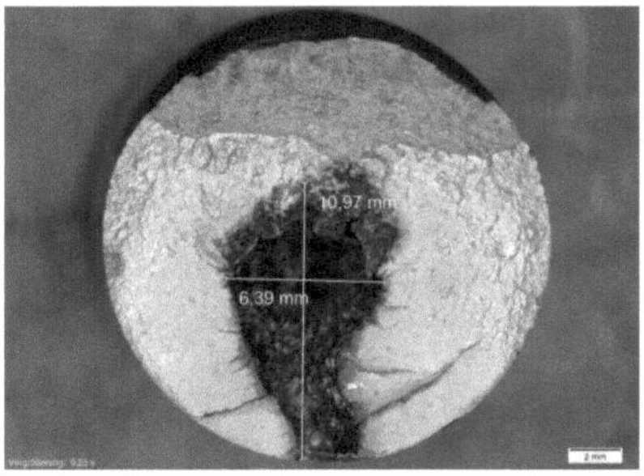


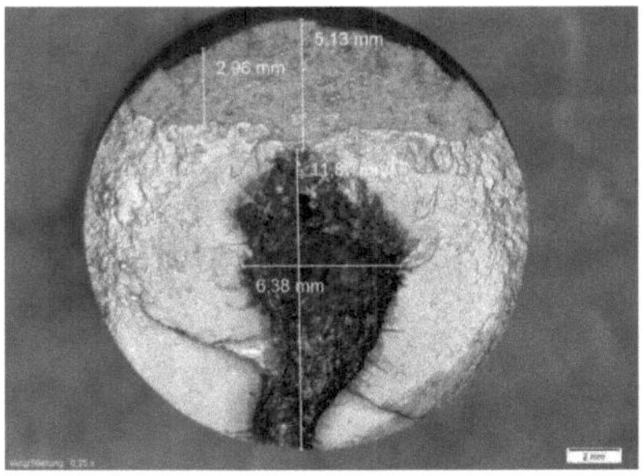

## U20M11

| $\Delta\sigma_{nom}$ [MPa] | $\Delta\sigma_{mic}$ [MPa] | cycles to failure [-] | foundry [-] | $\sigma_y$ [MPa] | $\sigma_u$ [MPa] | defect pos. | failure case |
|---|---|---|---|---|---|---|---|
| 220 | 262 | 194,253 | E | 409 | 597 | surf. | I |

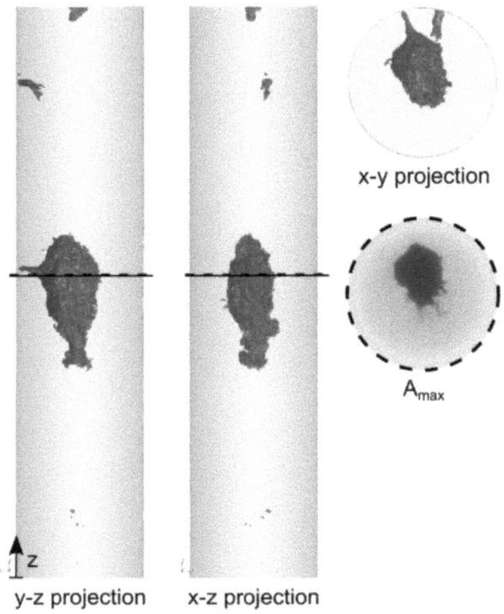


| $A_{proj,CT}$ [mm²]; [%] | $A_{max,CT}$ [mm²]; [%] | $z_{max,CT}$ [mm] |
|---|---|---|
| 1 | 33.5; 13 | 42.3 |

| $V_{max}/V_{tot}$ [%] | $A_{fail,mic}$ [mm]; [%] | $z_{fail}$ [mm] |
|---|---|---|
| 98 | 41.0; 16 | 42.3 |

| $\sigma_{max}/\sigma_y$ [%] | $\sigma_{max}/\sigma_u$ [%] |
|---|---|
| - | - |

Upper Fracture Surface

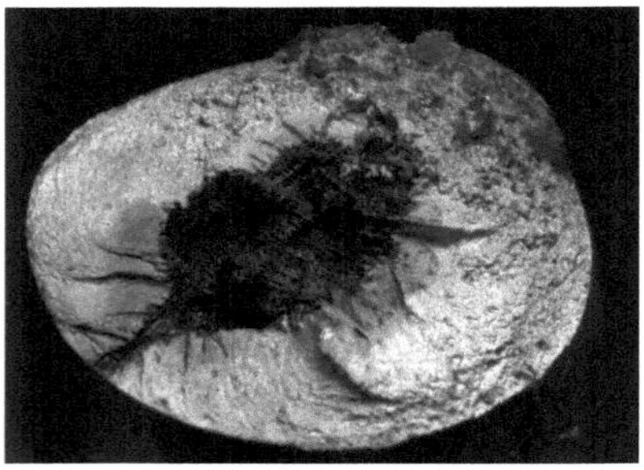

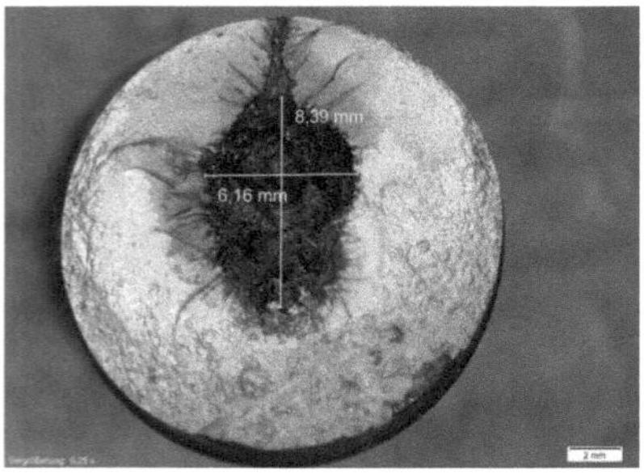


Lower Fracture Surface

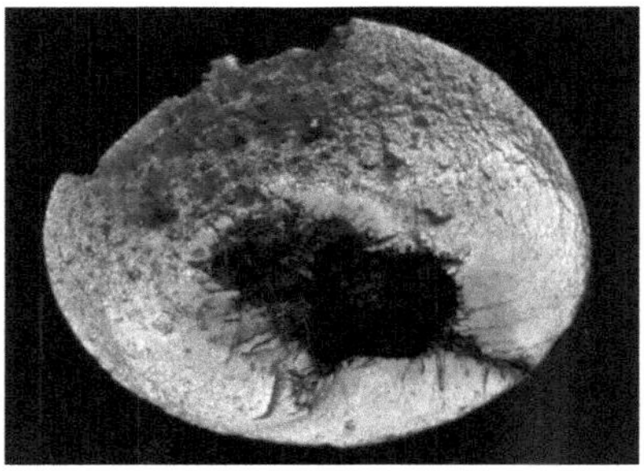

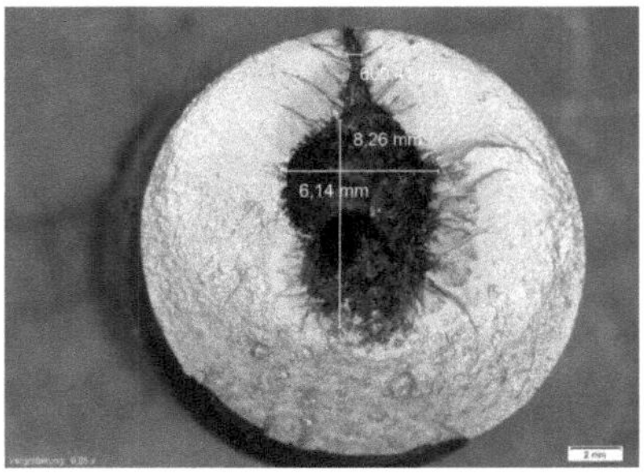

## U22O1

| $\Delta\sigma_{nom}$ [MPa] | $\Delta\sigma_{mic}$ [MPa] | cycles to failure [-] | foundry [-] | $\sigma_y$ [MPa] | $\sigma_u$ [MPa] | defect pos. | failure case |
|---|---|---|---|---|---|---|---|
| 310 | 361 | 78,784 | D | 889 | 997 | int. | III |

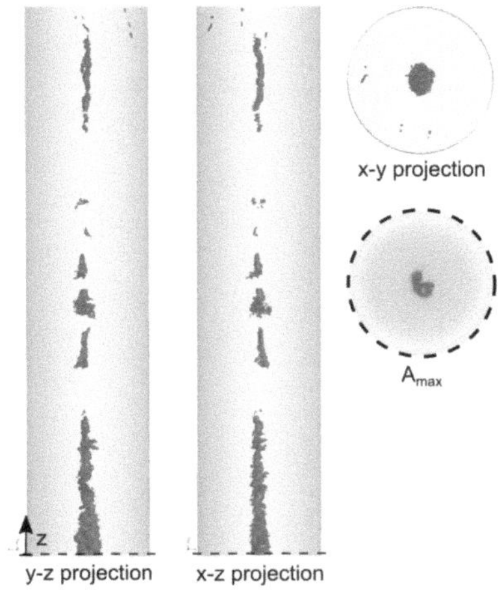


| $A_{proj,CT}$ [mm²]; [%] | $A_{max,CT}$ [mm²]; [%] | $z_{max,CT}$ [mm] |
|---|---|---|
| 1 | 5.6; 2 | -5.5 |

| $V_{max}/V_{tot}$ [%] | $A_{fail,mic}$ [mm]; [%] | $z_{fail}$ [mm] |
|---|---|---|
| 69 | 36.0; 14 | 0.0 |

| $\sigma_{max}/\sigma_y$ [%] | $\sigma_{max}/\sigma_u$ [%] |
|---|---|
| 105 | 94 |

Upper Fracture Surface

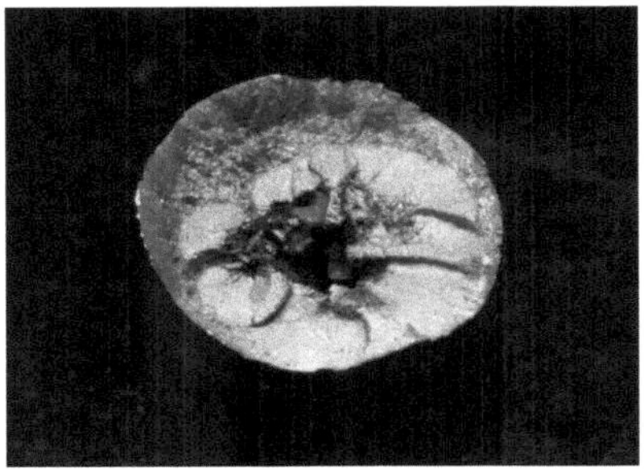

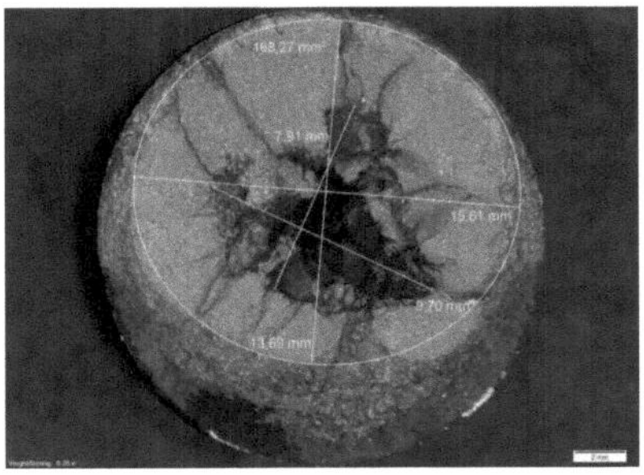

Lower Fracture Surface

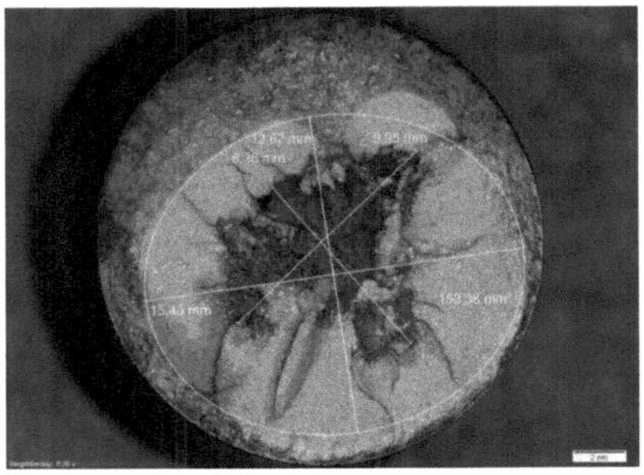

## U22O2

| $\Delta\sigma_{nom}$ [MPa] | $\Delta\sigma_{mic}$ [MPa] | cycles to failure [-] | foundry [-] | $\sigma_y$ [MPa] | $\sigma_u$ [MPa] | defect pos. | failure case |
|---|---|---|---|---|---|---|---|
| 400 | 451 | 45,327 | D | 889 | 997 | int. | III |

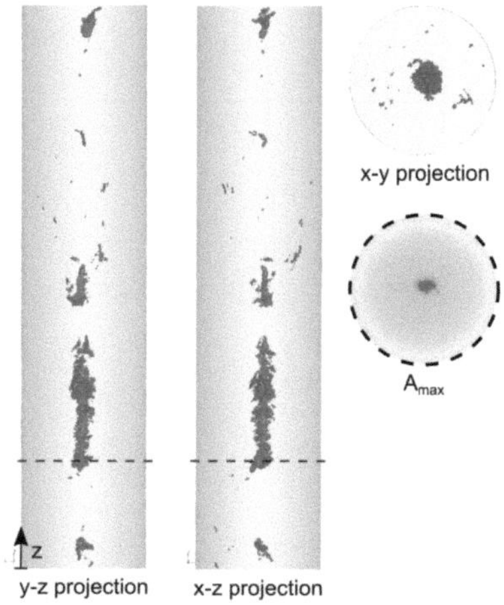


| $A_{proj,CT}$ [mm²]; [%] | $A_{max,CT}$ [mm²]; [%] | $z_{max,CT}$ [mm] |
|---|---|---|
| 1 | 2.9; 1 | 90.0 |

| $V_{max}/V_{tot}$ [%] | $A_{fail,mic}$ [mm]; [%] | $z_{fail}$ [mm] |
|---|---|---|
| 68 | 28.7; 11 | 15.2 |

| $\sigma_{max}/\sigma_y$ [%] | $\sigma_{max}/\sigma_u$ [%] |
|---|---|
| 143 | 127 |

Upper Fracture Surface

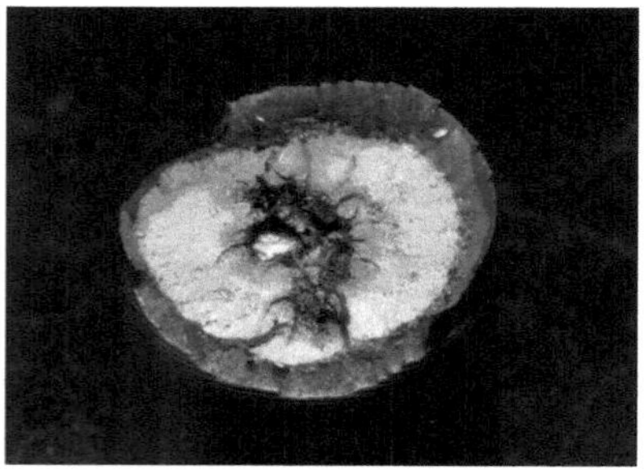

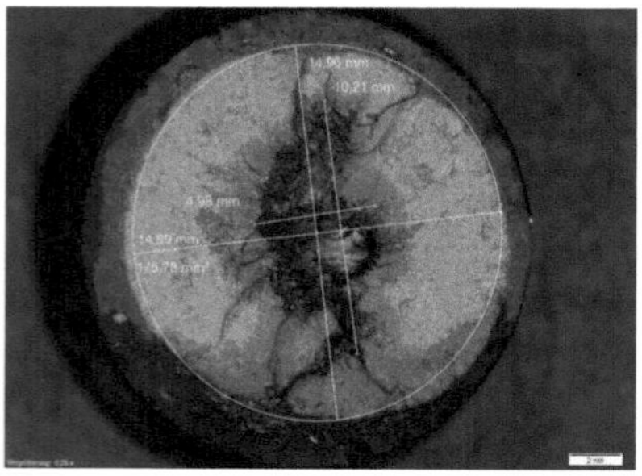

Lower Fracture Surface

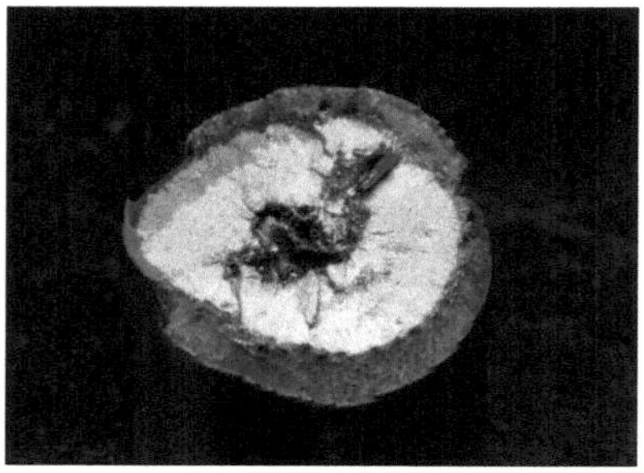

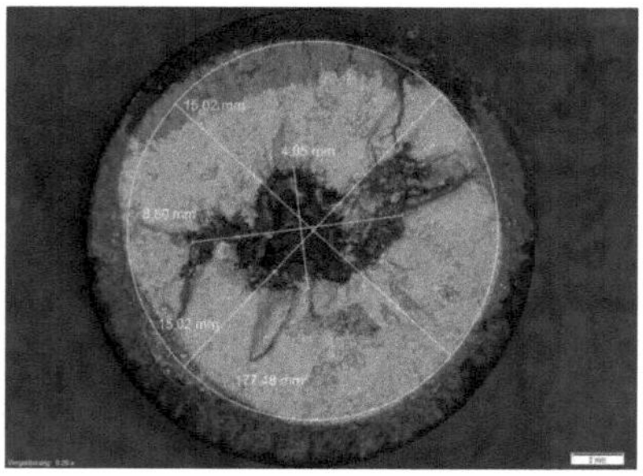

## U22O3

| $\Delta\sigma_{nom}$ [MPa] | $\Delta\sigma_{mic}$ [MPa] | cycles to failure [-] | foundry [-] | $\sigma_y$ [MPa] | $\sigma_u$ [MPa] | defect pos. | failure case |
|---|---|---|---|---|---|---|---|
| 310 | 346 | 522,615 | D | 889 | 997 | int. | I |

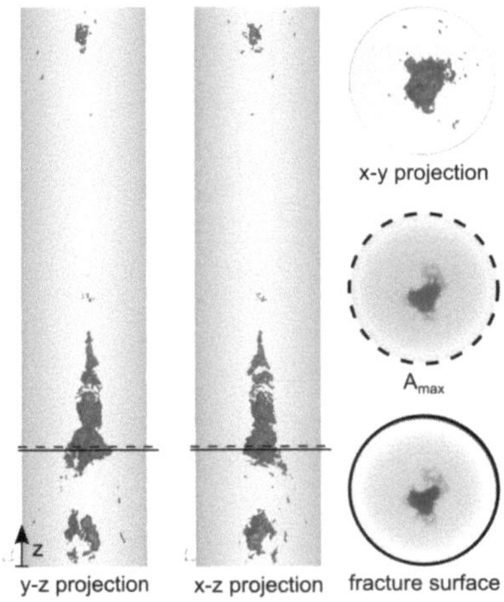


| $A_{proj,CT}$ [mm²]; [%] | $A_{max,CT}$ [mm²]; [%] | $z_{max,CT}$ [mm] |
|---|---|---|
| 1 | 12.9; 5 | 16.5 |

| $V_{max}/V_{tot}$ [%] | $A_{fail,mic}$ [mm]; [%] | $z_{fail}$ [mm] |
|---|---|---|
| 64 | 26.4; 10 | 16.5 |

| $\sigma_{max}/\sigma_y$ [%] | $\sigma_{max}/\sigma_u$ [%] |
|---|---|
| 78 | 70 |

Upper Fracture Surface

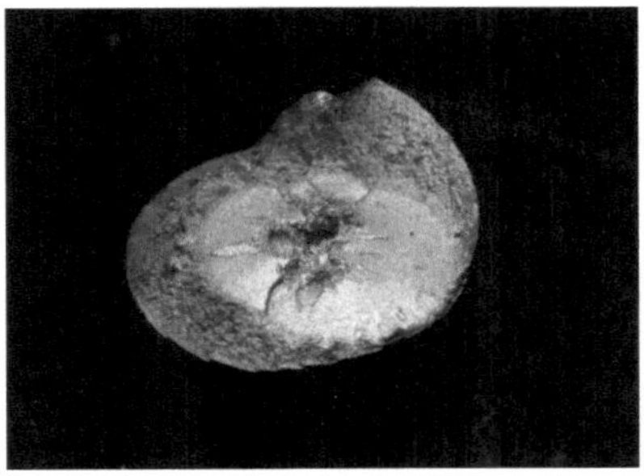

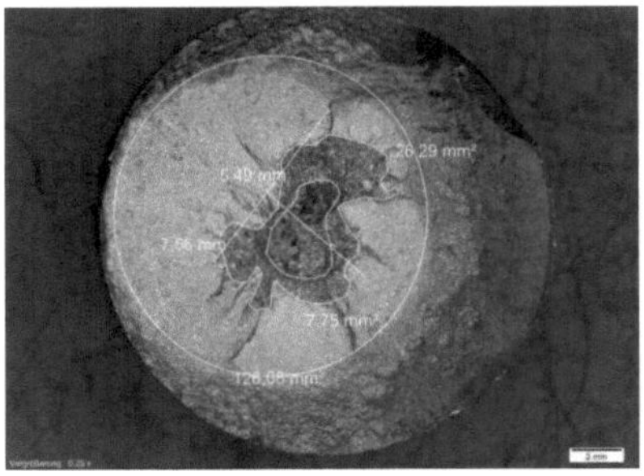

Lower Fracture Surface

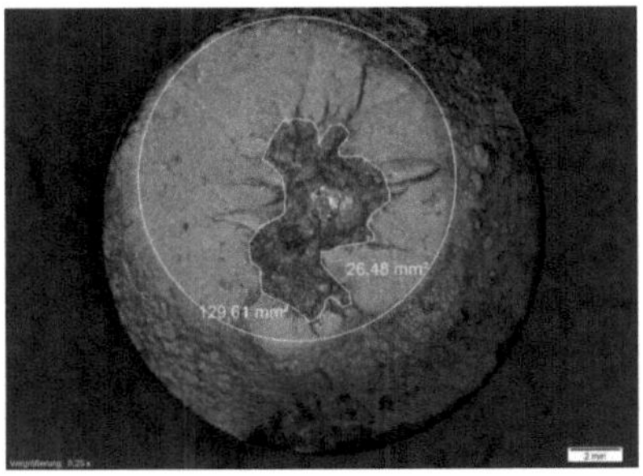

## U22O4

| $\Delta\sigma_{nom}$ [MPa] | $\Delta\sigma_{mic}$ [MPa] | cycles to failure [-] | foundry [-] | $\sigma_y$ [MPa] | $\sigma_u$ [MPa] | defect pos. | failure case |
|---|---|---|---|---|---|---|---|
| 230 | 254 | 1,677,207 | D | 889 | 997 | int. | III |

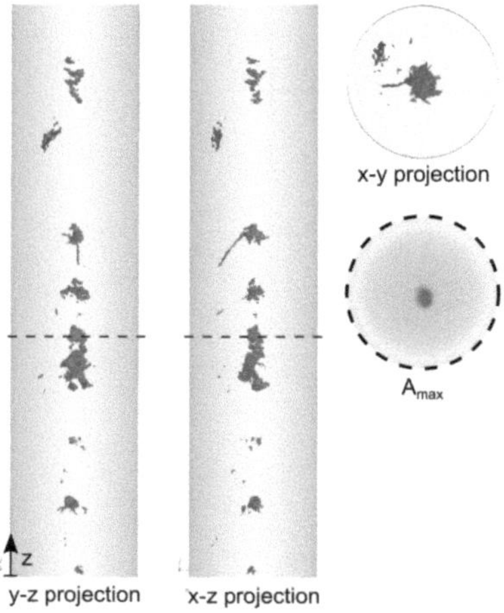


| $A_{proj,CT}$ [mm²]; [%] | $A_{max,CT}$ [mm²]; [%] | $z_{max,CT}$ [mm] |
|---|---|---|
| 1 | 4.3; 2 | -5.0 |

| $V_{max}/V_{tot}$ [%] | $A_{fail,mic}$ [mm]; [%] | $z_{fail}$ [mm] |
|---|---|---|
| 52 | 24.4; 10 | 33.4 |

| $\sigma_{max}/\sigma_y$ [%] | $\sigma_{max}/\sigma_u$ [%] |
|---|---|
| 83 | 74 |

Upper Fracture Surface

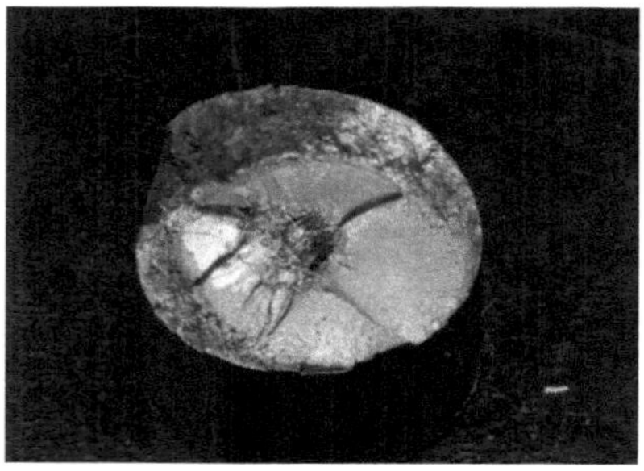

Lower Fracture Surface

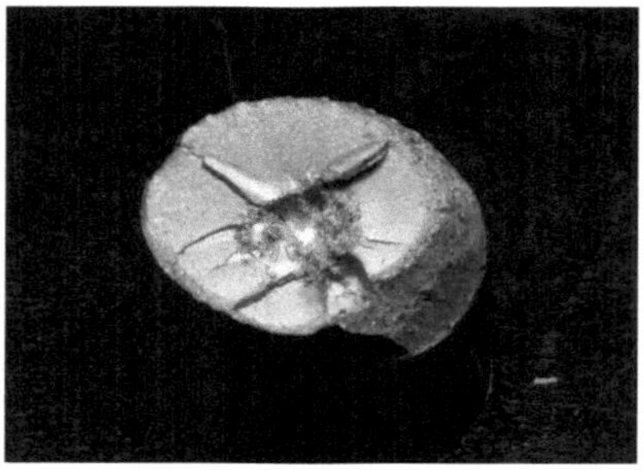

## U22O5

| $\Delta\sigma_{nom}$ [MPa] | $\Delta\sigma_{mic}$ [MPa] | cycles to failure [-] | foundry [-] | $\sigma_y$ [MPa] | $\sigma_u$ [MPa] | defect pos. | failure case |
|---|---|---|---|---|---|---|---|
| 240 | 263 | 1,922,456 | D | 889 | 997 | int. | I |

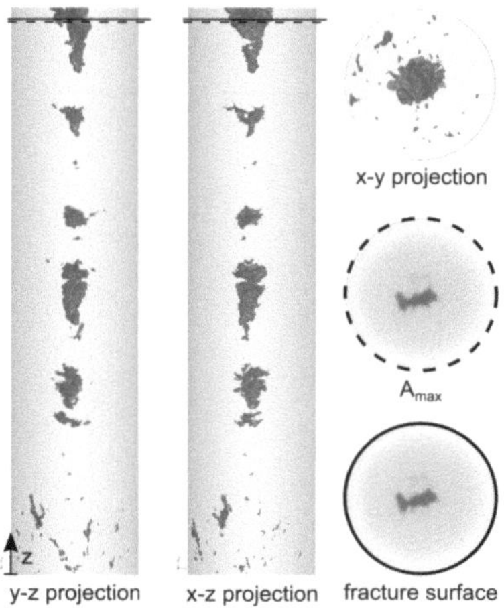


| $A_{proj,CT}$ [mm²]; [%] | $A_{max,CT}$ [mm²]; [%] | $z_{max,CT}$ [mm] |
|---|---|---|
| 1 | 7.0; 3 | 78.0 |

| $V_{max}/V_{tot}$ [%] | $A_{fail,mic}$ [mm]; [%] | $z_{fail}$ [mm] |
|---|---|---|
| 36 | 22.2; 9 | 77.9 |

| $\sigma_{max}/\sigma_y$ [%] | $\sigma_{max}/\sigma_u$ [%] |
|---|---|
| 84 | 75 |

Upper Fracture Surface

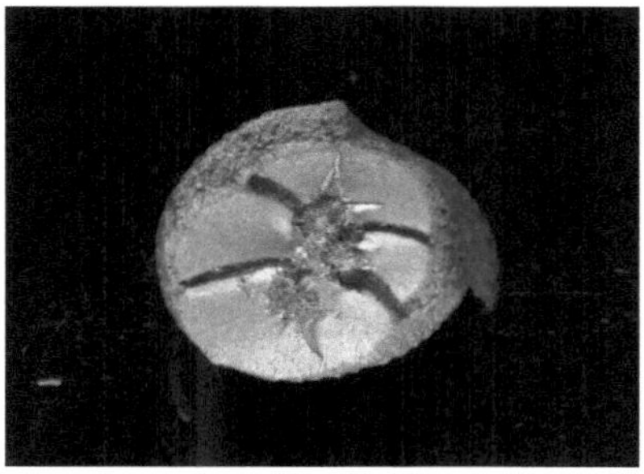

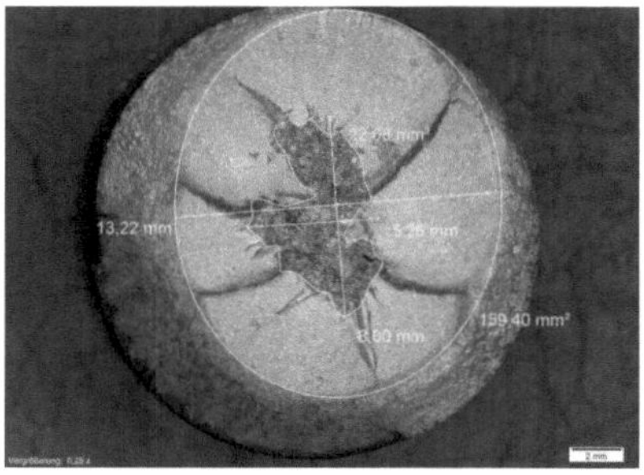


Lower Fracture Surface

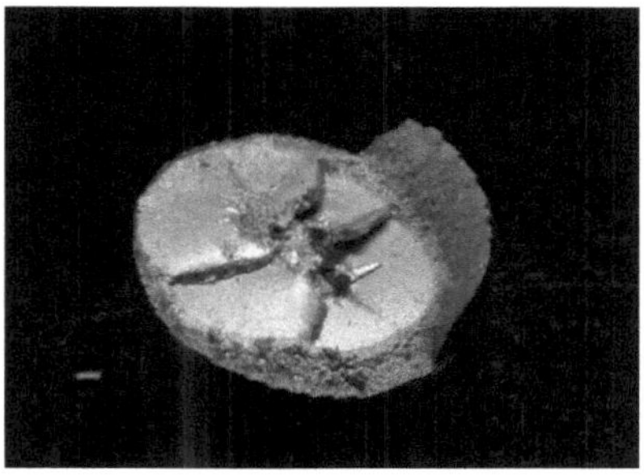

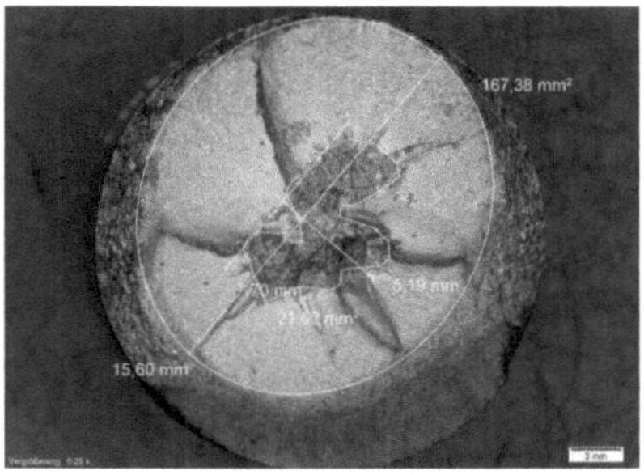

## U22O6

| $\Delta\sigma_{nom}$ [MPa] | $\Delta\sigma_{mic}$ [MPa] | cycles to failure [-] | foundry [-] | $\sigma_y$ [MPa] | $\sigma_u$ [MPa] | defect pos. | failure case |
|---|---|---|---|---|---|---|---|
| 320 | 366 | 248,296 | D | 889 | 997 | int. | III |

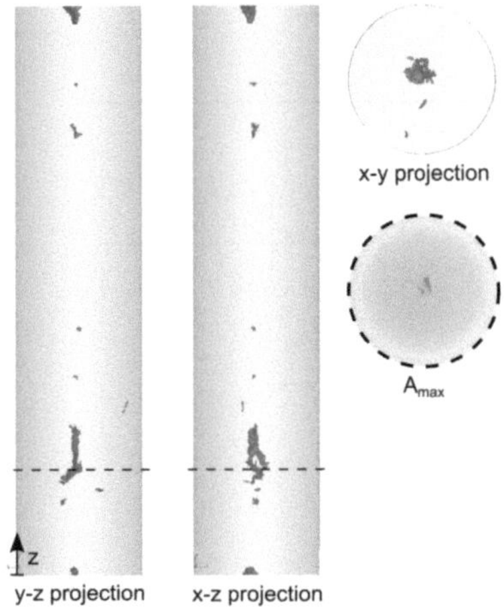


| $A_{proj,CT}$ [mm²]; [%] | $A_{max,CT}$ [mm²]; [%] | $z_{max,CT}$ [mm] |
|---|---|---|
| 1 | 1.4; 1 | 90.0 |

| $V_{max}/V_{tot}$ [%] | $A_{fail,mic}$ [mm]; [%] | $z_{fail}$ [mm] |
|---|---|---|
| 43 | 32.3; 13 | 14.6 |

| $\sigma_{max}/\sigma_y$ [%] | $\sigma_{max}/\sigma_u$ [%] |
|---|---|
| 78 | 70 |

Upper Fracture Surface

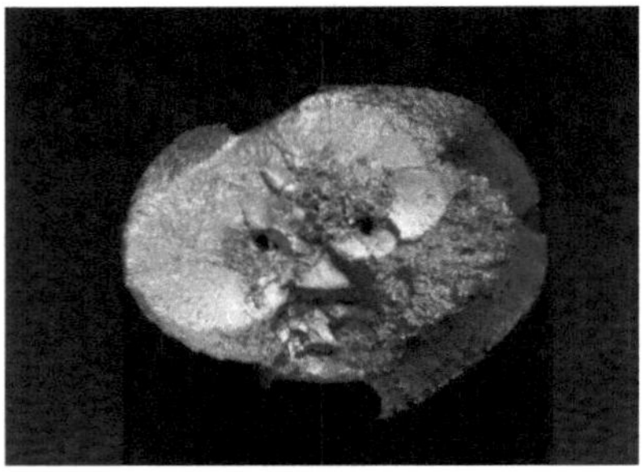

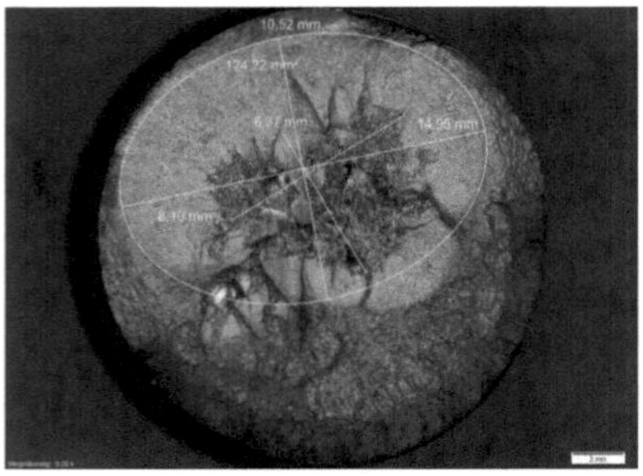

Lower Fracture Surface

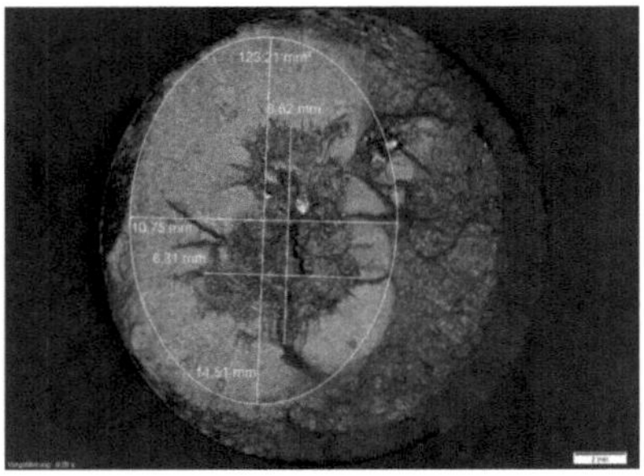

## U22O7

| $\Delta\sigma_{nom}$ [MPa] | $\Delta\sigma_{mic}$ [MPa] | cycles to failure [-] | foundry [-] | $\sigma_y$ [MPa] | $\sigma_u$ [MPa] | defect pos. | failure case |
|---|---|---|---|---|---|---|---|
| 500 | 550 | 51,507 | D | 889 | 997 | int. | II |

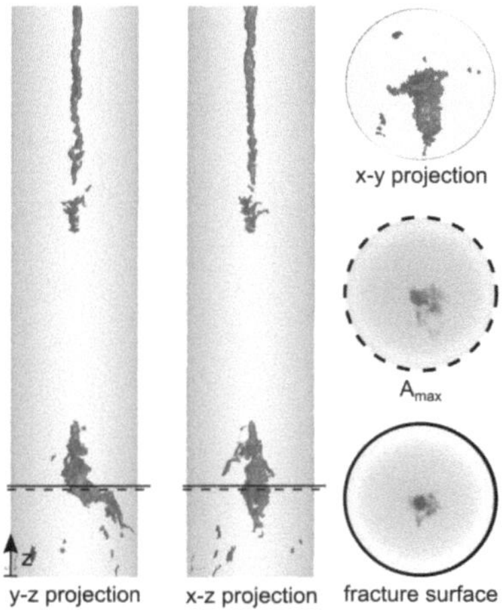


| $A_{proj,CT}$ [mm²]; [%] | $A_{max,CT}$ [mm²]; [%] | $z_{max,CT}$ [mm] |
|---|---|---|
| 1 | 10.0; 4 | 12.5 |

| $V_{max}/V_{tot}$ [%] | $A_{fail,mic}$ [mm]; [%] | $z_{fail}$ [mm] |
|---|---|---|
| 52 | 23.2; 9 | 12.0 |

| $\sigma_{max}/\sigma_y$ [%] | $\sigma_{max}/\sigma_u$ [%] |
|---|---|
| 103 | 92 |

Upper Fracture Surface

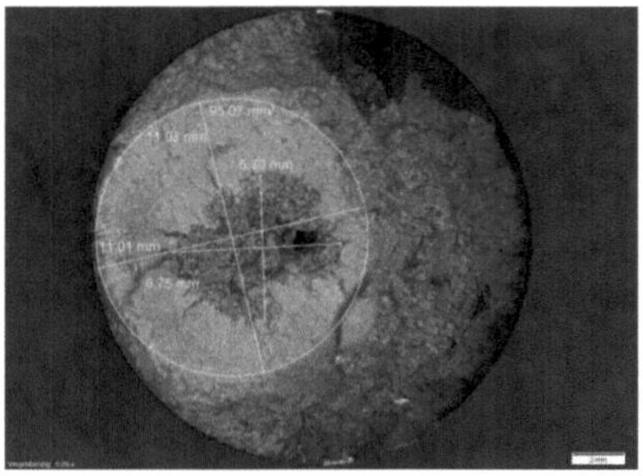

Lower Fracture Surface

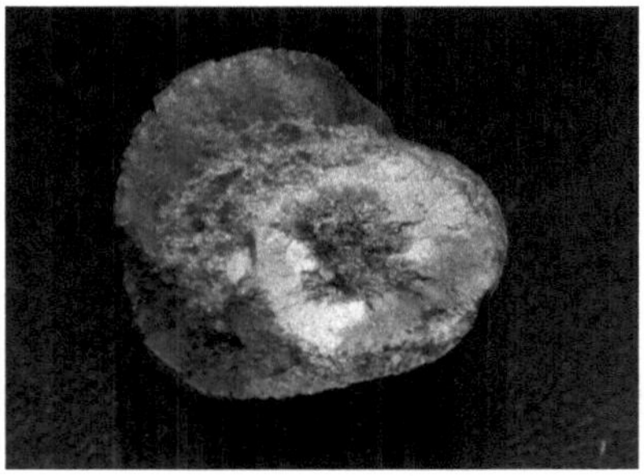

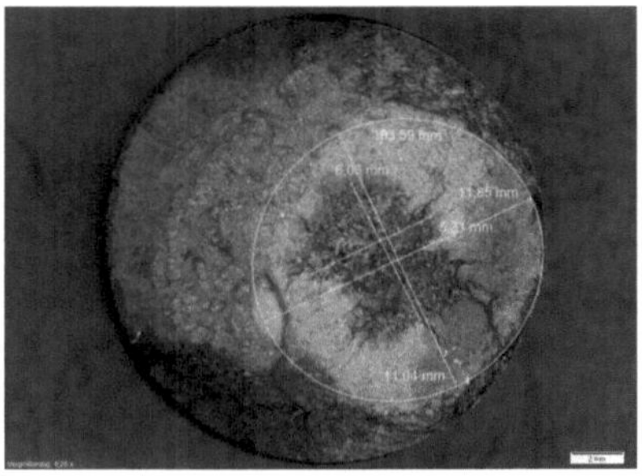

## U22O8

| $\Delta\sigma_{nom}$ [MPa] | $\Delta\sigma_{mic}$ [MPa] | cycles to failure [-] | foundry [-] | $\sigma_y$ [MPa] | $\sigma_u$ [MPa] | defect pos. | failure case |
|---|---|---|---|---|---|---|---|
| 280 | 281 | 2,212,114 | D | 889 | 997 | surf. | III |

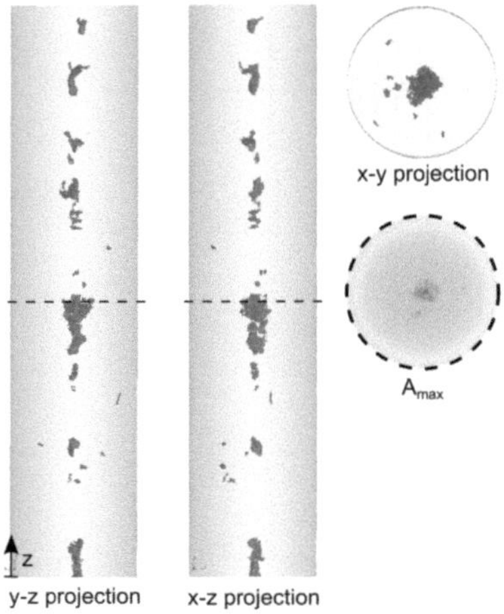


| $A_{proj,CT}$ [mm²]; [%] | $A_{max,CT}$ [mm²]; [%] | $z_{max,CT}$ [mm] |
|---|---|---|
| 1 | 4.3; 2 | 107.0 |

| $V_{max}/V_{tot}$ [%] | $A_{fail,mic}$ [mm]; [%] | $z_{fail}$ [mm] |
|---|---|---|
| 46 | 0.5; 0 | 38.4 |

| $\sigma_{max}/\sigma_y$ [%] | $\sigma_{max}/\sigma_u$ [%] |
|---|---|
| - | - |

Upper Fracture Surface

Lower Fracture Surface

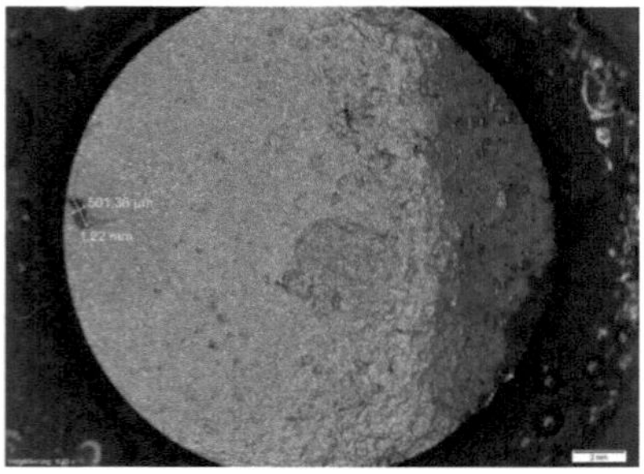

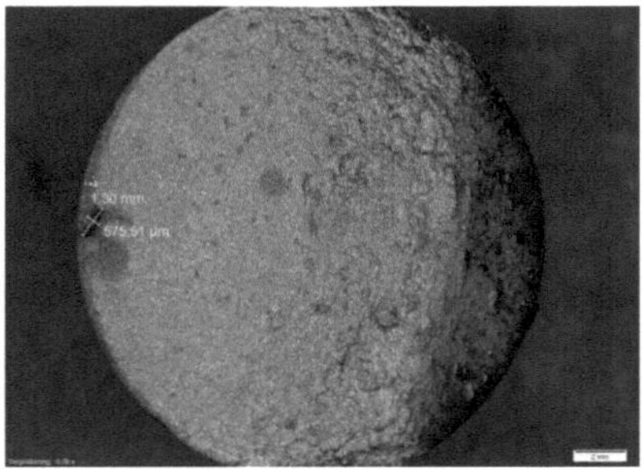

## U22O9

| $\Delta\sigma_{nom}$ [MPa] | $\Delta\sigma_{mic}$ [MPa] | cycles to failure [-] | foundry [-] | $\sigma_y$ [MPa] | $\sigma_u$ [MPa] | defect pos. | failure case |
|---|---|---|---|---|---|---|---|
| 320 | 346 | 457,191 | D | 889 | 997 | int. | III |

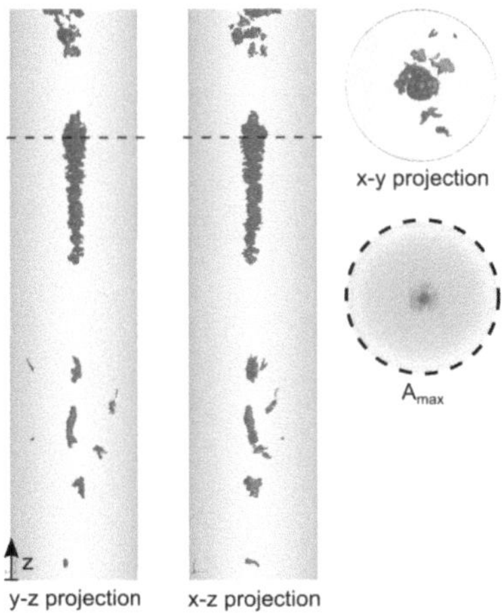


| $A_{proj,CT}$ [mm²]; [%] | $A_{max,CT}$ [mm²]; [%] | $z_{max,CT}$ [mm] |
|---|---|---|
| 1 | 6.2; 2 | 83.0 |

| $V_{max}/V_{tot}$ [%] | $A_{fail,mic}$ [mm]; [%] | $z_{fail}$ [mm] |
|---|---|---|
| 69 | 19.1; 7 | 61.9 |

| $\sigma_{max}/\sigma_y$ [%] | $\sigma_{max}/\sigma_u$ [%] |
|---|---|
| 62 | 55 |

Upper Fracture Surface

Lower Fracture Surface

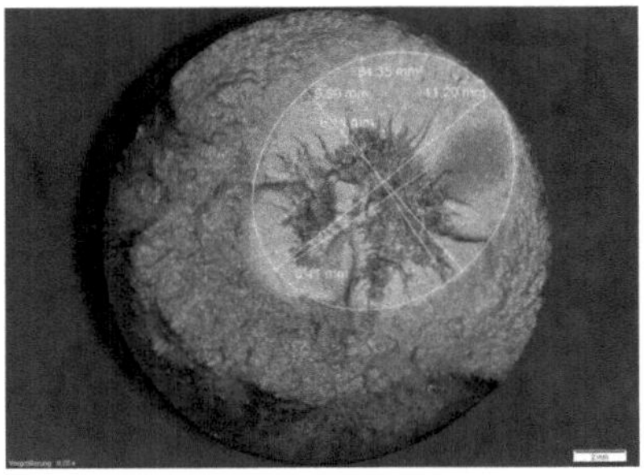

## U22O11

| $\Delta\sigma_{nom}$ [MPa] | $\Delta\sigma_{mic}$ [MPa] | cycles to failure [-] | foundry [-] | $\sigma_y$ [MPa] | $\sigma_u$ [MPa] | defect pos. | failure case |
|---|---|---|---|---|---|---|---|
| 400 | 429 | 227,264 | E | 889 | 997 | int. | II |

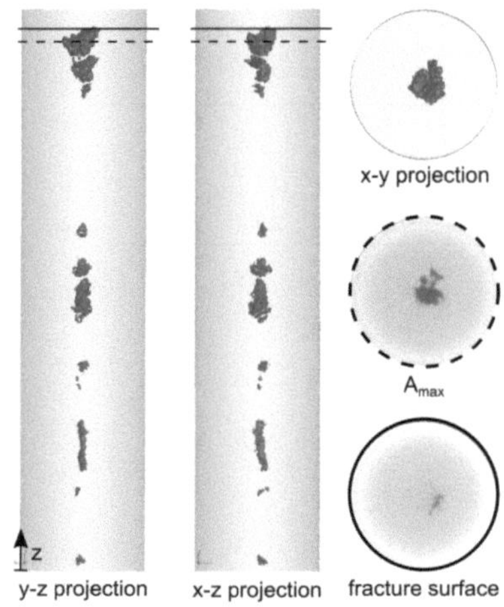


| $A_{proj,CT}$ [mm²]; [%] | $A_{max,CT}$ [mm²]; [%] | $z_{max,CT}$ [mm] |
|---|---|---|
| 1 | 8; 3 | 77.5 |

| $V_{max}/V_{tot}$ [%] | $A_{fail,mic}$ [mm]; [%] | $z_{fail}$ [mm] |
|---|---|---|
| 39 | 17.3; 7 | 75.2 |

| $\sigma_{max}/\sigma_y$ [%] | $\sigma_{max}/\sigma_u$ [%] |
|---|---|
| 89 | 79 |

Upper Fracture Surface

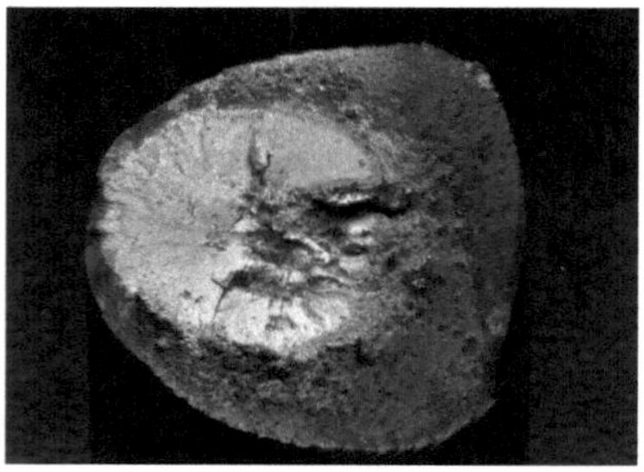

Lower Fracture Surface

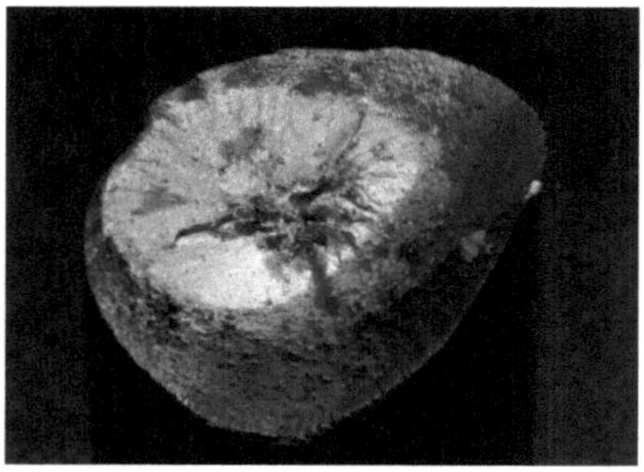

## U22O12

| $\Delta\sigma_{nom}$ [MPa] | $\Delta\sigma_{mic}$ [MPa] | cycles to failure [-] | foundry [-] | $\sigma_y$ [MPa] | $\sigma_u$ [MPa] | defect pos. | failure case |
|---|---|---|---|---|---|---|---|
| 450 | 475 | 214,567 | E | 889 | 997 | int. | III |

no data

| $A_{proj,CT}$ [mm²]; [%] | $A_{max,CT}$ [mm²]; [%] | $z_{max,CT}$ [mm] |
|---|---|---|
| 1 | -; - | 81.4 |

| $V_{max}/V_{tot}$ [%] | $A_{fail,mic}$ [mm]; [%] | $z_{fail}$ [mm] |
|---|---|---|
| - | 13.3; 5 | - |

| $\sigma_{max}/\sigma_y$ [%] | $\sigma_{max}/\sigma_u$ [%] |
|---|---|
| 104 | 93 |

Upper Fracture Surface

Lower Fracture Surface

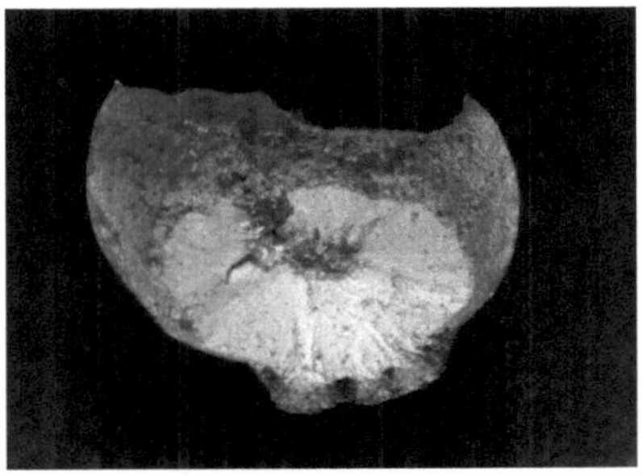

## U22O14

| $\Delta\sigma_{nom}$ [MPa] | $\Delta\sigma_{mic}$ [MPa] | cycles to failure [-] | foundry [-] | $\sigma_y$ [MPa] | $\sigma_u$ [MPa] | defect pos. | failure case |
|---|---|---|---|---|---|---|---|
| 260 | 268 | 4,947,995 | E | 889 | 997 | int. | III |

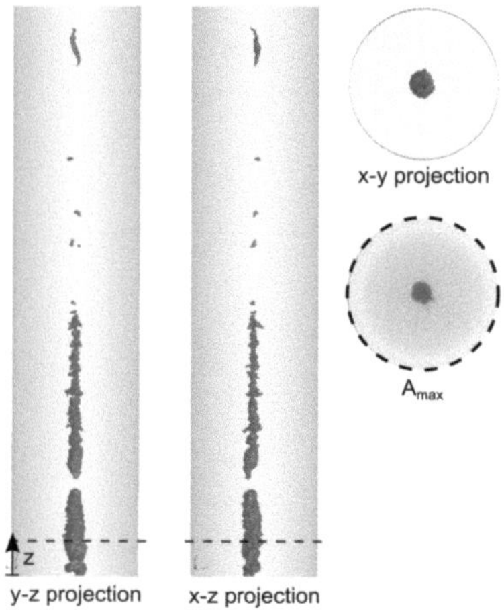


| $A_{proj,CT}$ [mm²]; [%] | $A_{max,CT}$ [mm²]; [%] | $z_{max,CT}$ [mm] |
|---|---|---|
| 1 | 5.6; 2 | -1.4 |

| $V_{max}/V_{tot}$ [%] | $A_{fail,mic}$ [mm]; [%] | $z_{fail}$ [mm] |
|---|---|---|
| 61 | 7.6; 3 | 4.7 |

| $\sigma_{max}/\sigma_y$ [%] | $\sigma_{max}/\sigma_u$ [%] |
|---|---|
| 72 | 65 |

Upper Fracture Surface

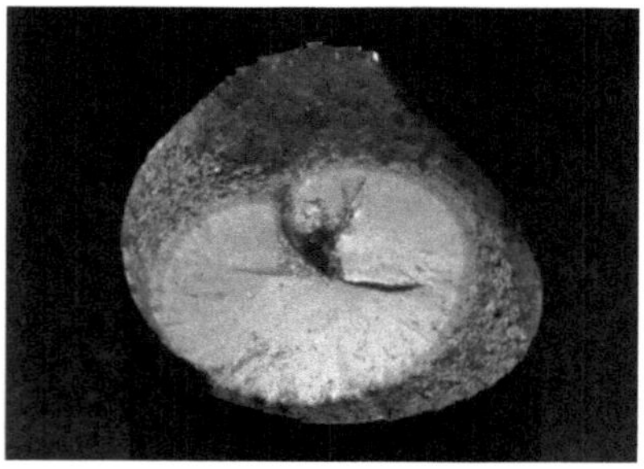

Lower Fracture Surface

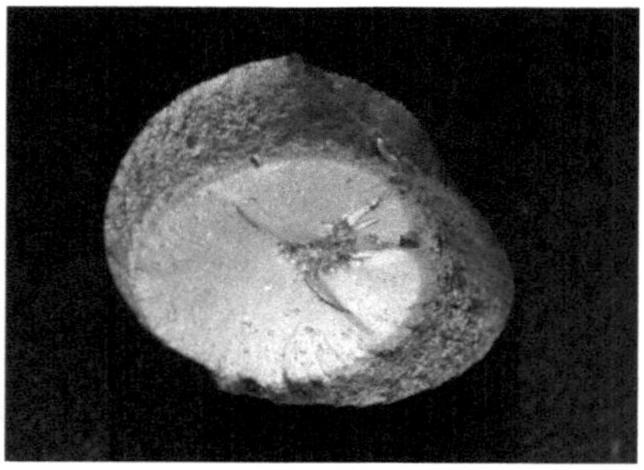

## U22O15

| $\Delta\sigma_{nom}$ [MPa] | $\Delta\sigma_{mic}$ [MPa] | cycles to failure [-] | foundry [-] | $\sigma_y$ [MPa] | $\sigma_u$ [MPa] | defect pos. | failure case |
|---|---|---|---|---|---|---|---|
| 300 | 315 | 1,304,602 | E | 889 | 997 | int. | III |

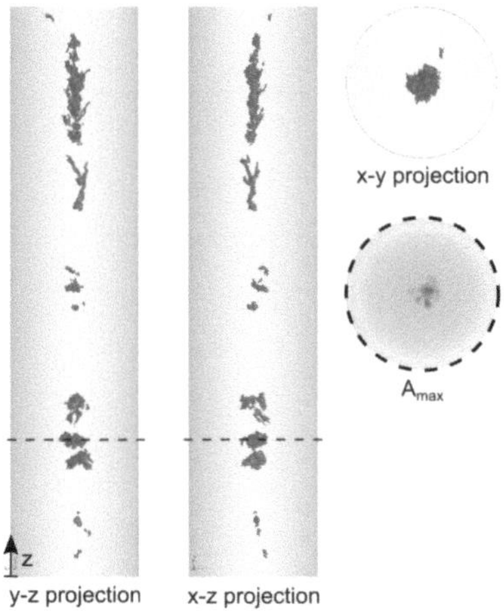


| $A_{proj,CT}$ [mm²]; [%] | $A_{max,CT}$ [mm²]; [%] | $z_{max,CT}$ [mm] |
|---|---|---|
| 1 | 5.6; 2 | -17.0 |

| $V_{max}/V_{tot}$ [%] | $A_{fail,mic}$ [mm]; [%] | $z_{fail}$ [mm] |
|---|---|---|
| 34 | 12.4; 5 | 19.1 |

| $\sigma_{max}/\sigma_y$ [%] | $\sigma_{max}/\sigma_u$ [%] |
|---|---|
| 72 | 64 |

Upper Fracture Surface

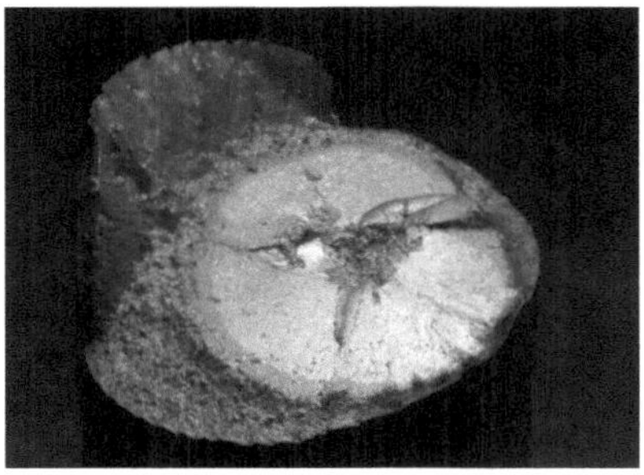

Lower Fracture Surface

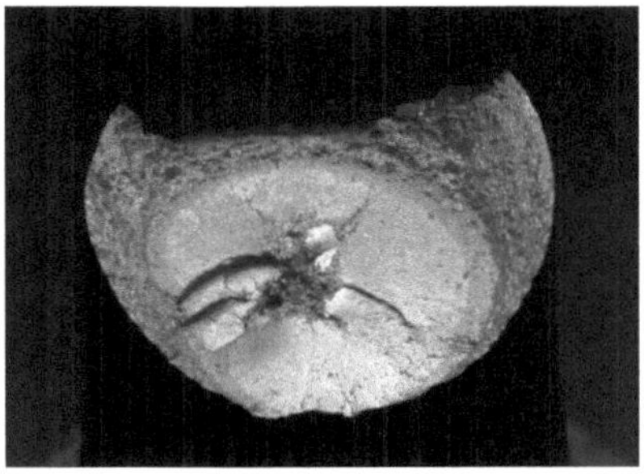

## U22O16

| $\Delta\sigma_{nom}$ [MPa] | $\Delta\sigma_{mic}$ [MPa] | cycles to failure [-] | foundry [-] | $\sigma_y$ [MPa] | $\sigma_u$ [MPa] | defect pos. | failure case |
|---|---|---|---|---|---|---|---|
| 260 | 273 | 2,062,405 | E | 889 | 997 | int. | III |

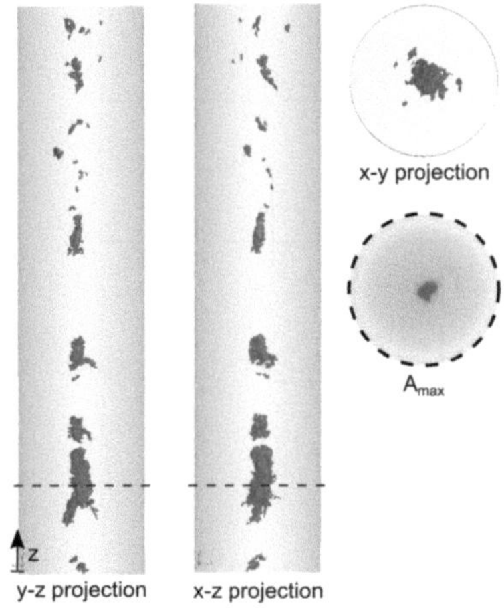


| $A_{proj,CT}$ [mm²]; [%] | $A_{max,CT}$ [mm²]; [%] | $z_{max,CT}$ [mm] |
|---|---|---|
| 1 | 4.2; 2 | -17.1 |

| $V_{max}/V_{tot}$ [%] | $A_{fail,mic}$ [mm]; [%] | $z_{fail}$ [mm] |
|---|---|---|
| 56 | 12.2; 5 | 12.0 |

| $\sigma_{max}/\sigma_y$ [%] | $\sigma_{max}/\sigma_u$ [%] |
|---|---|
| 64 | 57 |

Upper Fracture Surface

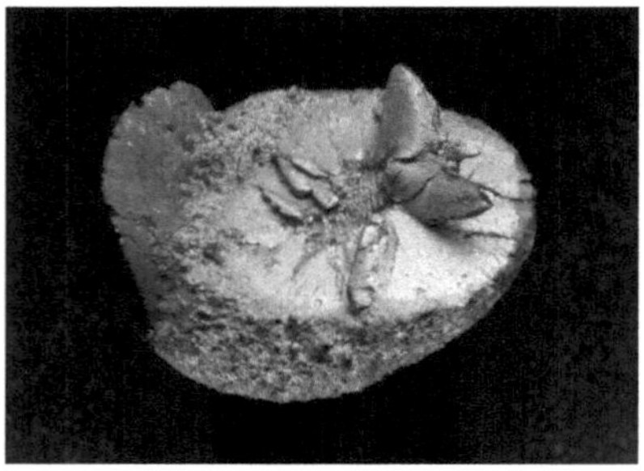

Lower Fracture Surface

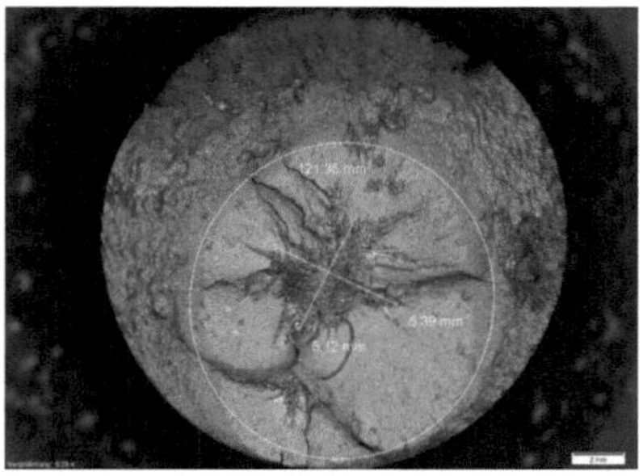

## U22O17

| $\Delta\sigma_{nom}$ [MPa] | $\Delta\sigma_{mic}$ [MPa] | cycles to failure [-] | foundry [-] | $\sigma_y$ [MPa] | $\sigma_u$ [MPa] | defect pos. | failure case |
|---|---|---|---|---|---|---|---|
| 360 | 361 | 262,987 | E | 889 | 997 | surf. | II |

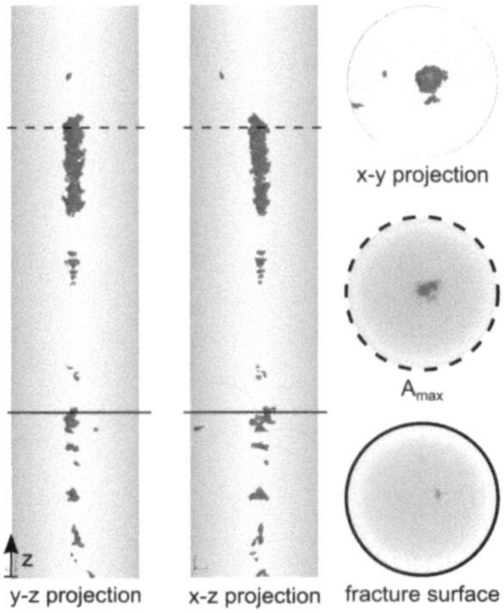


| $A_{proj,CT}$ [mm²]; [%] | $A_{max,CT}$ [mm²]; [%] | $z_{max,CT}$ [mm] |
|---|---|---|
| 1 | 3.7; 1 | 23.0 |

| $V_{max}/V_{tot}$ [%] | $A_{fail,mic}$ [mm]; [%] | $z_{fail}$ [mm] |
|---|---|---|
| 75 | 0.8; 0 | 62.7 |

| $\sigma_{max}/\sigma_y$ [%] | $\sigma_{max}/\sigma_u$ [%] |
|---|---|
| - | - |

Upper Fracture Surface

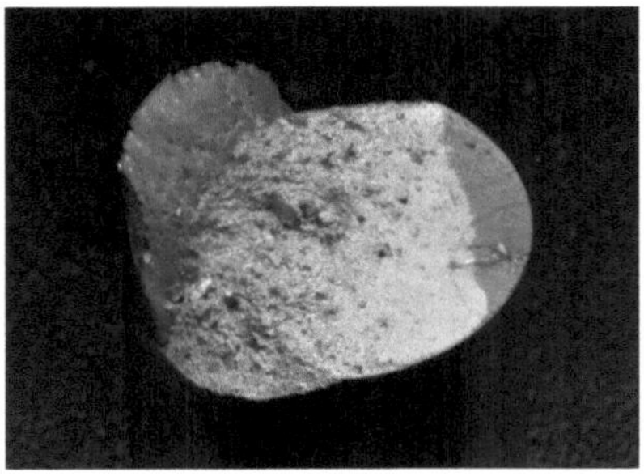

Lower Fracture Surface

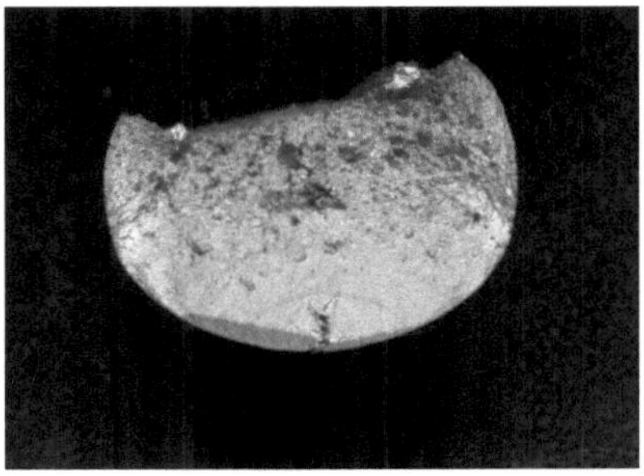

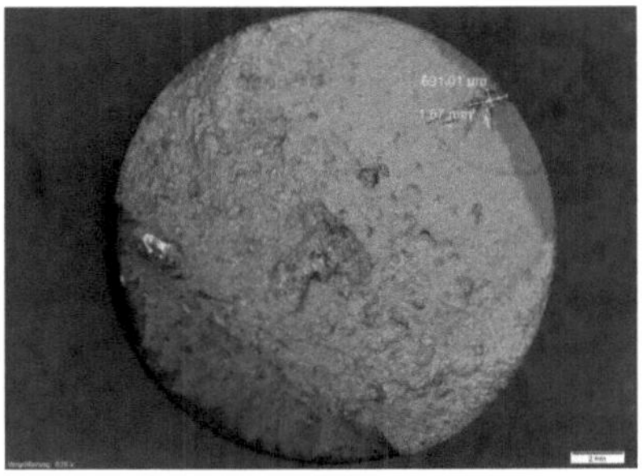

## U22O18

| $\Delta\sigma_{nom}$ [MPa] | $\Delta\sigma_{mic}$ [MPa] | cycles to failure [-] | foundry [-] | $\sigma_y$ [MPa] | $\sigma_u$ [MPa] | defect pos. | failure case |
|---|---|---|---|---|---|---|---|
| 300 | 318 | 1,009,273 | E | 889 | 997 | int. | III |

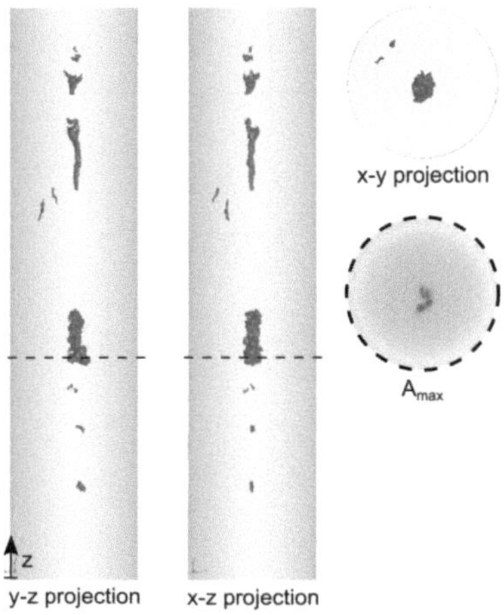


| $A_{proj,CT}$ [mm²]; [%] | $A_{max,CT}$ [mm²]; [%] | $z_{max,CT}$ [mm] |
|---|---|---|
| 1 | 3.3; 1 | -8.0 |

| $V_{max}/V_{tot}$ [%] | $A_{fail,mic}$ [mm]; [%] | $z_{fail}$ [mm] |
|---|---|---|
| 62 | 14.2; 6 | 30.7 |

| $\sigma_{max}/\sigma_y$ [%] | $\sigma_{max}/\sigma_u$ [%] |
|---|---|
| 64 | 57 |

Upper Fracture Surface

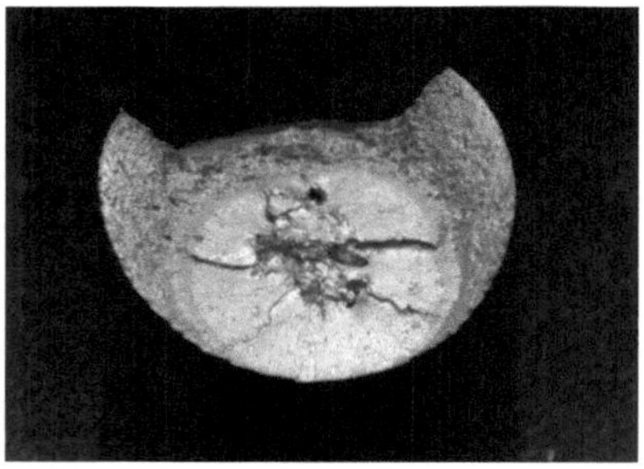

Lower Fracture Surface

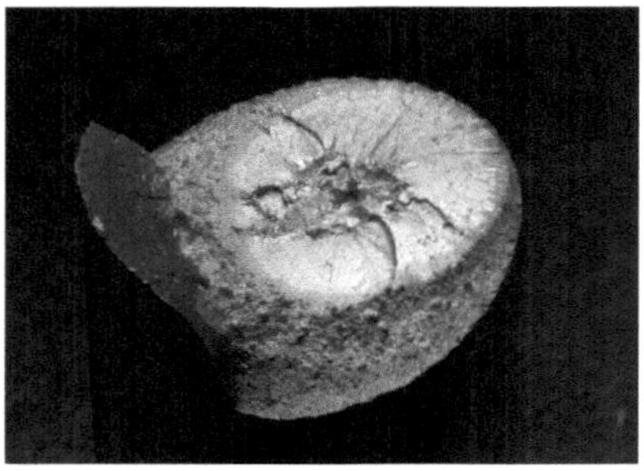

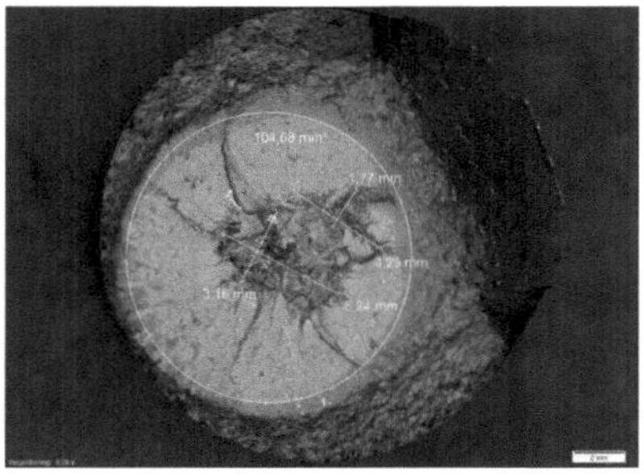

## U22O19

| $\Delta\sigma_{nom}$ [MPa] | $\Delta\sigma_{mic}$ [MPa] | cycles to failure [-] | foundry [-] | $\sigma_y$ [MPa] | $\sigma_u$ [MPa] | defect pos. | failure case |
|---|---|---|---|---|---|---|---|
| 360 | 362 | 130,824 | E | 889 | 997 | surf. | II |

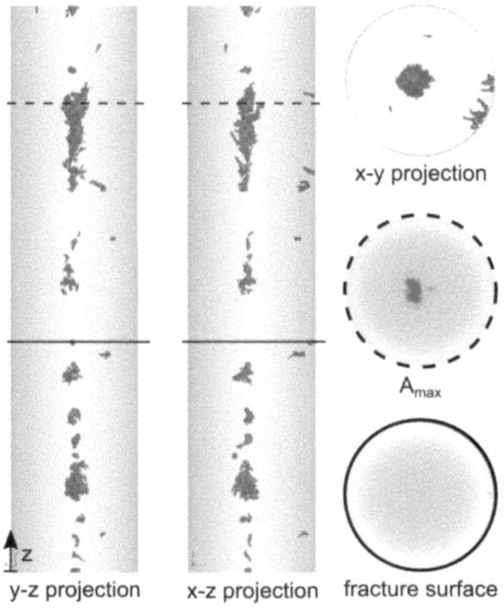


| $A_{proj,CT}$ [mm²]; [%] | $A_{max,CT}$ [mm²]; [%] | $z_{max,CT}$ [mm] |
|---|---|---|
| 1 | 4.3; 2 | 32.3 |

| $V_{max}/V_{tot}$ [%] | $A_{fail,mic}$ [mm]; [%] | $z_{fail}$ [mm] |
|---|---|---|
| 44 | 1.5; 1 | 66 |

| $\sigma_{max}/\sigma_y$ [%] | $\sigma_{max}/\sigma_u$ [%] |
|---|---|
| - | - |

Upper Fracture Surface

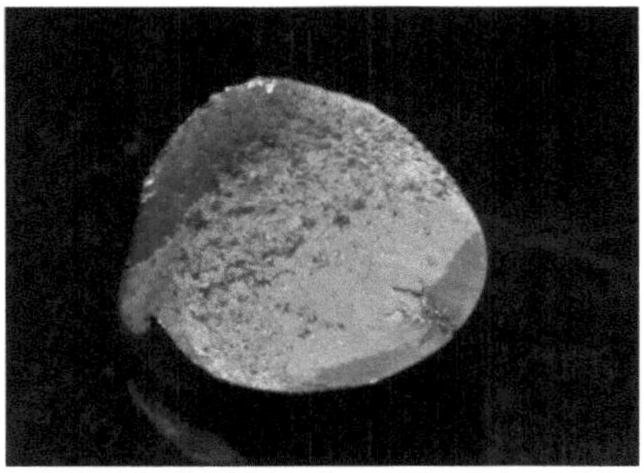

Lower Fracture Surface

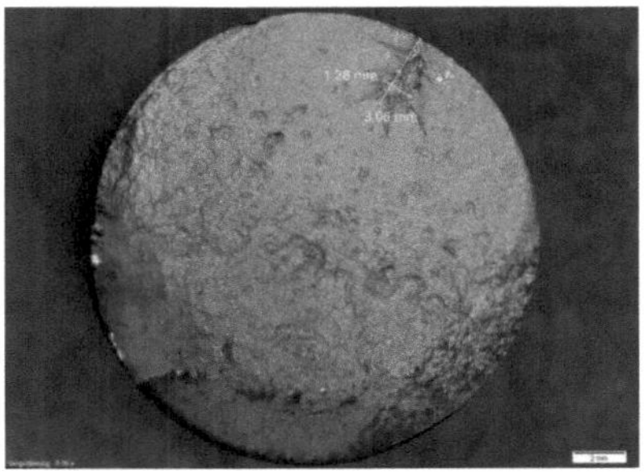

### U22O20

| $\Delta\sigma_{nom}$ [MPa] | $\Delta\sigma_{mic}$ [MPa] | cycles to failure [-] | foundry [-] | $\sigma_y$ [MPa] | $\sigma_u$ [MPa] | defect pos. | failure case |
|---|---|---|---|---|---|---|---|
| 280 | 302 | 860,620 | E | 889 | 997 | int. | III |

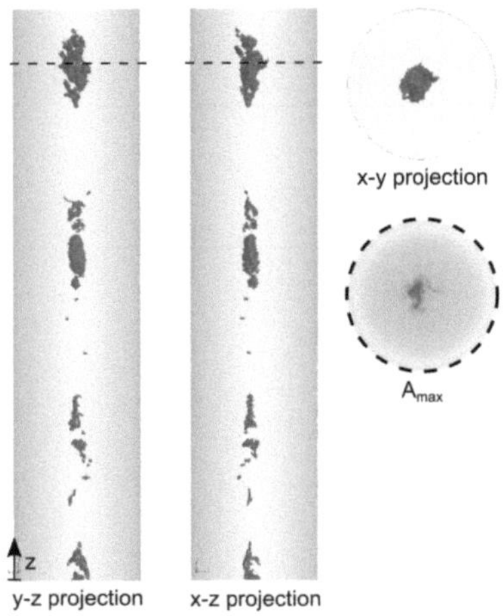


| $A_{proj,CT}$ [mm²]; [%] | $A_{max,CT}$ [mm²]; [%] | $z_{max,CT}$ [mm] |
|---|---|---|
| 1 | 8.3; 3 | 97.7 |

| $V_{max}/V_{tot}$ [%] | $A_{fail,mic}$ [mm]; [%] | $z_{fail}$ [mm] |
|---|---|---|
| - | 18.9; 7 | 72.3 |

| $\sigma_{max}/\sigma_y$ [%] | $\sigma_{max}/\sigma_u$ [%] |
|---|---|
| 66 | 58 |

Upper Fracture Surface

Lower Fracture Surface

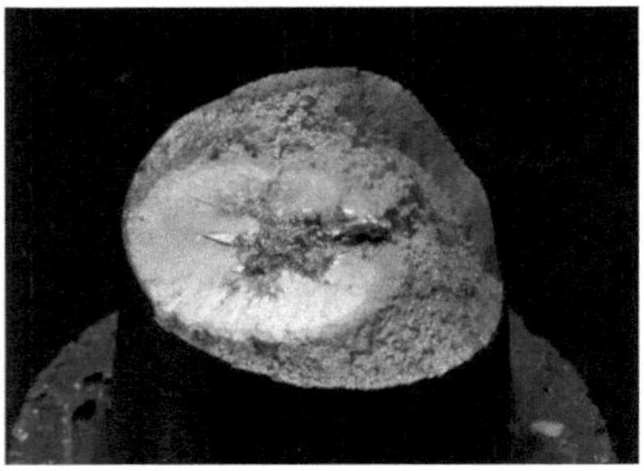

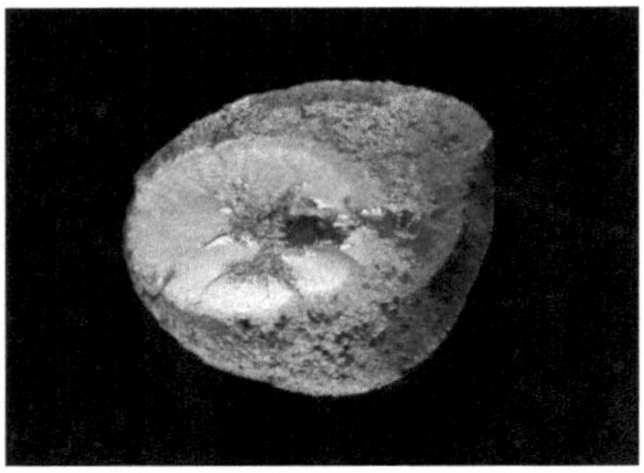

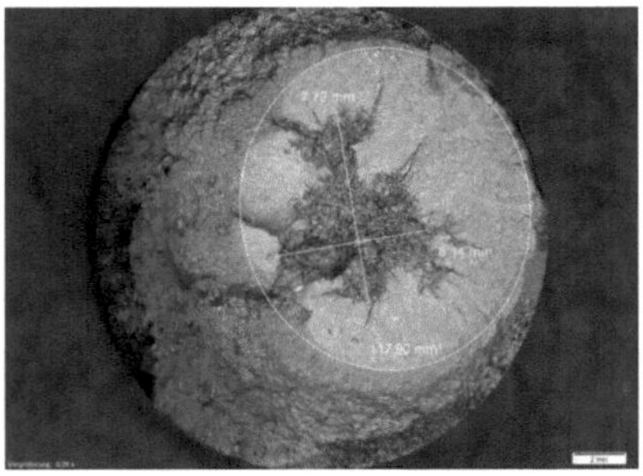

## U22M2

| $\Delta\sigma_{nom}$ [MPa] | $\Delta\sigma_{mic}$ [MPa] | cycles to failure [-] | foundry [-] | $\sigma_y$ [MPa] | $\sigma_u$ [MPa] | defect pos. | failure case |
|---|---|---|---|---|---|---|---|
| 260 | 297 | 113,745 | D | 889 | 997 | surf. | II |

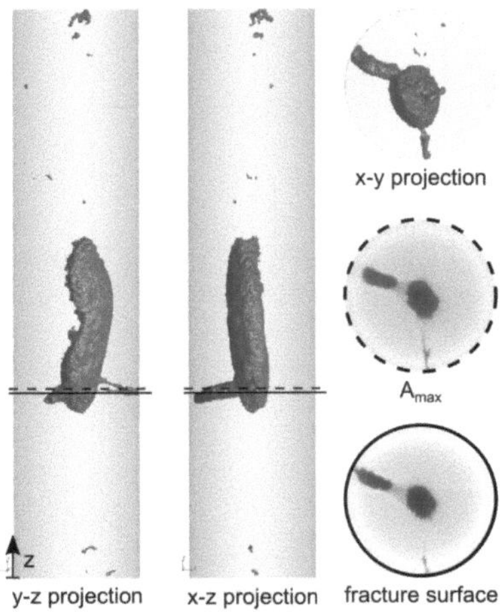


| $A_{proj,CT}$ [mm²]; [%] | $A_{max,CT}$ [mm²]; [%] | $z_{max,CT}$ [mm] |
|---|---|---|
| 1 | 21.4; 8 | 25.8 |

| $V_{max}/V_{tot}$ [%] | $A_{fail,mic}$ [mm]; [%] | $z_{fail}$ [mm] |
|---|---|---|
| 99 | 32.0; 12 | 26.6 |

| $\sigma_{max}/\sigma_y$ [%] | $\sigma_{max}/\sigma_u$ [%] |
|---|---|
| - | - |

Upper Fracture Surface

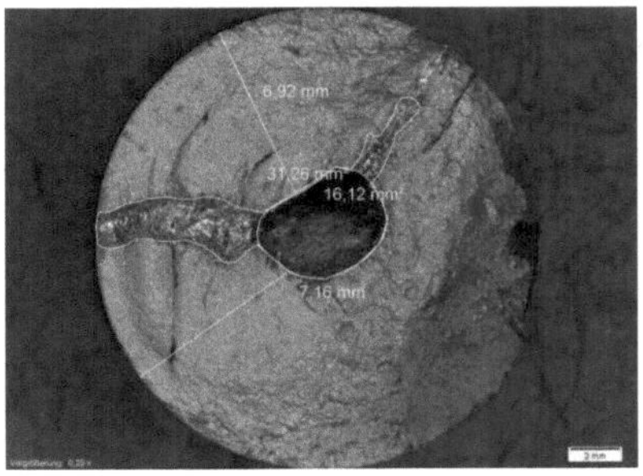

Lower Fracture Surface

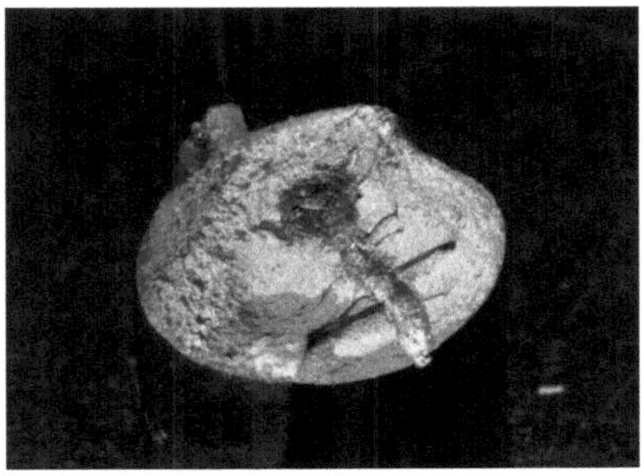

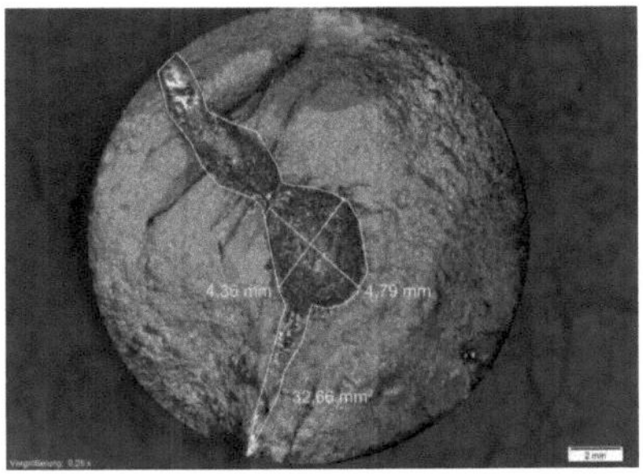

## U22M3

| $\Delta\sigma_{nom}$ [MPa] | $\Delta\sigma_{mic}$ [MPa] | cycles to failure [-] | foundry [-] | $\sigma_y$ [MPa] | $\sigma_u$ [MPa] | defect pos. | failure case |
|---|---|---|---|---|---|---|---|
| 280 | 341 | 370,442 | D | 889 | 997 | int. | III |

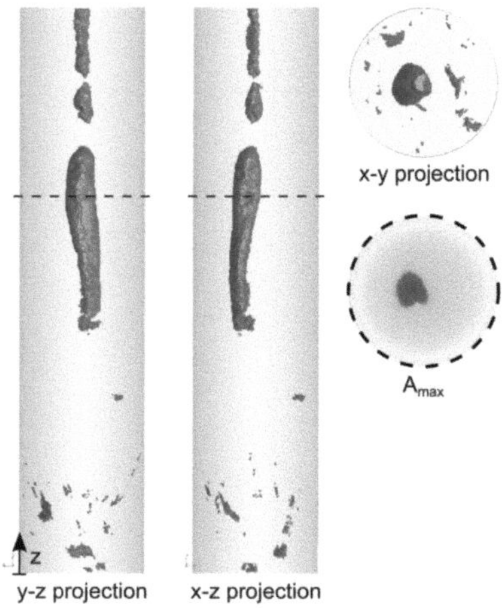


| $A_{proj,CT}$ [mm²]; [%] | $A_{max,CT}$ [mm²]; [%] | $z_{max,CT}$ [mm] |
|---|---|---|
| 1 | 11.9; 5 | 91.0 |

| $V_{max}/V_{tot}$ [%] | $A_{fail,mic}$ [mm]; [%] | $z_{fail}$ [mm] |
|---|---|---|
| 83 | 46.1; 18 | 53.0 |

| $\sigma_{max}/\sigma_y$ [%] | $\sigma_{max}/\sigma_u$ [%] |
|---|---|
| 78 | 69 |

Upper Fracture Surface

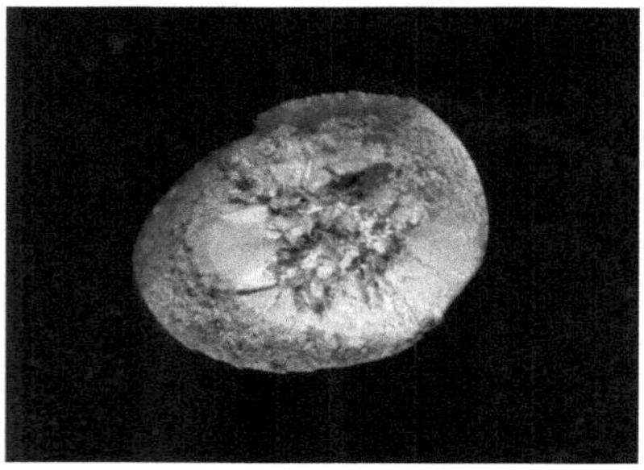

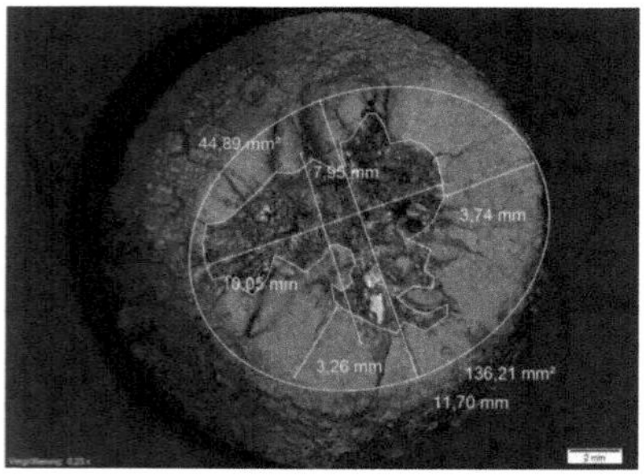


Lower Fracture Surface

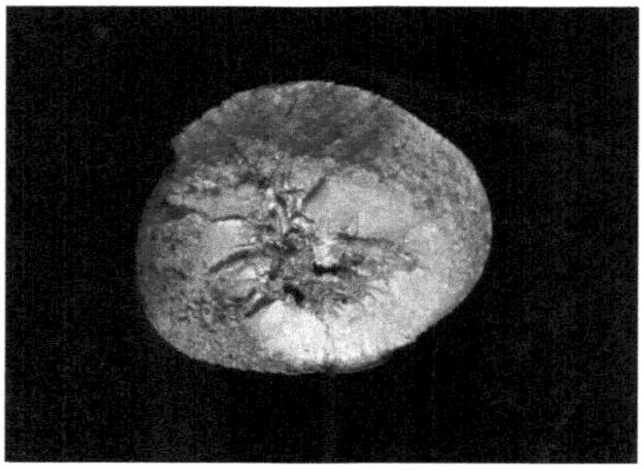

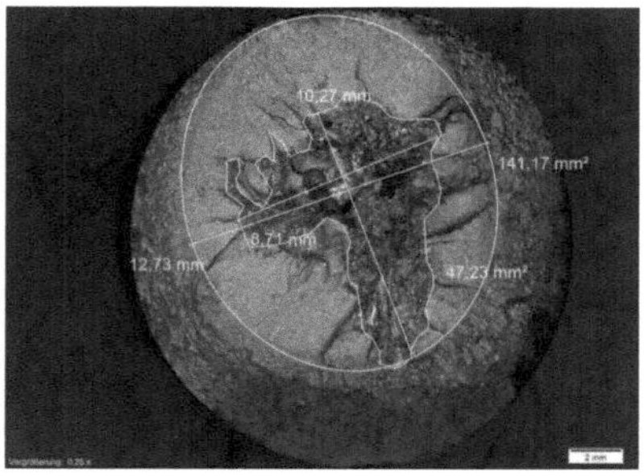

## U22M4

| $\Delta\sigma_{nom}$ [MPa] | $\Delta\sigma_{mic}$ [MPa] | cycles to failure [-] | foundry [-] | $\sigma_y$ [MPa] | $\sigma_u$ [MPa] | defect pos. | failure case |
|---|---|---|---|---|---|---|---|
| 220 | 239 | 195,017 | D | 889 | 997 | surf. | II |

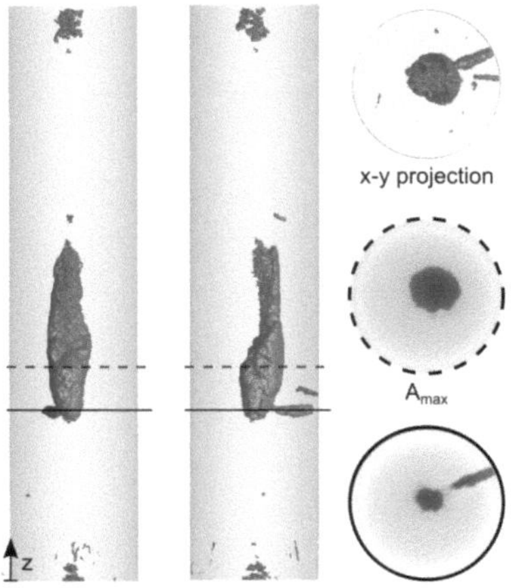


| $A_{proj,CT}$ [mm²]; [%] | $A_{max,CT}$ [mm²]; [%] | $z_{max,CT}$ [mm] |
|---|---|---|
| 1 | 24.7; 10 | 23.8 |

| $V_{max}/V_{tot}$ [%] | $A_{fail,mic}$ [mm]; [%] | $z_{fail}$ [mm] |
|---|---|---|
| 95 | 20.8; 8 | 29.8 |

| $\sigma_{max}/\sigma_y$ [%] | $\sigma_{max}/\sigma_u$ [%] |
|---|---|
| - | - |

Upper Fracture Surface

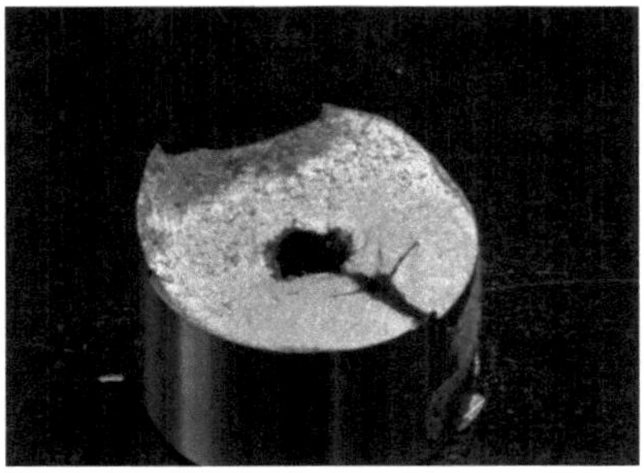

Lower Fracture Surface

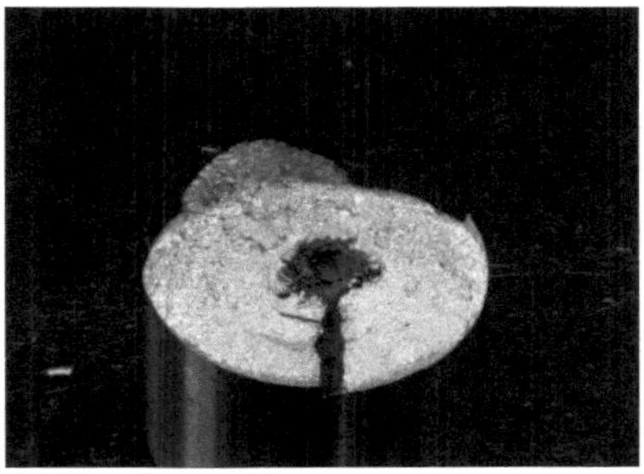

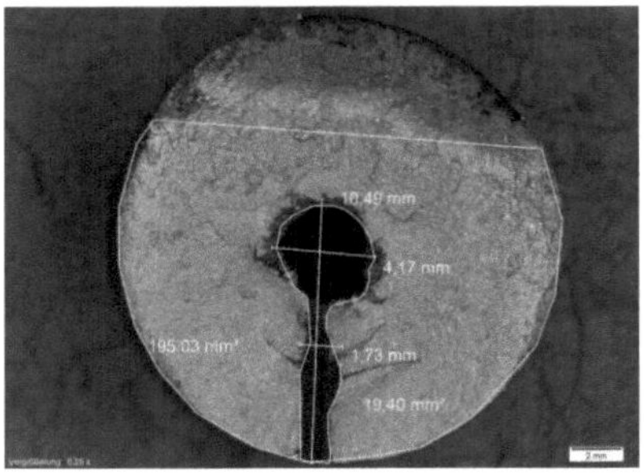


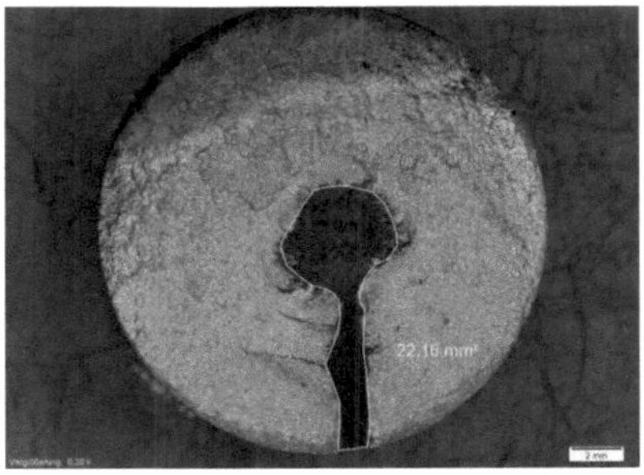

## U22M5

| $\Delta\sigma_{nom}$ [MPa] | $\Delta\sigma_{mic}$ [MPa] | cycles to failure [-] | foundry [-] | $\sigma_y$ [MPa] | $\sigma_u$ [MPa] | defect pos. | failure case |
|---|---|---|---|---|---|---|---|
| 220 | 220 | 2,578,188 | D | 889 | 997 | surf. | II |

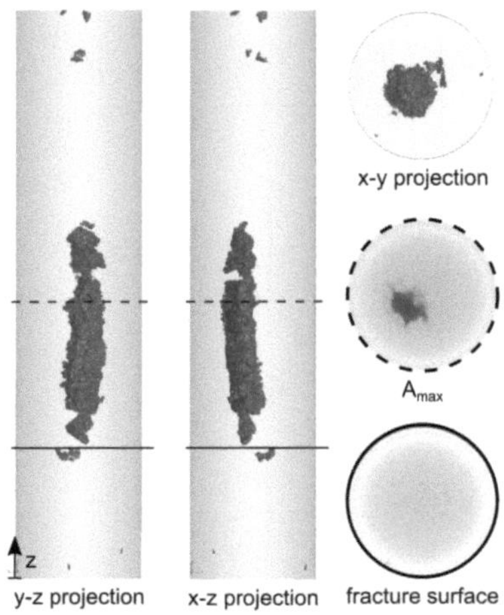


| $A_{proj,CT}$ [mm²]; [%] | $A_{max,CT}$ [mm²]; [%] | $z_{max,CT}$ [mm] |
|---|---|---|
| 1 | 11.2; 4 | 18.7 |

| $V_{max}/V_{tot}$ [%] | $A_{fail,mic}$ [mm]; [%] | $z_{fail}$ [mm] |
|---|---|---|
| 98 | 0.4; 0 | 38.8 |

| $\sigma_{max}/\sigma_y$ [%] | $\sigma_{max}/\sigma_u$ [%] |
|---|---|
| - | - |

Upper Fracture Surface

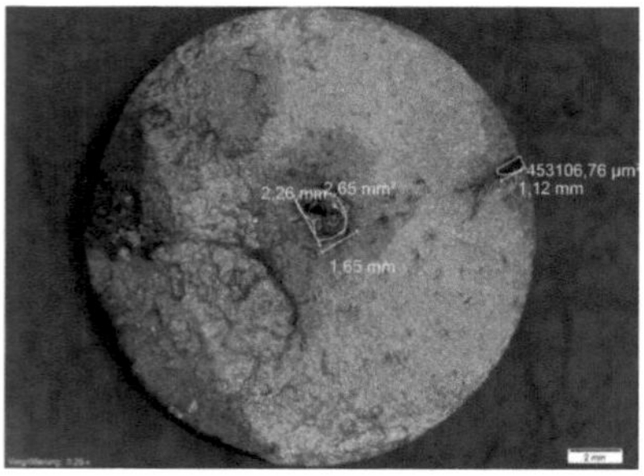

Lower Fracture Surface

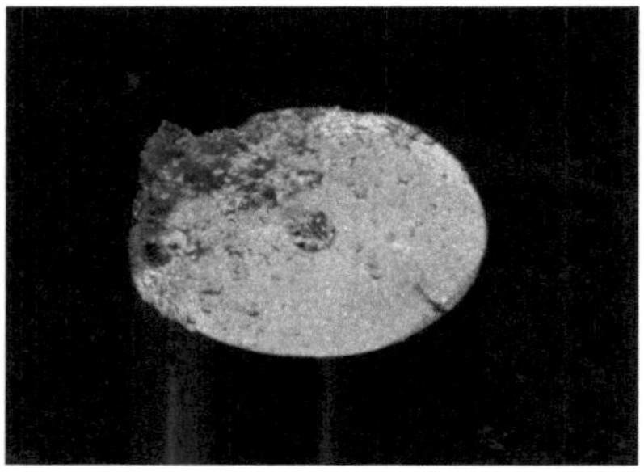

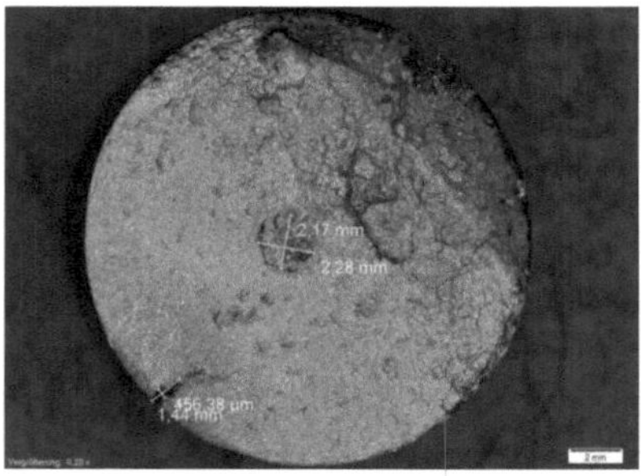

## U22M6

| $\Delta\sigma_{\mathrm{nom}}$ [MPa] | $\Delta\sigma_{\mathrm{mic}}$ [MPa] | cycles to failure [-] | foundry [-] | $\sigma_{\mathrm{y}}$ [MPa] | $\sigma_{\mathrm{u}}$ [MPa] | defect pos. | failure case |
|---|---|---|---|---|---|---|---|
| 280 | 312 | 161,991 | D | 889 | 997 | int. | III |

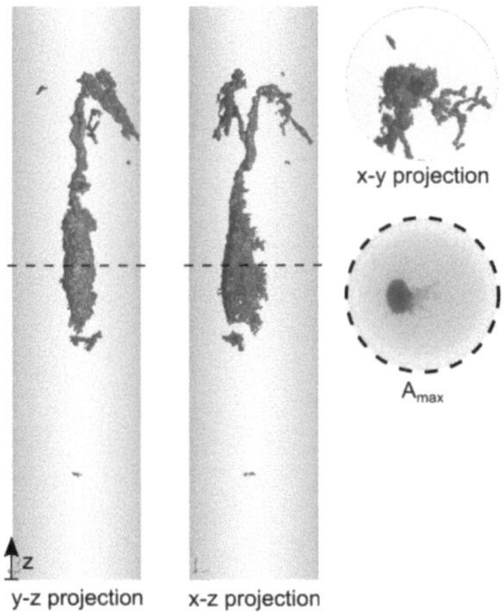


| $A_{\mathrm{proj,CT}}$ [mm²]; [%] | $A_{\mathrm{max,CT}}$ [mm²]; [%] | $z_{\mathrm{max,CT}}$ [mm] |
|---|---|---|
| 1 | 13.2; 5 | -12.0 |

| $V_{\mathrm{max}}/V_{\mathrm{tot}}$ [%] | $A_{\mathrm{fail,mic}}$ [mm]; [%] | $z_{\mathrm{fail}}$ [mm] |
|---|---|---|
| 88 | 26.4; 10 | 43.7 |

| $\sigma_{\mathrm{max}}/\sigma_{\mathrm{y}}$ [%] | $\sigma_{\mathrm{max}}/\sigma_{\mathrm{u}}$ [%] |
|---|---|
| 71 | 63 |

Upper Fracture Surface

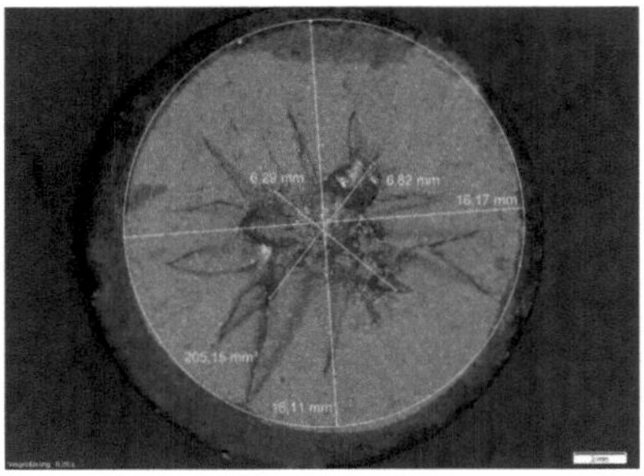

Lower Fracture Surface

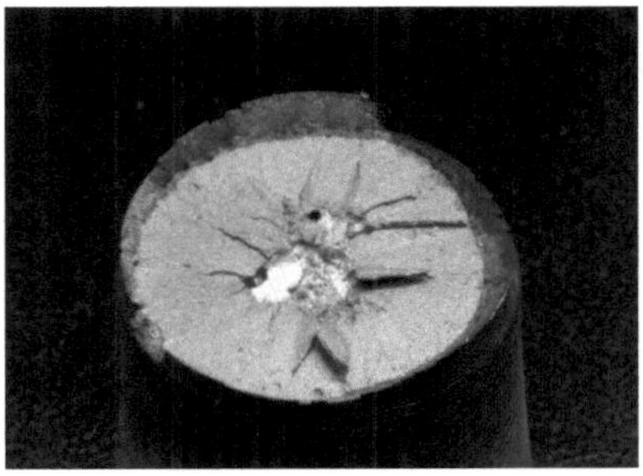

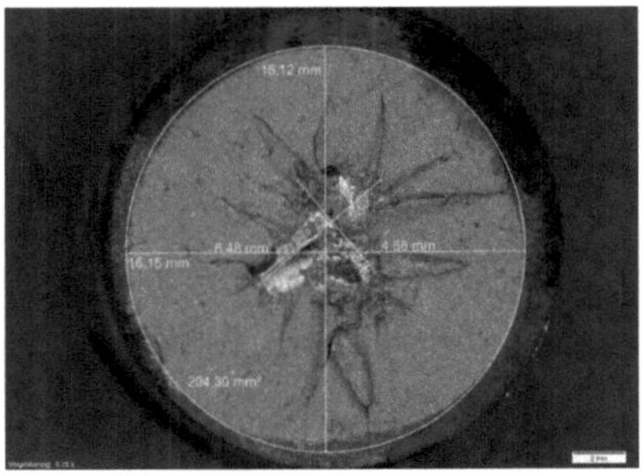

## U22M7

| $\Delta\sigma_{nom}$ [MPa] | $\Delta\sigma_{mic}$ [MPa] | cycles to failure [-] | foundry [-] | $\sigma_y$ [MPa] | $\sigma_u$ [MPa] | defect pos. | failure case |
|---|---|---|---|---|---|---|---|
| 400 | 477 | 93,317 | E | 889 | 997 | int. | II |

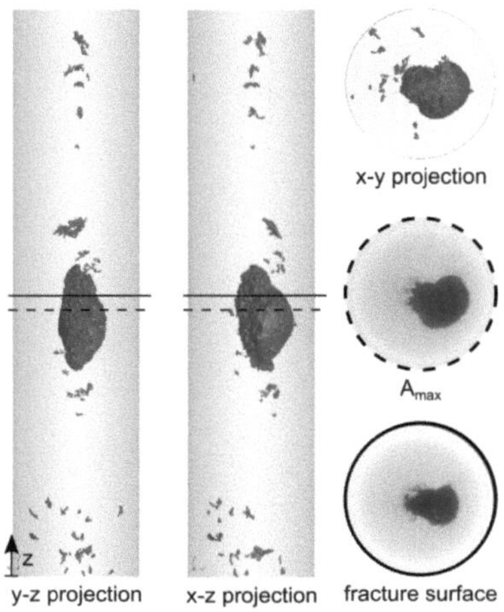


| $A_{proj,CT}$ [mm²]; [%] | $A_{max,CT}$ [mm²]; [%] | $z_{max,CT}$ [mm] |
|---|---|---|
| 1 | 35.6; 14 | 39.4 |

| $V_{max}/V_{tot}$ [%] | $A_{fail,mic}$ [mm]; [%] | $z_{fail}$ [mm] |
|---|---|---|
| 96 | 41.6; 16 | 37.4 |

| $\sigma_{max}/\sigma_y$ [%] | $\sigma_{max}/\sigma_u$ [%] |
|---|---|
| 92 | 82 |

Upper Fracture Surface

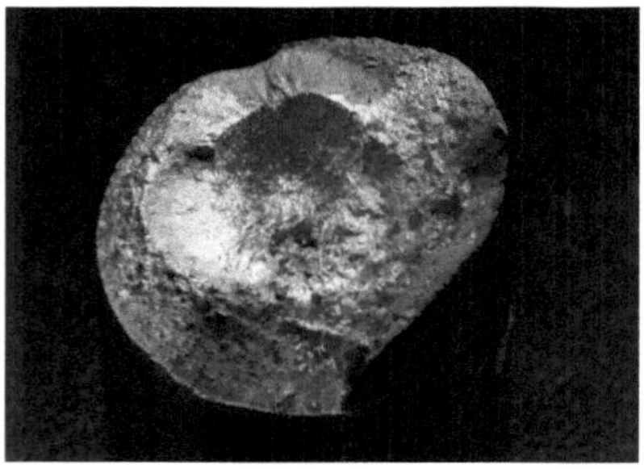

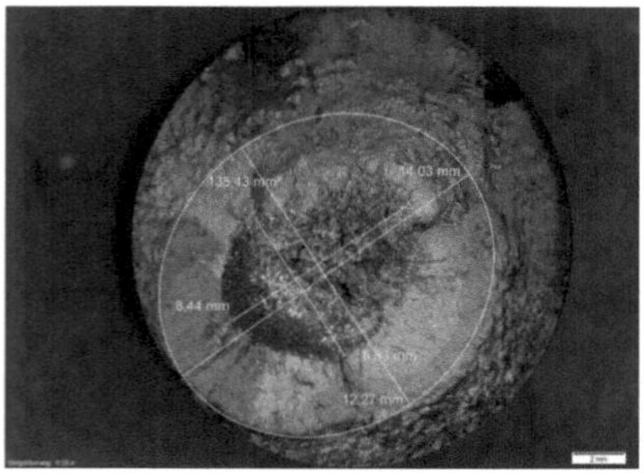

Lower Fracture Surface

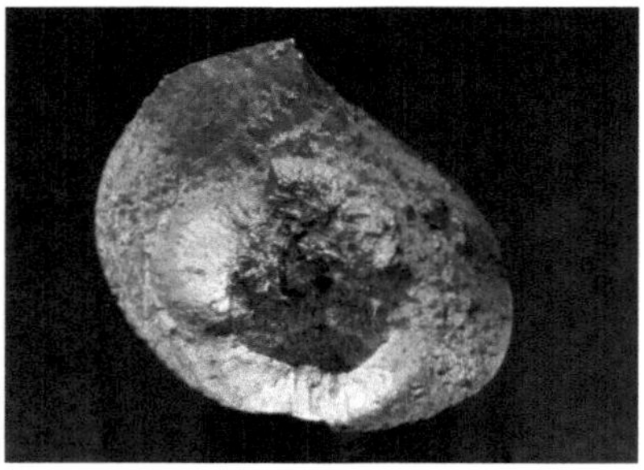

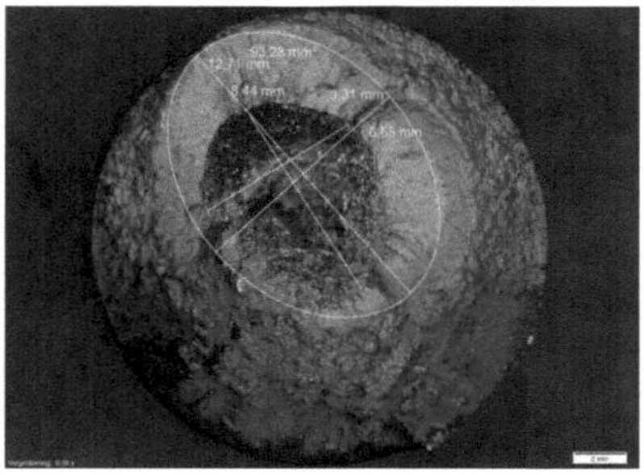

## U22M8

| $\Delta\sigma_{nom}$ [MPa] | $\Delta\sigma_{mic}$ [MPa] | cycles to failure [-] | foundry [-] | $\sigma_y$ [MPa] | $\sigma_u$ [MPa] | defect pos. | failure case |
|---|---|---|---|---|---|---|---|
| 260 | 305 | 939,729 | E | 889 | 997 | int. | I |

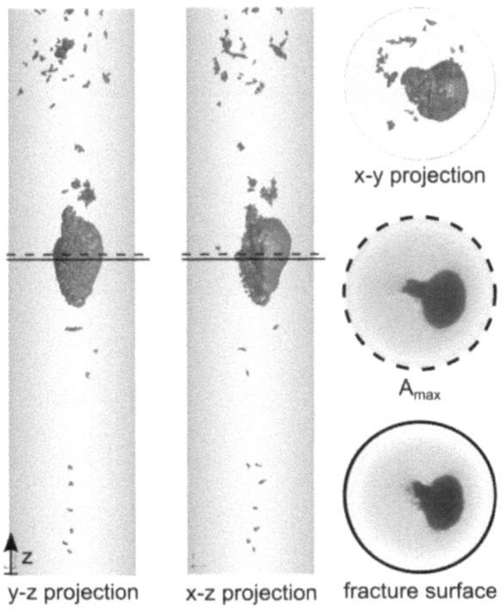


| $A_{proj,CT}$ [mm²]; [%] | $A_{max,CT}$ [mm²]; [%] | $z_{max,CT}$ [mm] |
|---|---|---|
| 1 | 33.8; 13 | 44.4 |

| $V_{max}/V_{tot}$ [%] | $A_{fail,mic}$ [mm]; [%] | $z_{fail}$ [mm] |
|---|---|---|
| 96 | 37.6; 15 | 45.1 |

| $\sigma_{max}/\sigma_y$ [%] | $\sigma_{max}/\sigma_u$ [%] |
|---|---|
| 86 | 77 |

Upper Fracture Surface

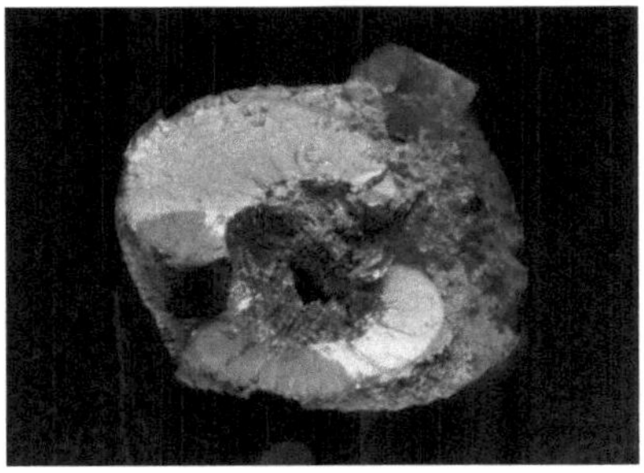

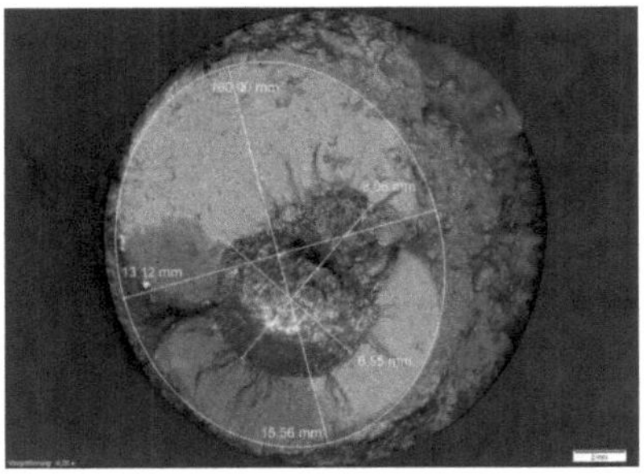

Lower Fracture Surface

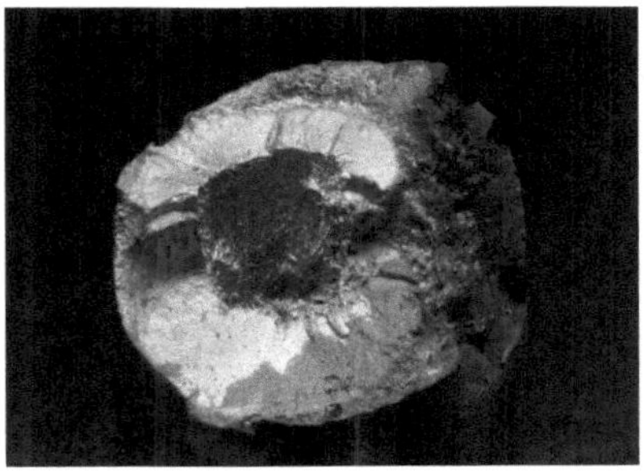

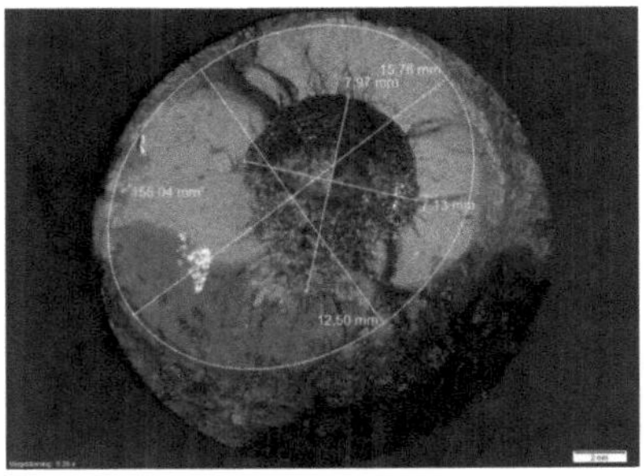

## U22M9

| $\Delta\sigma_{nom}$ [MPa] | $\Delta\sigma_{mic}$ [MPa] | cycles to failure [-] | foundry [-] | $\sigma_y$ [MPa] | $\sigma_u$ [MPa] | defect pos. | failure case |
|---|---|---|---|---|---|---|---|
| 320 | 322 | 332,607 | E | 889 | 997 | surf. | II |

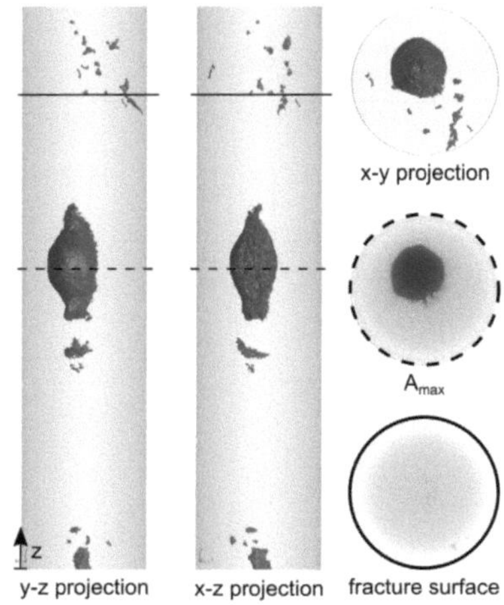


| $A_{proj,CT}$ [mm²]; [%] | $A_{max,CT}$ [mm²]; [%] | $z_{max,CT}$ [mm] |
|---|---|---|
| 1 | 35.7; 14 | 67.3 |

| $V_{max}/V_{tot}$ [%] | $A_{fail,mic}$ [mm]; [%] | $z_{fail}$ [mm] |
|---|---|---|
| 94 | 1.3; 0 | 42.5 |

| $\sigma_{max}/\sigma_y$ [%] | $\sigma_{max}/\sigma_u$ [%] |
|---|---|
| - | - |

Upper Fracture Surface

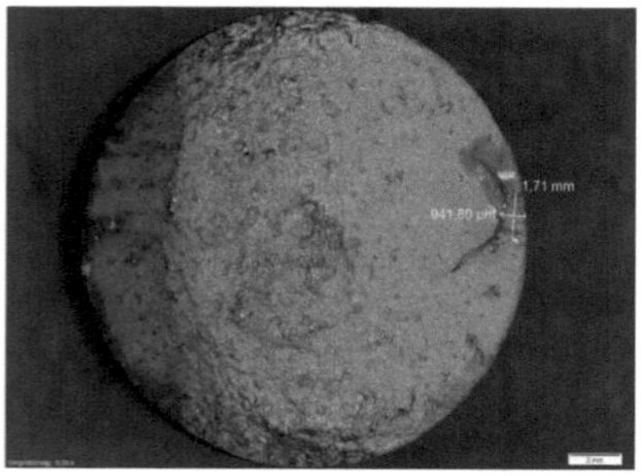

Lower Fracture Surface

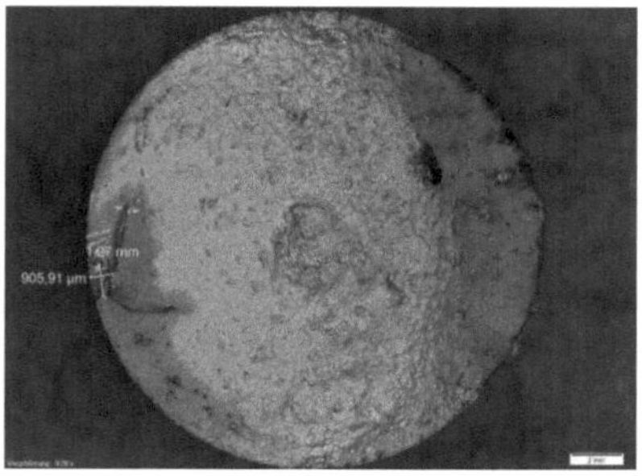

## U22M12

| $\Delta\sigma_{nom}$ [MPa] | $\Delta\sigma_{mic}$ [MPa] | cycles to failure [-] | foundry [-] | $\sigma_y$ [MPa] | $\sigma_u$ [MPa] | defect pos. | failure case |
|---|---|---|---|---|---|---|---|
| 240 | 256 | 3,779,906 | E | 889 | 997 | int. | III |

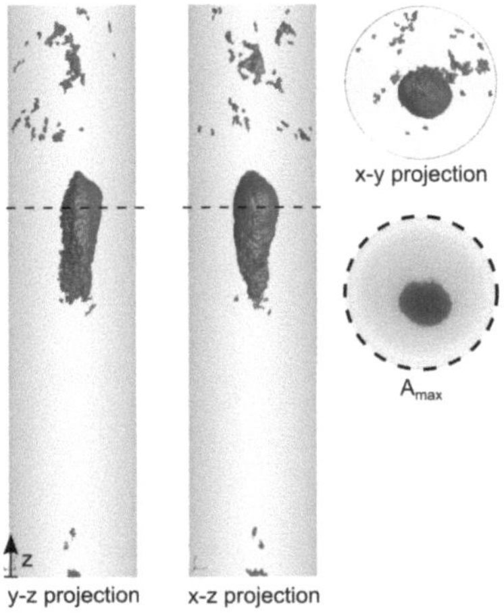


| $A_{proj,CT}$ [mm²]; [%] | $A_{max,CT}$ [mm²]; [%] | $z_{max,CT}$ [mm] |
|---|---|---|
| 1 | 26.5; 10 | -17.3 |

| $V_{max}/V_{tot}$ [%] | $A_{fail,mic}$ [mm]; [%] | $z_{fail}$ [mm] |
|---|---|---|
| 94 | 15.8; 6 | 51.6 |

| $\sigma_{max}/\sigma_y$ [%] | $\sigma_{max}/\sigma_u$ [%] |
|---|---|
| 79 | 70 |

Upper Fracture Surface

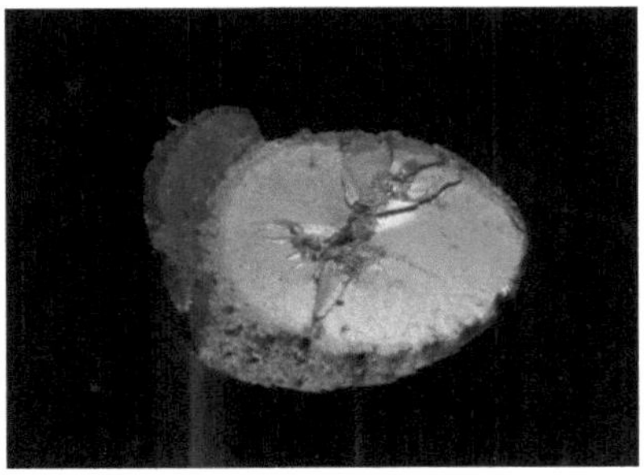

Lower Fracture Surface

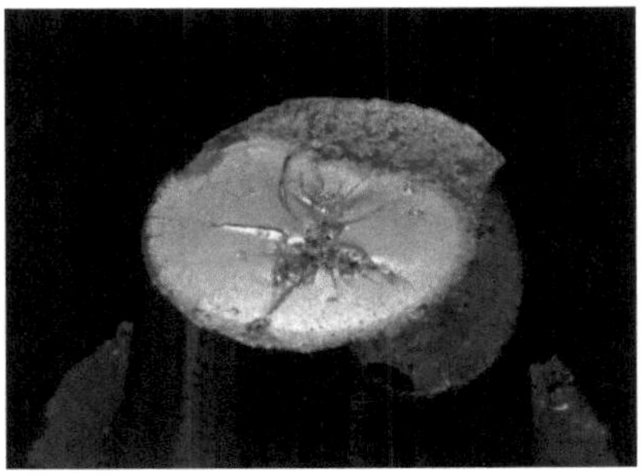

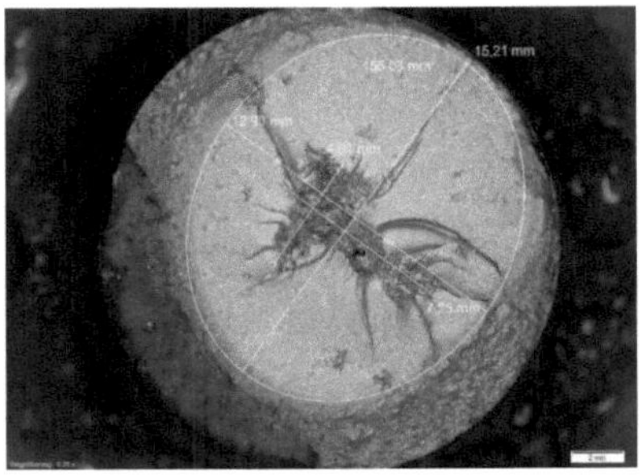

# D Details of Component Tests

This annex summarizes all component tests described in Section 3.3. Figure D.1 defines the geometric parameters and the applied coordinate systems. The structure of the data sheets is described in Table D.1 along with the notation used. An overview of the specimens including their nominal dimensions is given in Figure D.2.

The specimens are identified using the following concept.

G-(size)-(foundry)-(n)

| | |
|---|---|
| G | identifier for component tests |
| (size) | K = small; M = medium; G = large |
| (foundry) | A, B or C, according to Annex A |
| (n) | consecutive number |

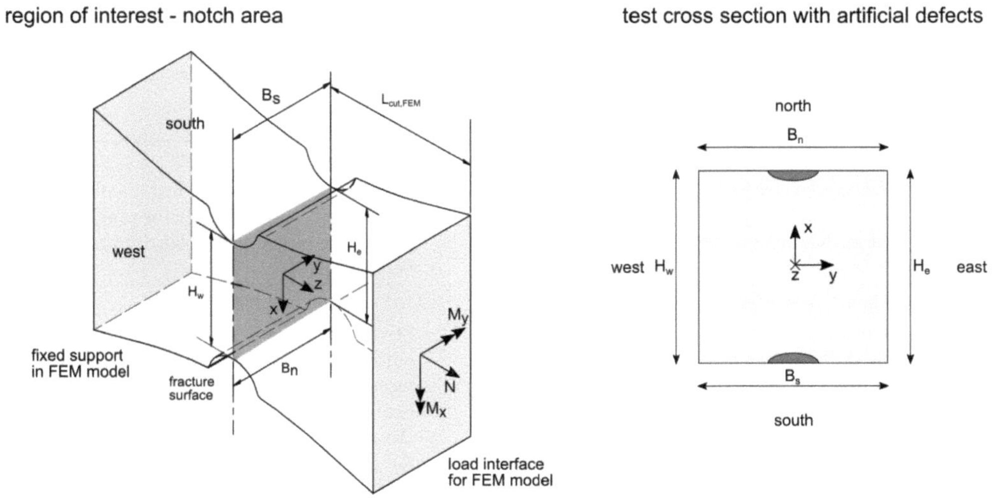


**Figure D.1:** Definition of geometric parameters and load directions

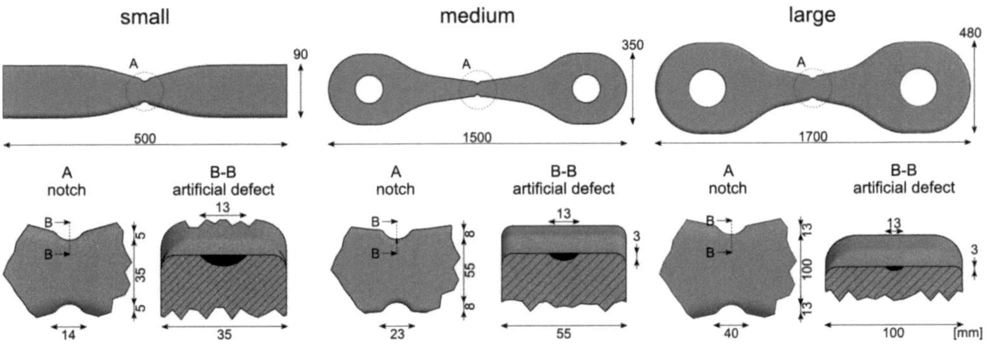


**Figure D.2:** Geometry of component test specimens

**Table D.1:** structure of the data sheets - component tests

**Table with Major Parameters**

| | |
|---|---|
| $\Delta\sigma$ | = nominal stress range related to the actual stressed cross section. Beach mark tests are marked by an additional index B |
| $\sigma_{max}/\sigma_y$ | = utilization factor in the fracture state (upper load) with material characteristics according to Eq. (3.4) |
| $H_w$, $H_e$, $B_n$, $B_s$ | = geometry parameters according to Figure D.1 |
| N, $M_x$, $M_y$ | = substitute loads to represent a realistic situation |
| SCF | = $\Delta\sigma_{max}/\Delta\sigma$ |

| | |
|---|---|
| **Fracture Surface**<br>Measured fracture surfaces with an indication of the spatial orientation and position of EMAT. The marked area corresponds to the residual fracture surface. | **Positions of Strain Gauges**<br>The label, type and exact position of all strain gauges is shown separately by view. |
| **Strain - Cycle Curves**<br>Strain ranges are plotted over the test cycles. The evaluation range for the initial situation as well as $\Delta N$ according to Section 3.3.2.3 is highlighted. | **Deviation - Cycle Curves**<br>Visualization of the deviations of the current strain range from the strain range at the beginning of the test according to Section 3.3.2.3. |
| **Initial Strain Ranges**<br>Measured srain ranges are compared with surface strains calculated by FEM under the substitute loads according to Section 4.2.2. The value of the strain gaue correspond to the mean values of the evaluation range marked in the "strain - cycle diagram" The size of the artificial defect as well as the width of the strain gauges are markt in relation to the stress curve. | **Initial 3D Stress Field**<br>Stress fields resulting from FEM with and without consideration of the substitute moments in the initial state according to Section 4.2.2.<br>$\Delta\sigma_{max}$ = maximum stress range;<br>$\Delta\sigma_{max,N\text{-}only}$ = maximum stress range without substitute moments;<br>$\Delta\sigma_{max,defect}$ = maximum stress ramge at the defect location under consideration of moments |

## GKA1

| $\Delta\sigma$ [MPa] | cycles to failure [-] | $\sigma_{max}/\sigma_y$ [%] | $H_w$ [mm] | $H_e$ [mm] | $B_n$ [mm] | $B_s$ [mm] | $F_x$ [kN] | $M_x$ [Nm] | $M_y$ [Nm] | SCF [-] |
|---|---|---|---|---|---|---|---|---|---|---|
| 220 | 212,567 | 234 | 35.0 | 35.0 | 34.6 | 34.5 | 256 | -70 | -180 | 2.22 |

Note: This specimen was tested at $\Delta\sigma = 85\,\text{MPa}$, stopped at $10^7$ cycles and retested at $\Delta\sigma = 220\,\text{MPa}$

### Fracture Surface

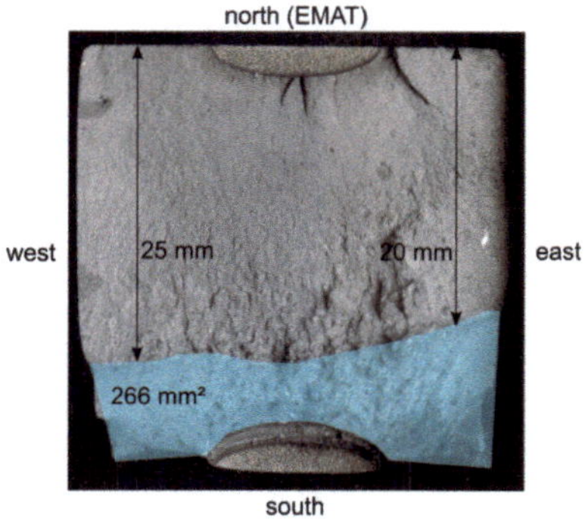


### Positions of Strain Gauges

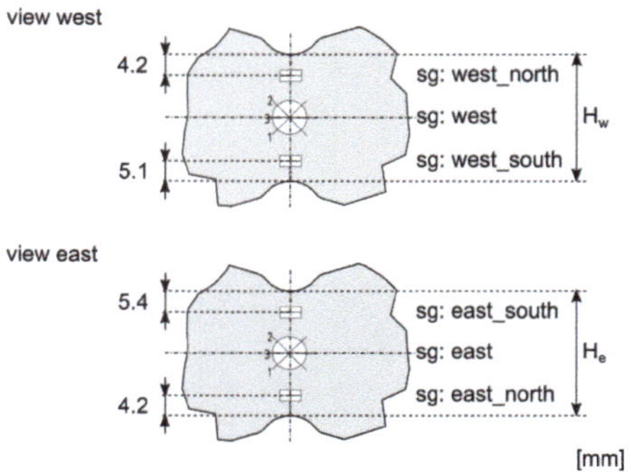


### Strain Ranges

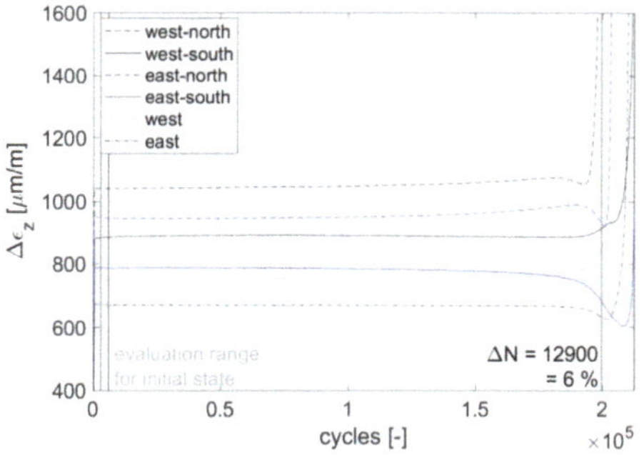


### Deviation from Initial Strain Ranges

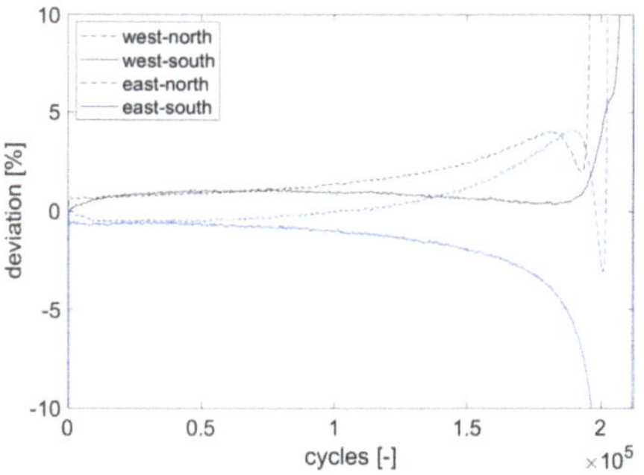


### Initial Strain Ranges

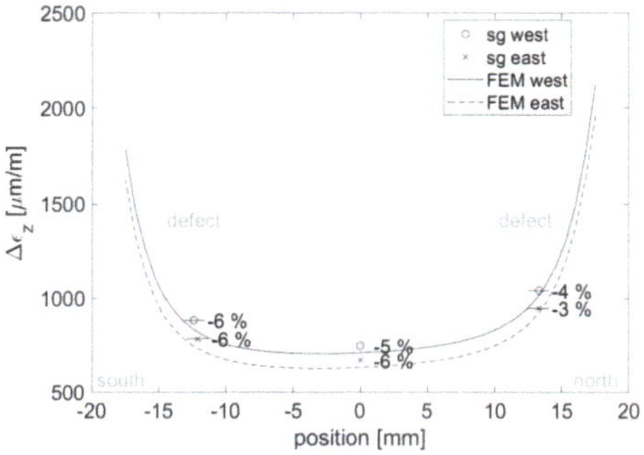


### Initial 3D Stress Field

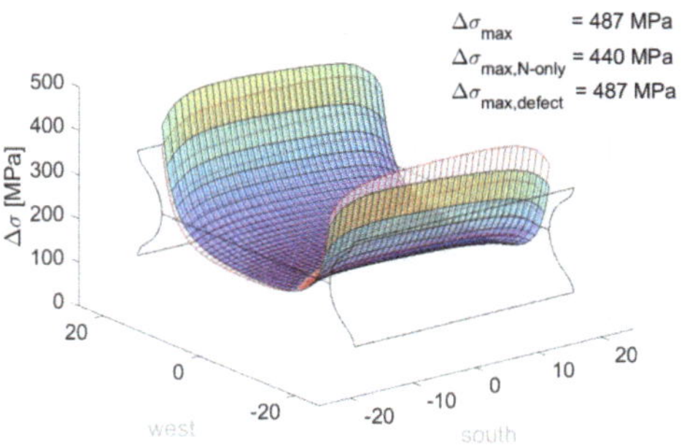

## GKA2

| $\Delta\sigma$ [MPa] | cycles to failure [-] | $\sigma_{max}/\sigma_y$ [%] | $H_w$ [mm] | $H_e$ [mm] | $B_n$ [mm] | $B_s$ [mm] | $F_x$ [kN] | $M_x$ [Nm] | $M_y$ [Nm] | SCF [-] |
|---|---|---|---|---|---|---|---|---|---|---|
| B150 | 1,255,246 | 198 | 35.1 | 35.1 | 34.4 | 34.3 | 174 | -20 | 100 | 2.16 |

Note: The strain gauges recording stopped during testing

### Fracture Surface

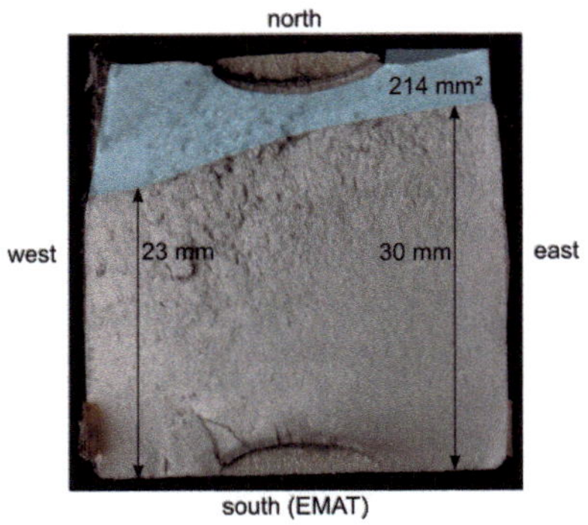


### Positions of Strain Gauges

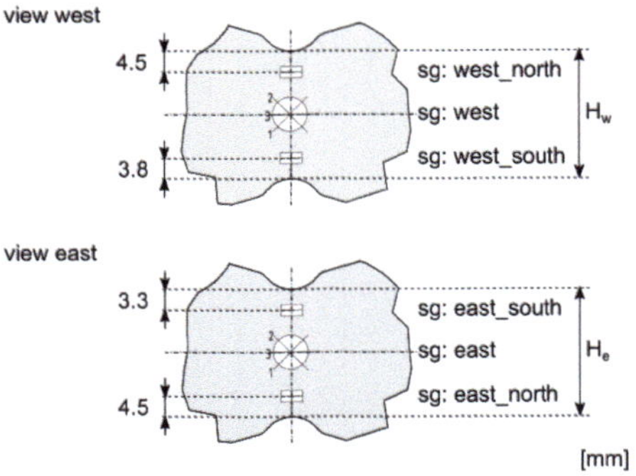


### Strain Ranges

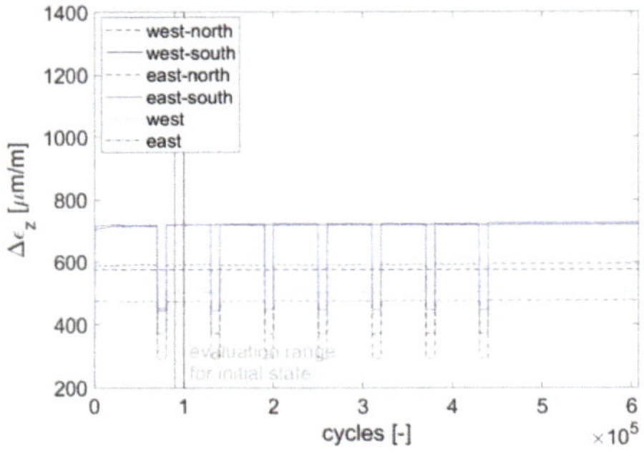


### Deviation from Initial Strain Ranges

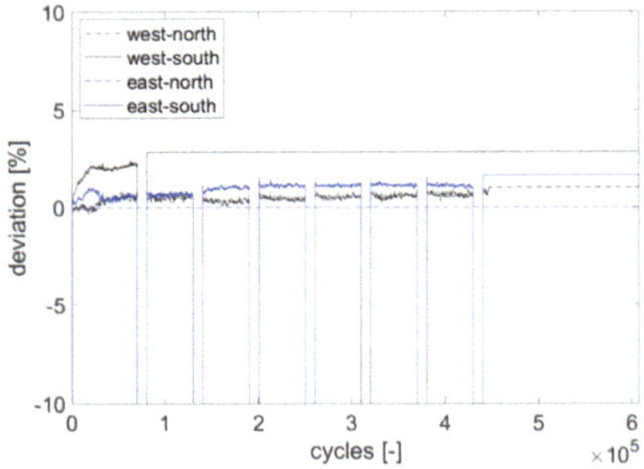


### Initial Strain Ranges

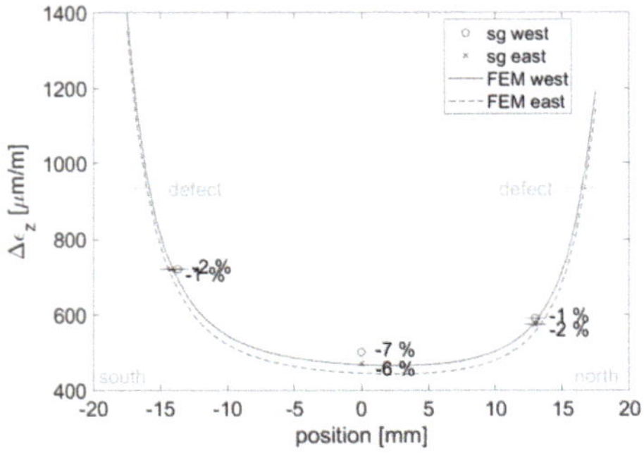


### Initial 3D Stress Field

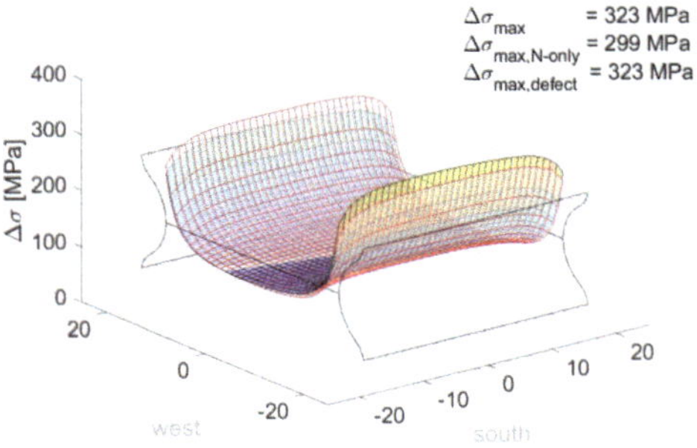

## GKB1

| $\Delta\sigma$ [MPa] | cycles to failure [-] | $\sigma_{max}/\sigma_y$ [%] | $H_w$ [mm] | $H_e$ [mm] | $B_n$ [mm] | $B_s$ [mm] | $F_x$ [kN] | $M_x$ [Nm] | $M_y$ [Nm] | SCF [-] |
|---|---|---|---|---|---|---|---|---|---|---|
| 154 | 452,670 | 221 | 35.7 | 35.6 | 35.0 | 35.0 | - | - | - | - |

Note: A calculation of the initial load condition was not possible with this strain gauge configuration. The strain measurement was disturbed in the beginning, due to a wire of the EMAT that scratched on the soldering base.

### Fracture Surface

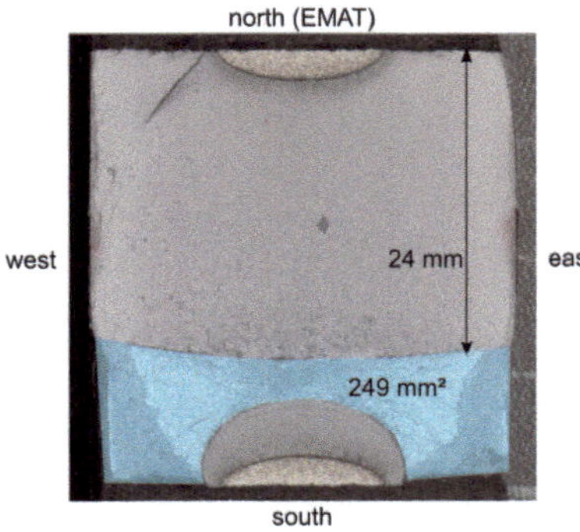


### Positions of Strain Gauges

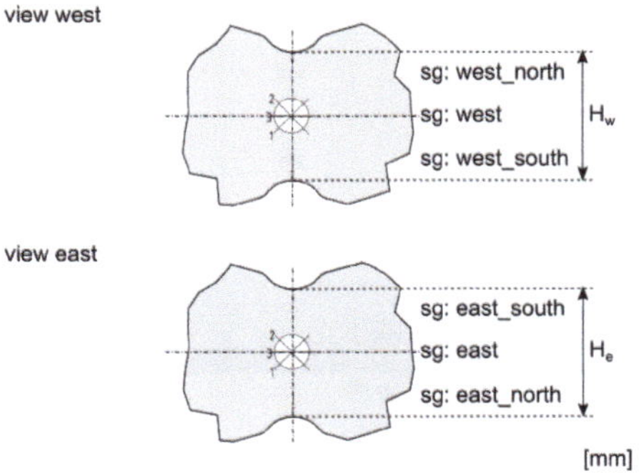


### Strain Ranges

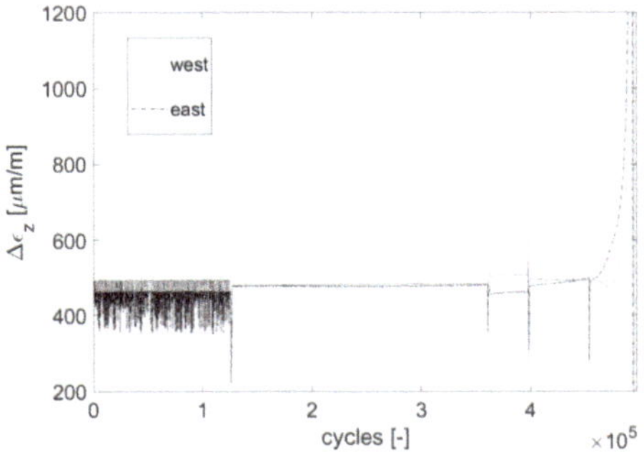


### Deviation from Initial Strain Ranges

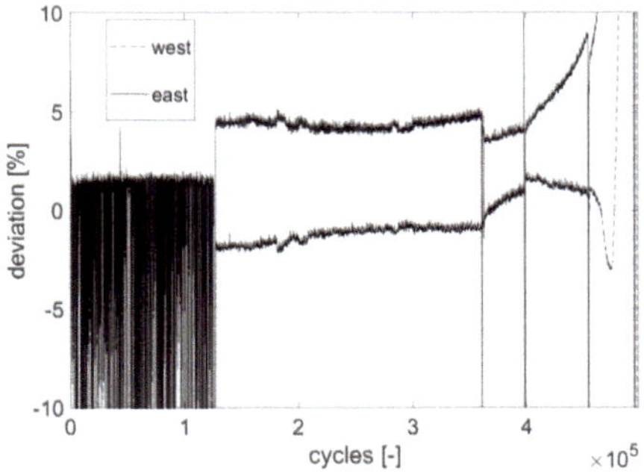


### Initial Strain Ranges

no data

### Initial 3D Stress Field

no data

## GKB2

| $\Delta\sigma$ [MPa] | cycles to failure [-] | $\sigma_{max}/\sigma_y$ [%] | $H_w$ [mm] | $H_e$ [mm] | $B_n$ [mm] | $B_s$ [mm] | $F_x$ [kN] | $M_x$ [Nm] | $M_y$ [Nm] | SCF [-] |
|---|---|---|---|---|---|---|---|---|---|---|
| 120 | 767,308 | 280 | 35.5 | 35.5 | 35.0 | 35.0 | 139 | -18 | -200 | 2.41 |

Note: -

### Fracture Surface

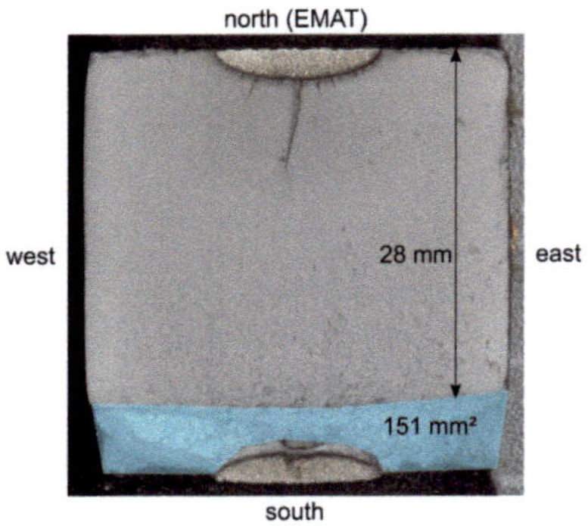


### Positions of Strain Gauges

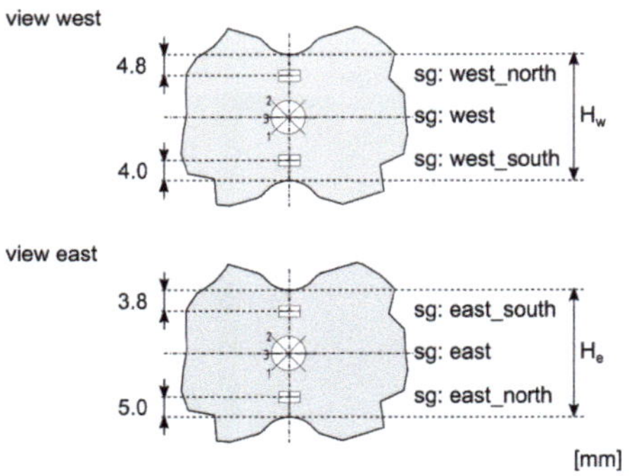


### Strain Ranges

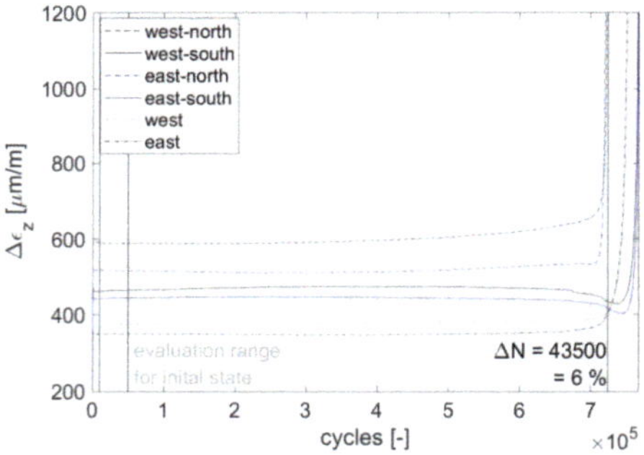


### Deviation from Initial Strain Ranges

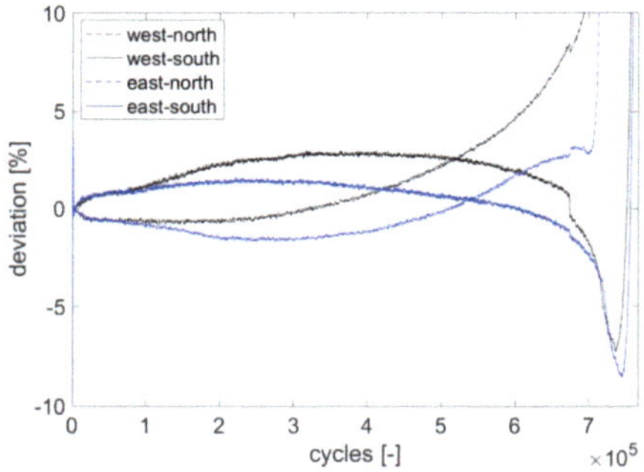


### Initial Strain Ranges

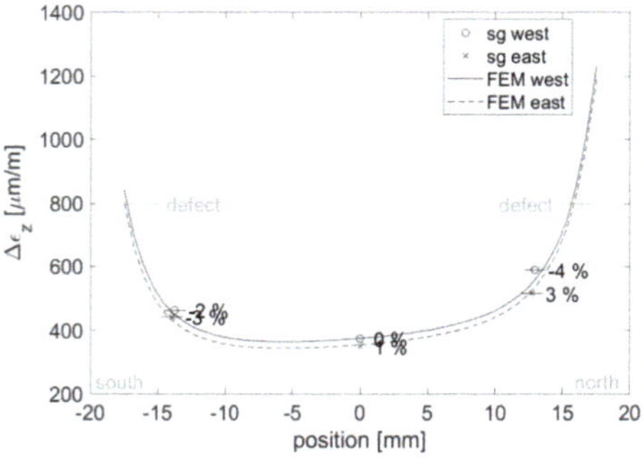


### Initial 3D Stress Field

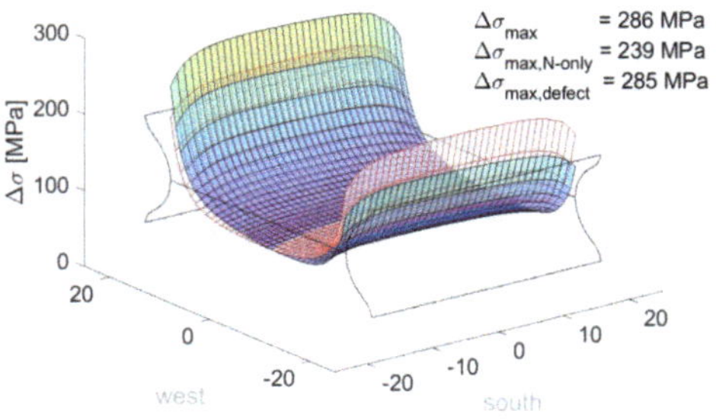

## GKB3

| $\Delta\sigma$ [MPa] | cycles to failure [-] | $\sigma_{max}/\sigma_y$ [%] | $H_w$ [mm] | $H_e$ [mm] | $B_n$ [mm] | $B_s$ [mm] | $F_x$ [kN] | $M_x$ [Nm] | $M_y$ [Nm] | SCF [-] |
|---|---|---|---|---|---|---|---|---|---|---|
| 220 | 107,430 | 252 | 36.7 | 36.8 | 35.0 | 35.0 | - | - | - | - |

Note: An overload of the testing machine at approx. 35,000 cycles led to severe plastic deformations in the notch area. Afterwards no more measurement was possible. This data point is questionable for a statistical evaluation and was considered carefully.

### Fracture Surface

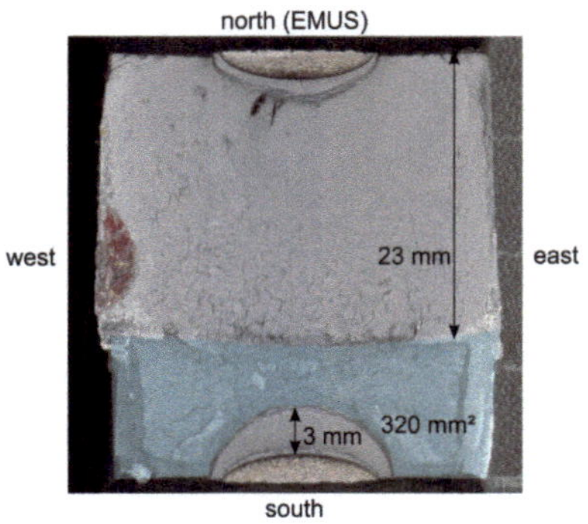


### Positions of Strain Gauges

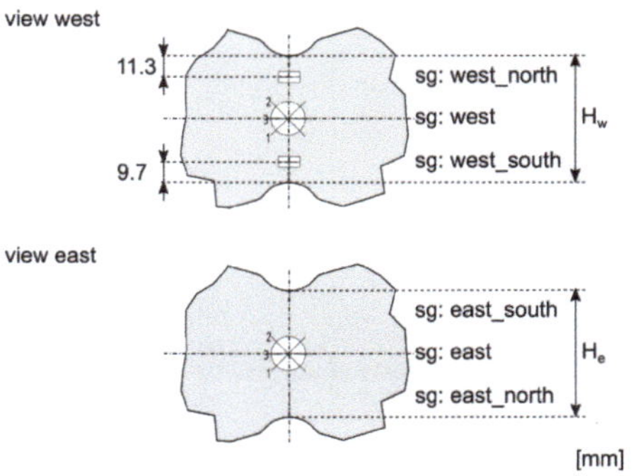


### Strain Ranges

no data

### Deviation from Initial Strain Ranges

no data

### Initial Strain Ranges

no data

### Initial 3D Stress Field

no data

## GKC1

| $\Delta\sigma$ [MPa] | cycles to failure [-] | $\sigma_{max}/\sigma_y$ [%] | $H_w$ [mm] | $H_e$ [mm] | $B_n$ [mm] | $B_s$ [mm] | $F_x$ [kN] | $M_x$ [Nm] | $M_y$ [Nm] | SCF [-] |
|---|---|---|---|---|---|---|---|---|---|---|
| 220 | 103,852 | 227 | 34.9 | 35.0 | 32.8 | 32.6 | 255 | -6 | 120 | 2.13 |

Note: -

### Fracture Surface

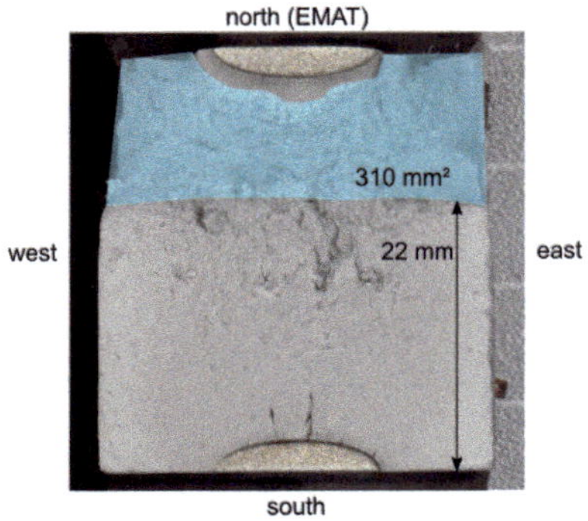


### Positions of Strain Gauges

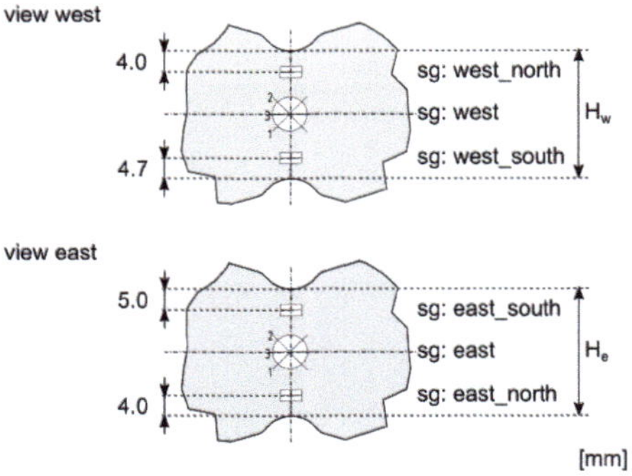


### Strain Ranges

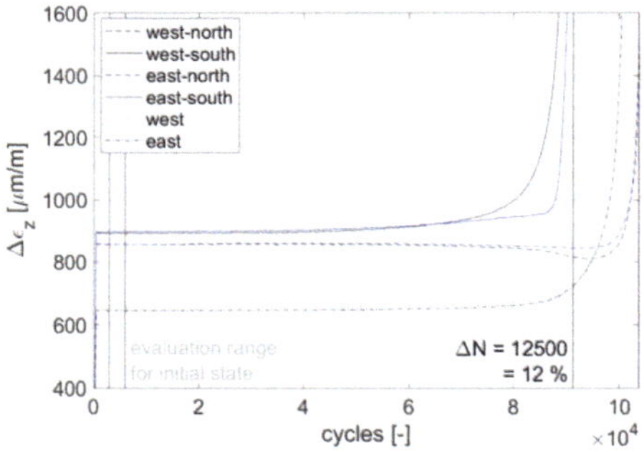


### Deviation from Initial Strain Ranges

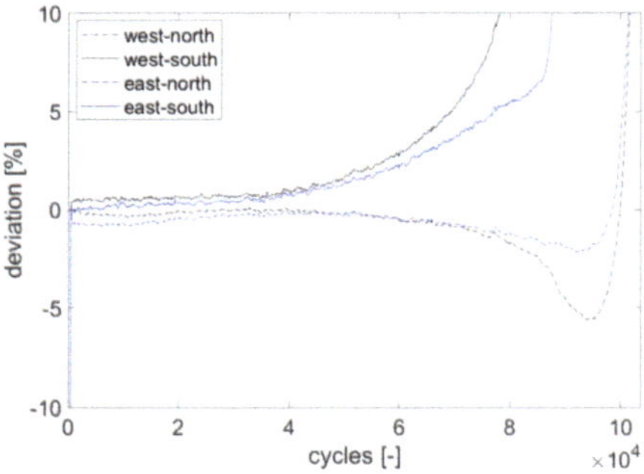


### Initial Strain Ranges

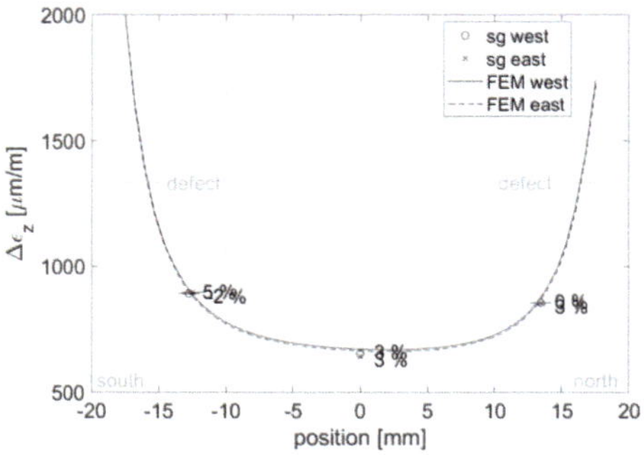


### Initial 3D Stress Field

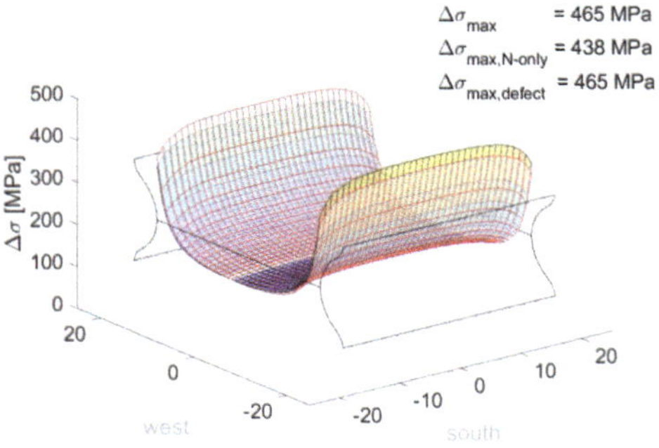

## GKC2

| $\Delta\sigma$ [MPa] | cycles to failure [-] | $\sigma_{max}/\sigma_y$ [%] | $H_w$ [mm] | $H_e$ [mm] | $B_n$ [mm] | $B_s$ [mm] | $F_x$ [kN] | $M_x$ [Nm] | $M_y$ [Nm] | SCF [-] |
|---|---|---|---|---|---|---|---|---|---|---|
| 100 | 1,117,568 | 244 | 34.9 | 34.9 | 32.7 | 32.6 | 115 | -27 | -60 | 2.16 |

Note: -

### Fracture Surface

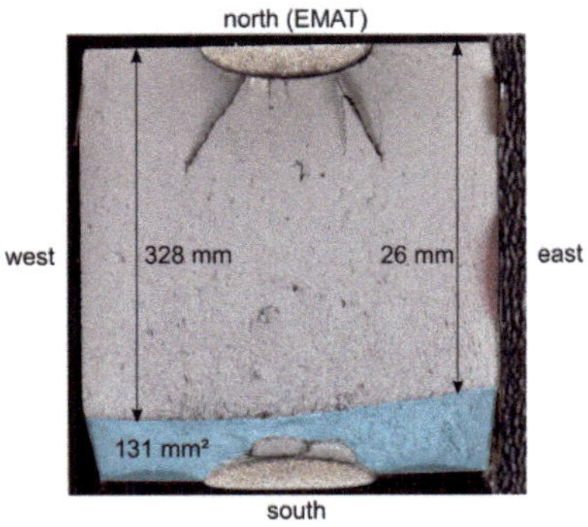


### Positions of Strain Gauges

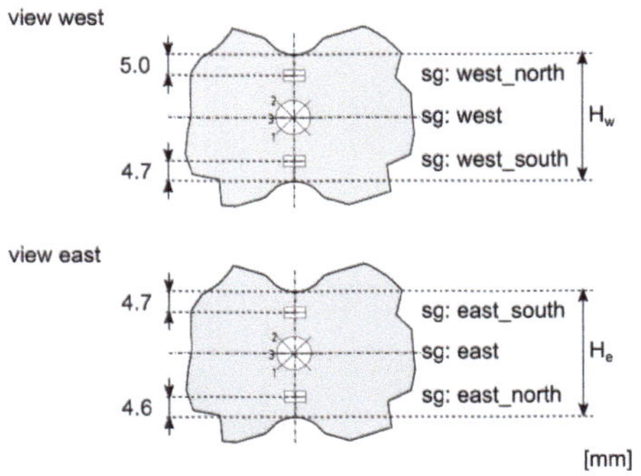


### Strain Ranges

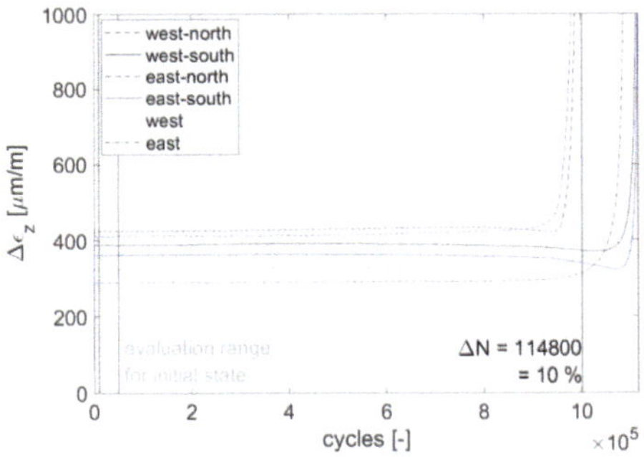


### Deviation from Initial Strain Ranges

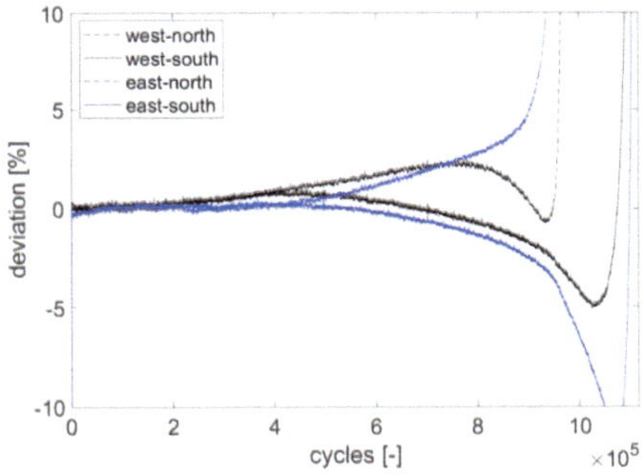


### Initial Strain Ranges

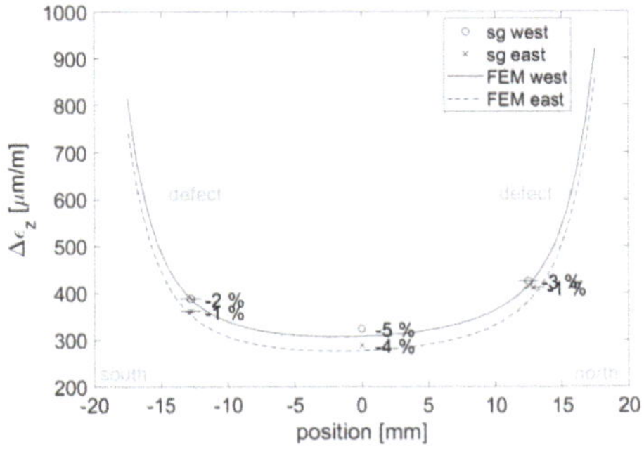


### Initial 3D Stress Field

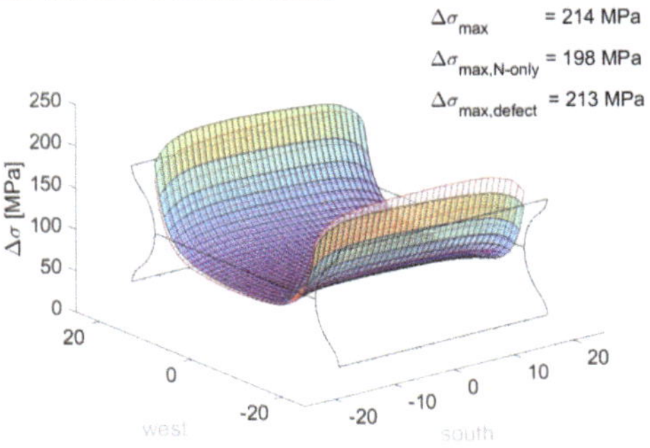

## GKC3

| $\Delta\sigma$ [MPa] | cycles to failure [-] | $\sigma_{max}/\sigma_y$ [%] | $H_w$ [mm] | $H_e$ [mm] | $B_n$ [mm] | $B_s$ [mm] | $F_x$ [kN] | $M_x$ [Nm] | $M_y$ [Nm] | SCF [-] |
|---|---|---|---|---|---|---|---|---|---|---|
| B150 | 348,279 | 200 | 35.0 | 35.0 | 32.6 | 32.4 | 174 | -50 | -100 | 2.18 |

Note: Strain data only exist within the first cycles

### Fracture Surface

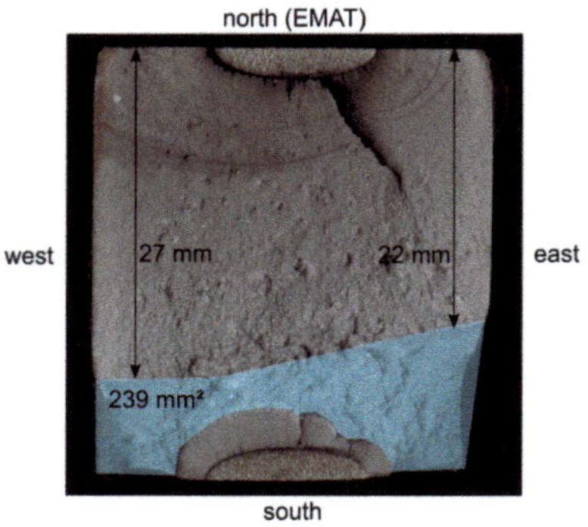


### Positions of Strain Gauges

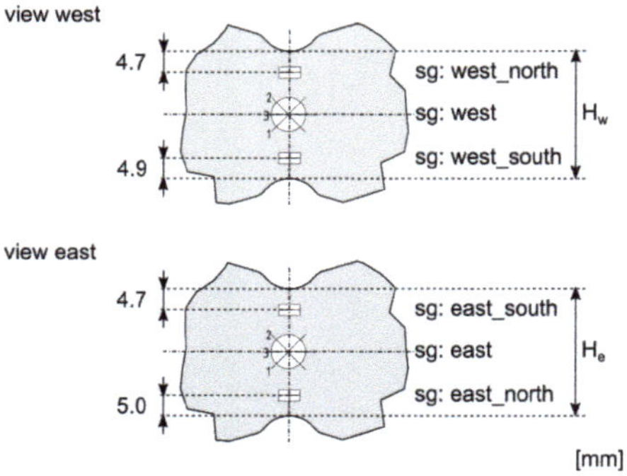


### Strain Ranges

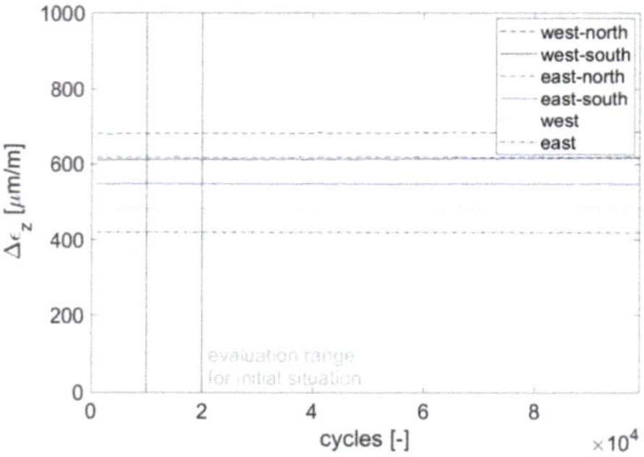


### Deviation from Initial Strain Ranges

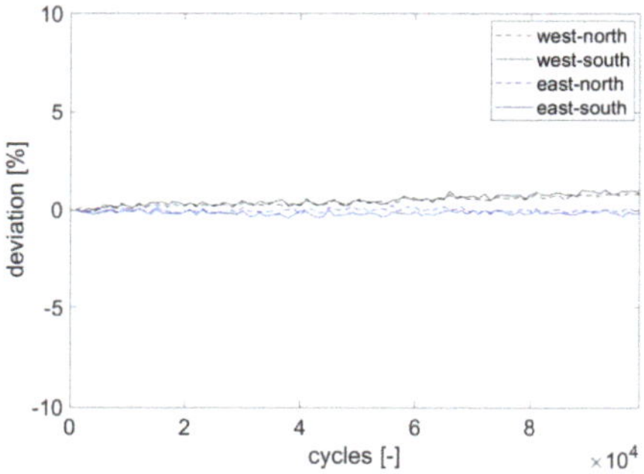


### Initial Strain Ranges

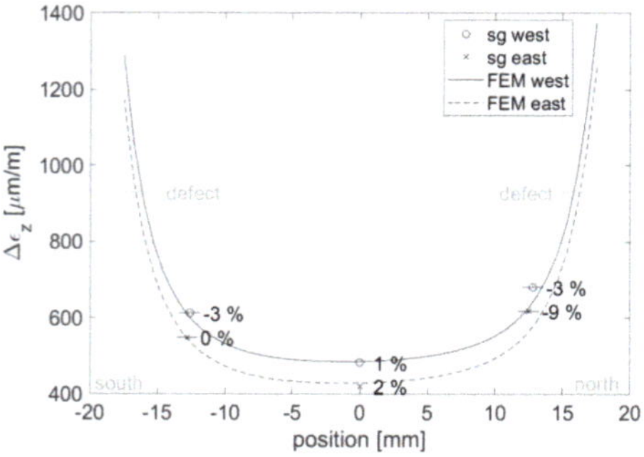


### Initial 3D Stress Field

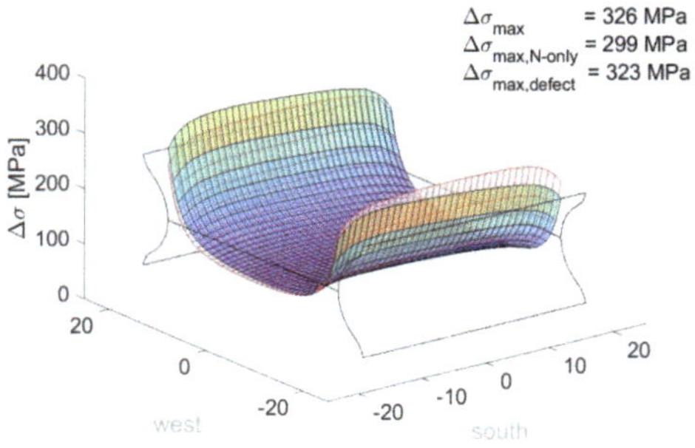

## GMA1

| $\Delta\sigma$ [MPa] | cycles to failure [-] | $\sigma_{max}/\sigma_y$ [%] | $H_w$ [mm] | $H_e$ [mm] | $B_n$ [mm] | $B_s$ [mm] | $F_x$ [kN] | $M_x$ [Nm] | $M_y$ [Nm] | SCF [-] |
|---|---|---|---|---|---|---|---|---|---|---|
| 180 | 284,668 | 224 | 55.1 | 54.8 | 55.5 | 55.6 | 536 | 70 | 500 | 2.17 |

Note: -

### Fracture Surface

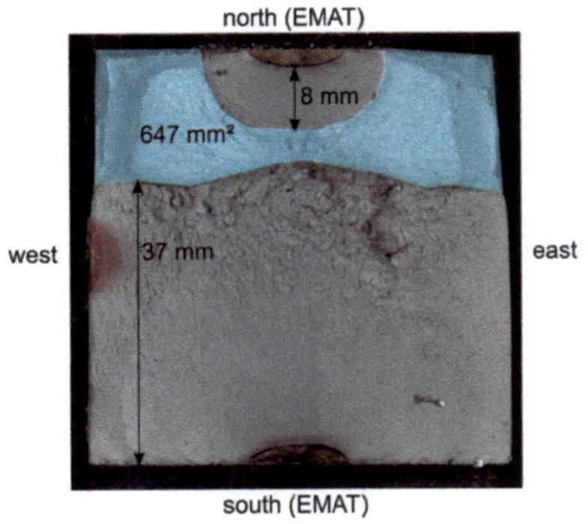


### Positions of Strain Gauges

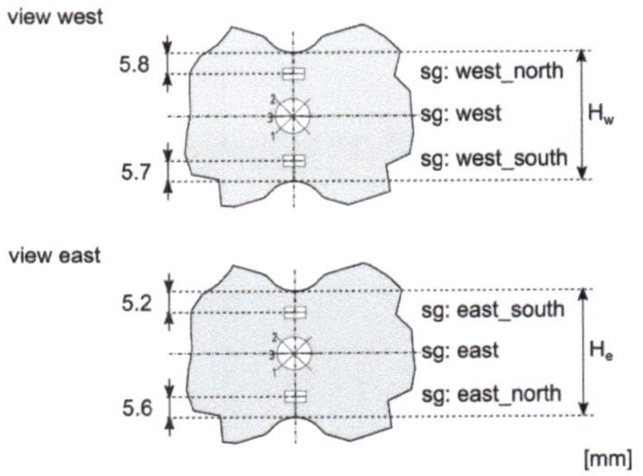


### Strain Ranges

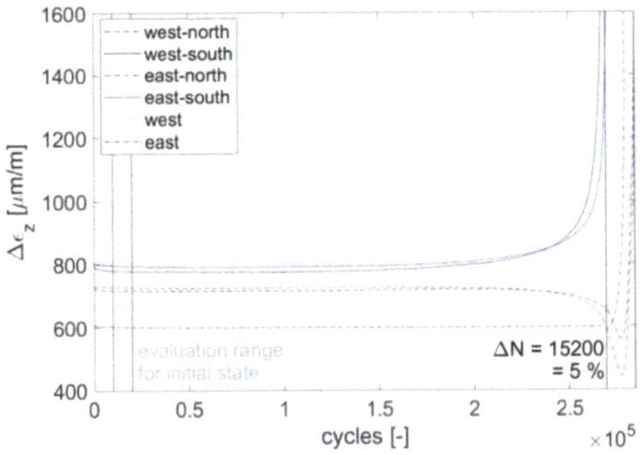


### Deviation from Initial Strain Ranges

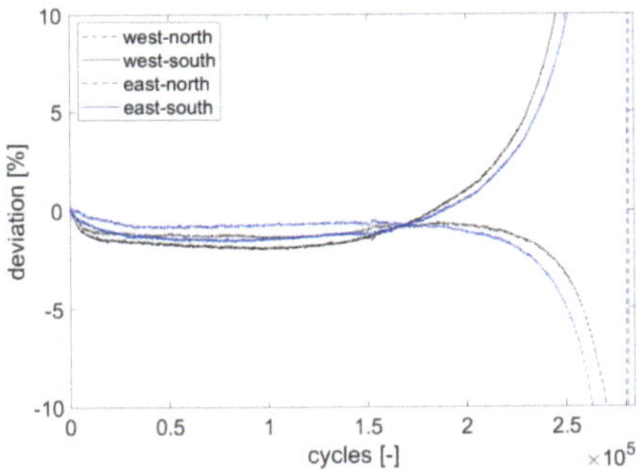


### Initial Strain Ranges

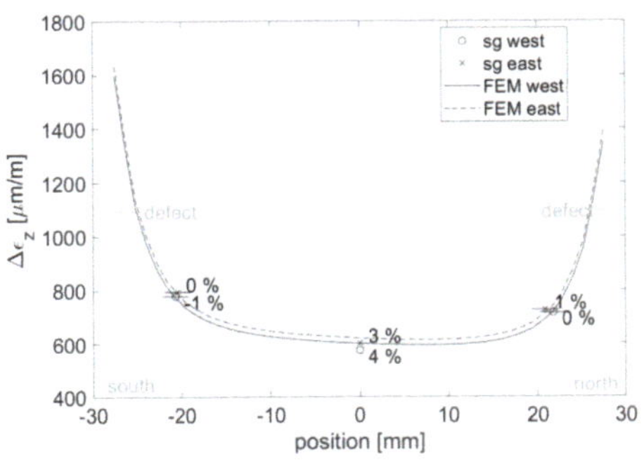


### Initial 3D Stress Field

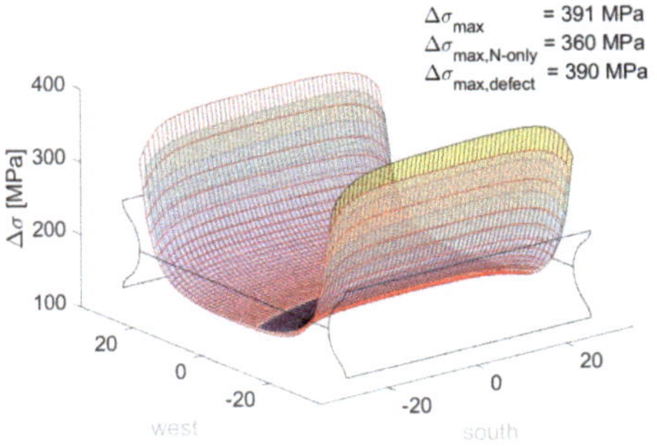

## GMA2

| $\Delta\sigma$ [MPa] | cycles to failure [-] | $\sigma_{max}$/ $\sigma_y$ [%] | $H_w$ [mm] | $H_e$ [mm] | $B_n$ [mm] | $B_s$ [mm] | $F_x$ [kN] | $M_x$ [Nm] | $M_y$ [Nm] | SCF [-] |
|---|---|---|---|---|---|---|---|---|---|---|
| 120 | 3,558,974 | - | 55.0 | 54.9 | 55.4 | 55.2 | - | - | - | - |

Note: Test was terminated without failure. After 691,482 cycles a sudden crack appeared which did not propagagate afterwards. Failure was enforced at a high load level. The reason for this is assumed to be the interaction of residual stresses (see Appendix F) and other effects. However, the exact background could not be clarified.

Fracture Surface

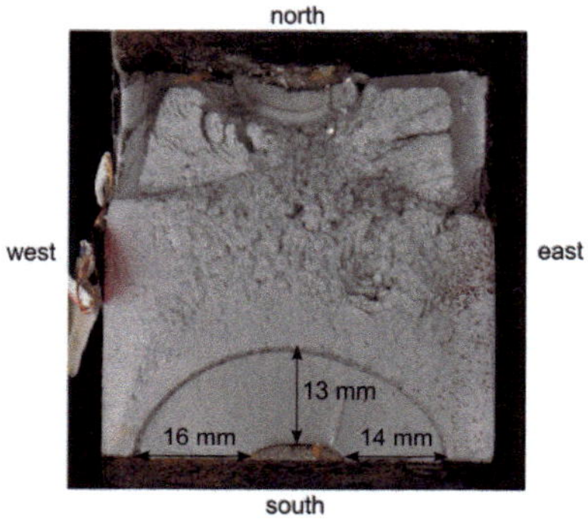


Positions of Strain Gauges

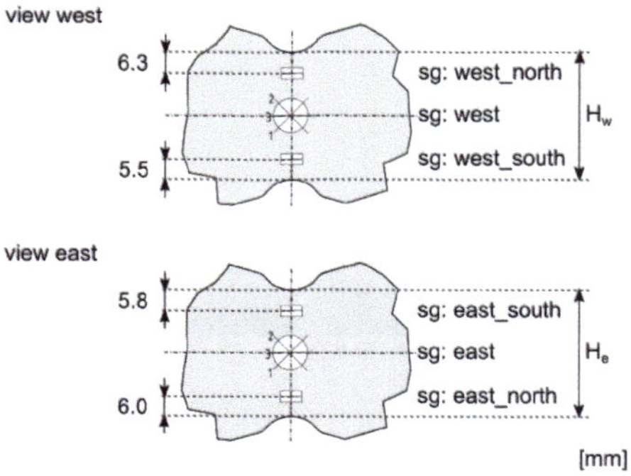


Strain Ranges

no data

Deviation from Initial Strain Ranges

no data

Initial Strain Ranges

no data

Initial 3D Stress Field

no data

## GMA3

| $\Delta\sigma$ [MPa] | cycles to failure [-] | $\sigma_{max}/\sigma_y$ [%] | $H_w$ [mm] | $H_e$ [mm] | $B_n$ [mm] | $B_s$ [mm] | $F_x$ [kN] | $M_x$ [Nm] | $M_y$ [Nm] | SCF [-] |
|---|---|---|---|---|---|---|---|---|---|---|
| 250 | 84,526 | 215 | 54.9 | 55.0 | 55.2 | 55.4 | 740 | -252 | 350 | 2.10 |

Note: -

### Fracture Surface

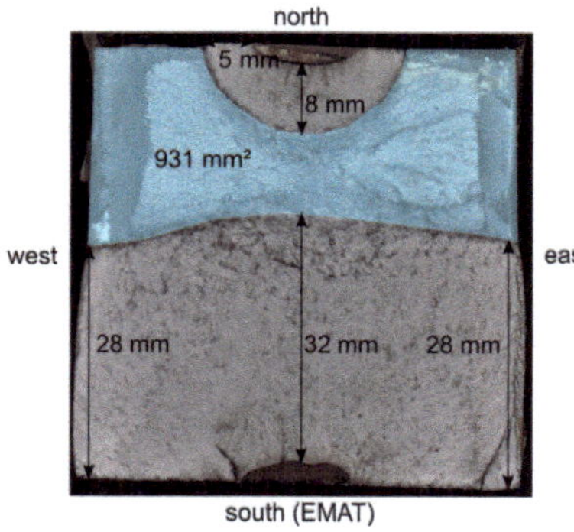


### Positions of Strain Gauges

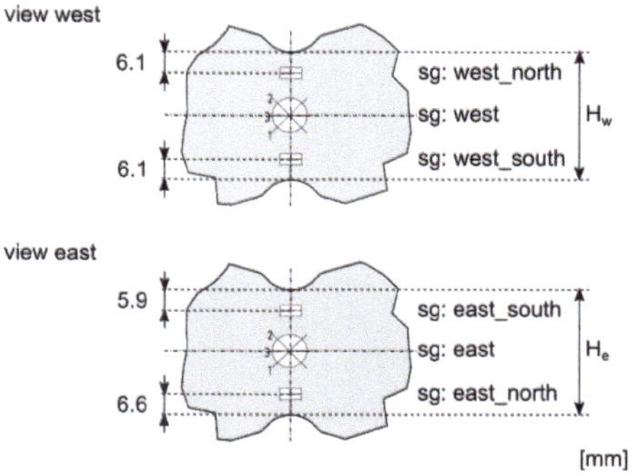


### Strain Ranges

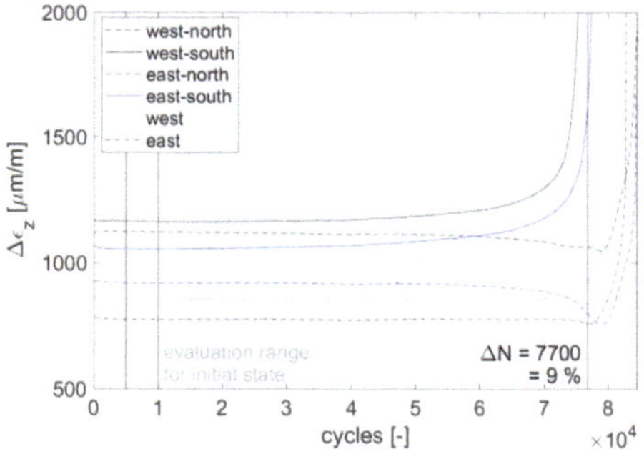


### Deviation from Initial Strain Ranges

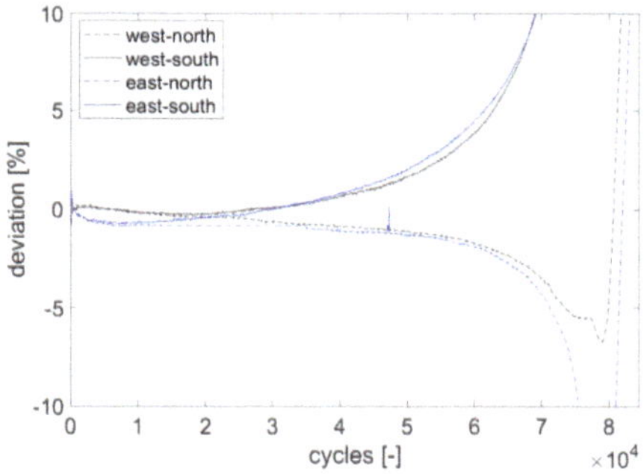


### Initial Strain Ranges

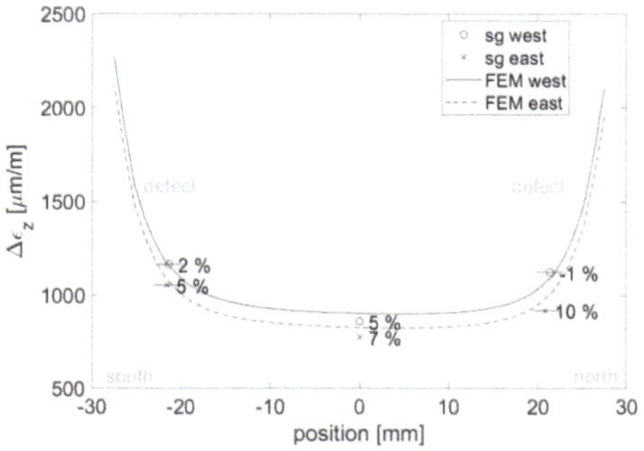


### Initial 3D Stress Field

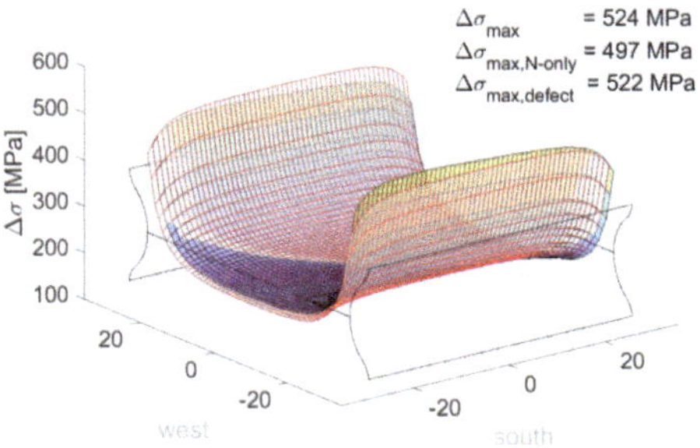

## GMB1

| $\Delta\sigma$ [MPa] | cycles to failure [-] | $\sigma_{max}/\sigma_y$ [%] | $H_w$ [mm] | $H_e$ [mm] | $B_n$ [mm] | $B_s$ [mm] | $F_x$ [kN] | $M_x$ [Nm] | $M_y$ [Nm] | SCF [-] |
|---|---|---|---|---|---|---|---|---|---|---|
| 93 | 1,882,787 | 219 | 55.5 | 56.4 | 54.7 | 54.8 | 277 | -180 | 600 | 2.44 |

Note: -

### Fracture Surface

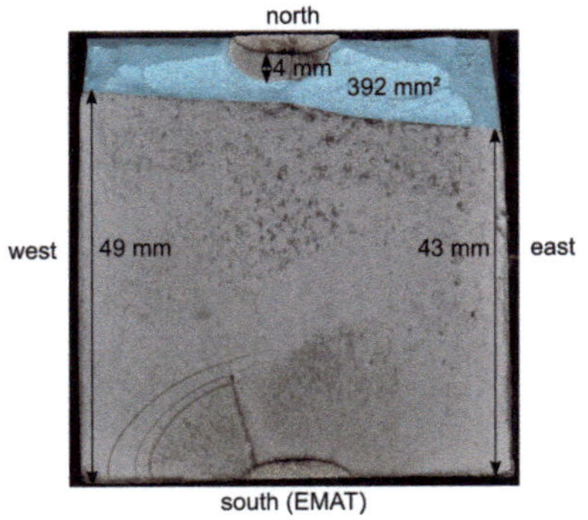


### Positions of Strain Gauges

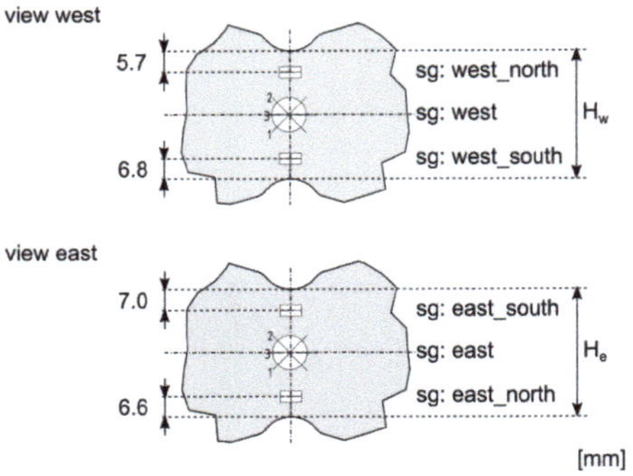


### Strain Ranges

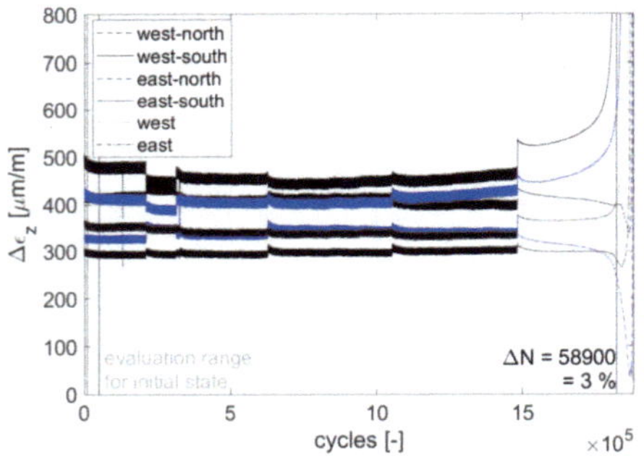


### Deviation from Initial Strain Ranges

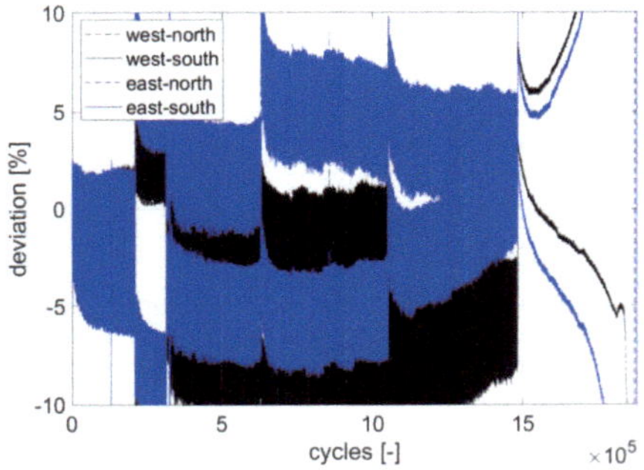


### Initial Strain Ranges

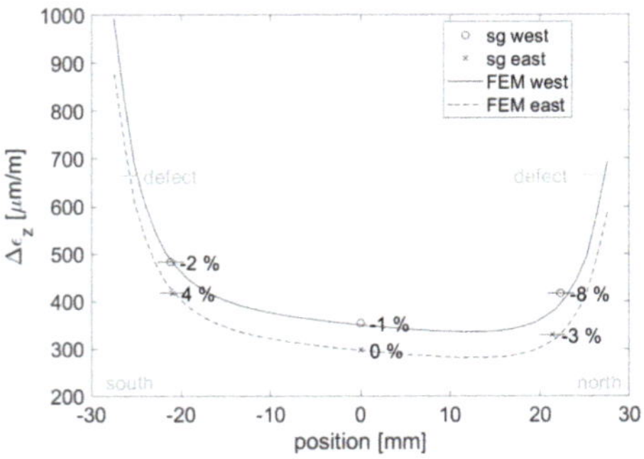


### Initial 3D Stress Field

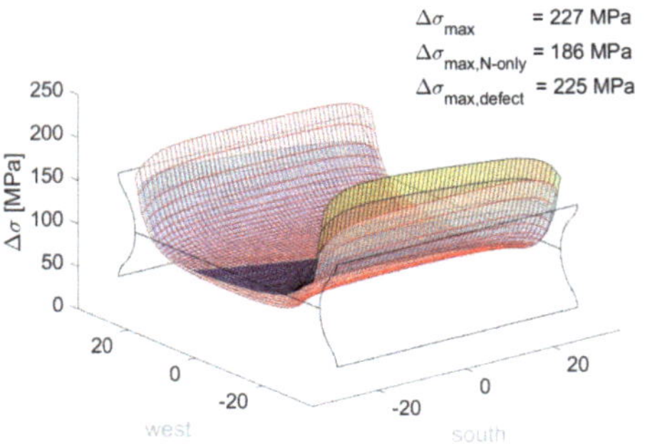

## GMB3

| $\Delta\sigma$ [MPa] | cycles to failure [-] | $\sigma_{max}/\sigma_y$ [%] | $H_w$ [mm] | $H_e$ [mm] | $B_n$ [mm] | $B_s$ [mm] | $F_x$ [kN] | $M_x$ [Nm] | $M_y$ [Nm] | SCF [-] |
|---|---|---|---|---|---|---|---|---|---|---|
| 220 | 169,522 | 225 | 55.6 | 55.1 | 53.6 | 53.4 | 651 | 704 | -500 | 2.25 |

Note: -

### Fracture Surface

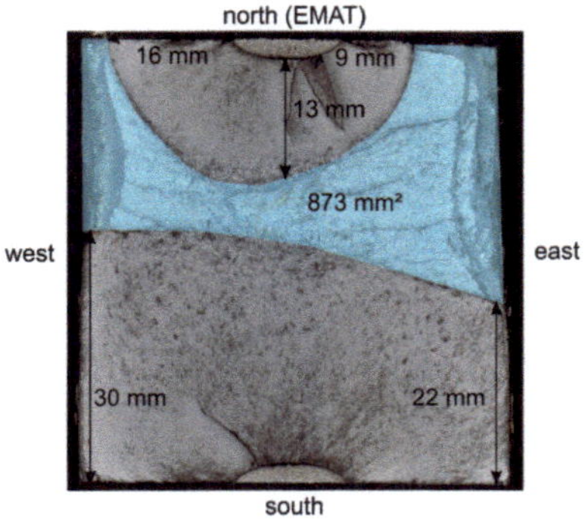


### Positions of Strain Gauges

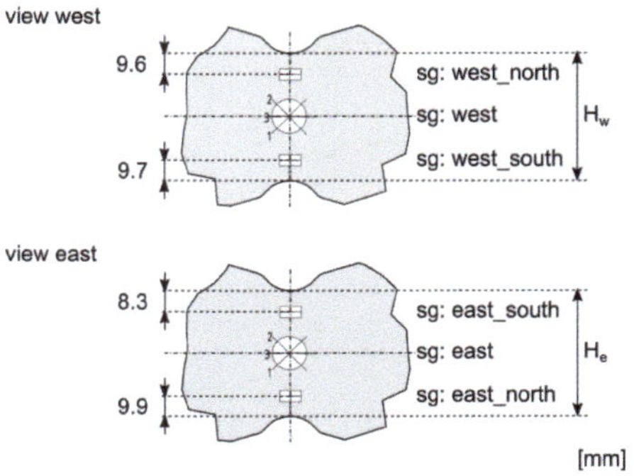


### Strain Ranges

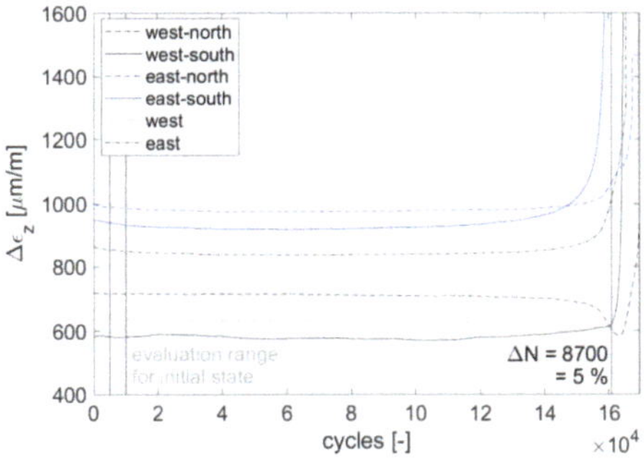


### Deviation from Initial Strain Ranges

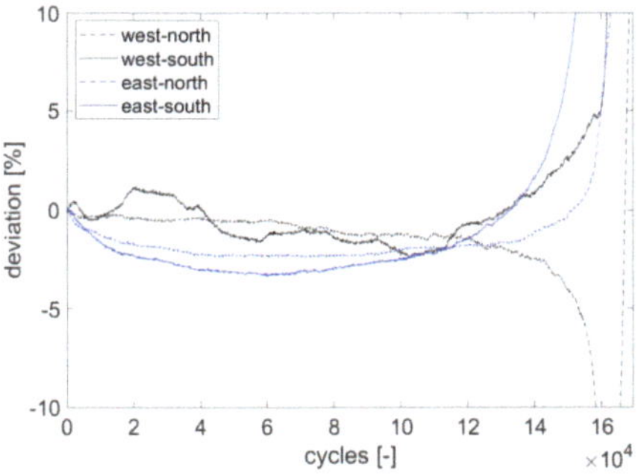


### Initial Strain Ranges

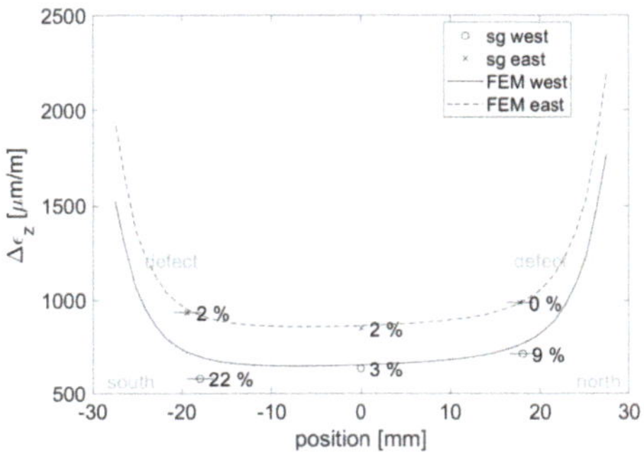


### Initial 3D Stress Field

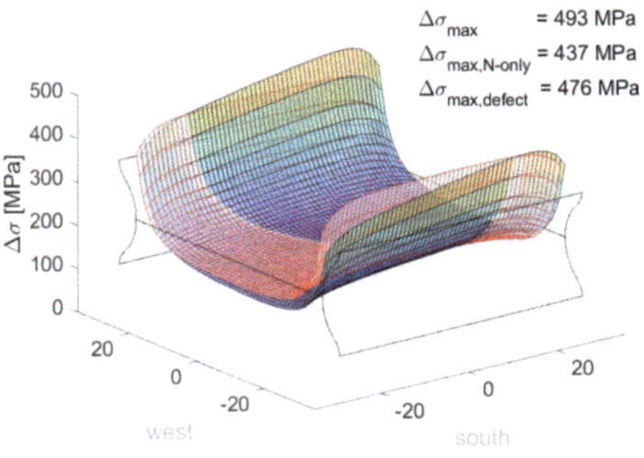

## GGA1

| $\Delta\sigma$ [MPa] | cycles to failure [-] | $\sigma_{max}/\sigma_y$ [%] | $H_w$ [mm] | $H_e$ [mm] | $B_n$ [mm] | $B_s$ [mm] | $F_x$ [kN] | $M_x$ [Nm] | $M_y$ [Nm] | SCF [-] |
|---|---|---|---|---|---|---|---|---|---|---|
| 220 | 511,009 | 235 | 100.0 | 100.4 | 99.9 | 99.7 | 2,200 | -4,500 | 500 | 2.21 |

Note: -

### Fracture Surface

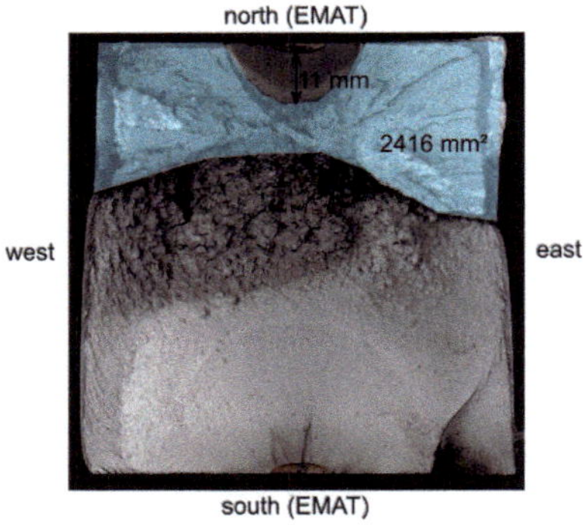


### Positions of Strain Gauges

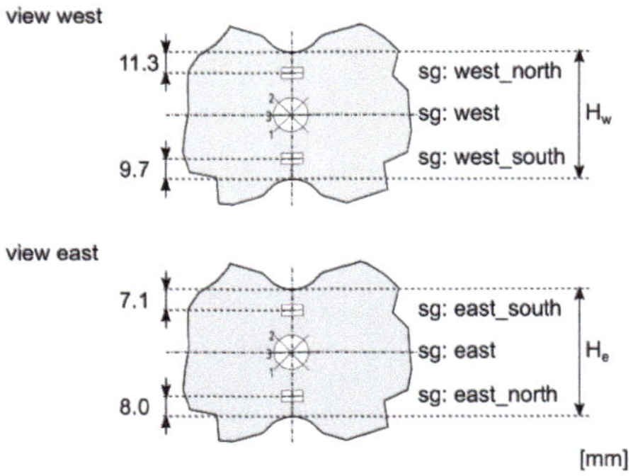


### Strain Ranges

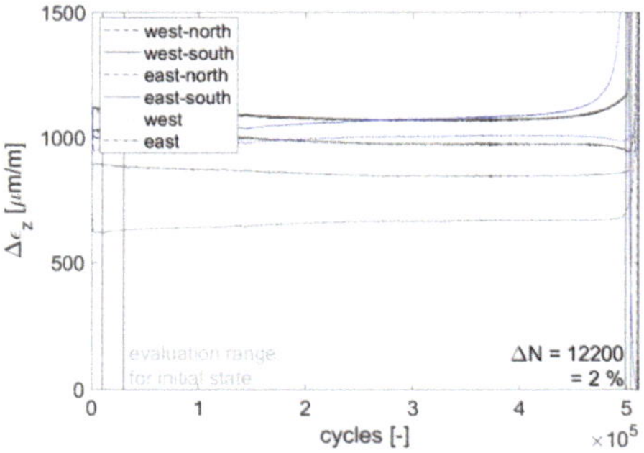


### Deviation from Initial Strain Ranges

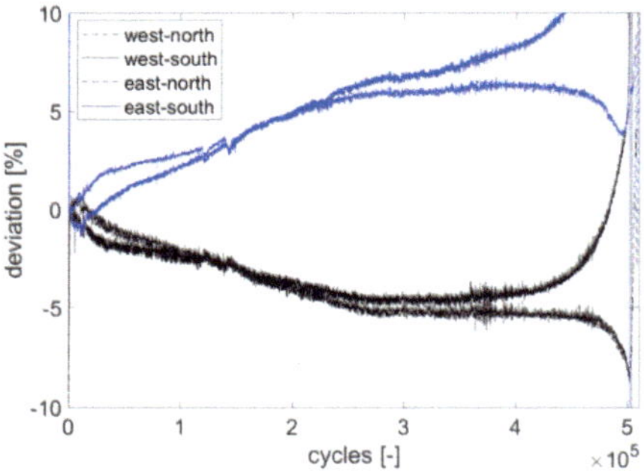


### Initial Strain Ranges

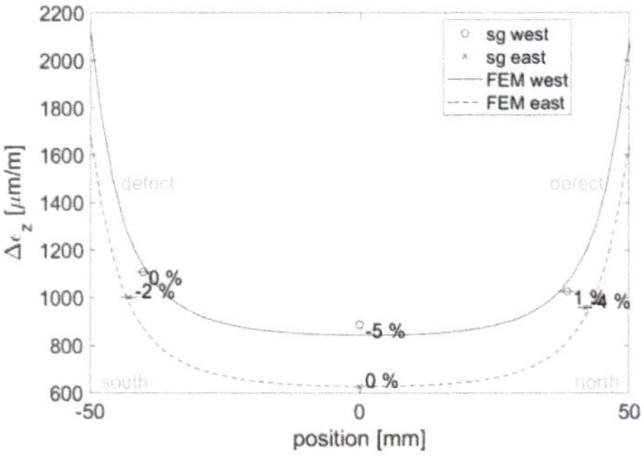


### Initial 3D Stress Field

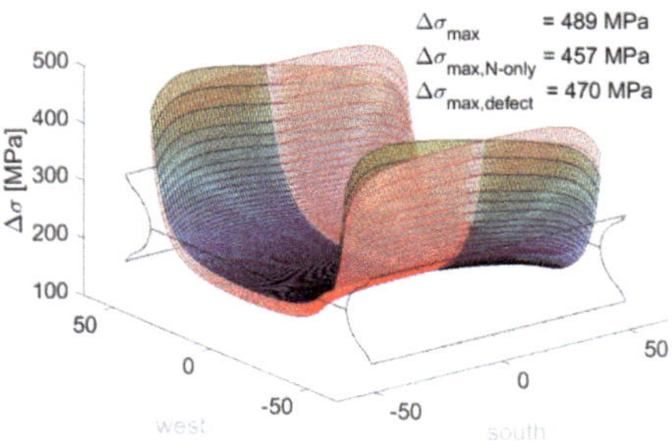

## GGA2

| Δσ [MPa] | cycles to failure [-] | $\sigma_{max}/\sigma_y$ [%] | $H_w$ [mm] | $H_e$ [mm] | $B_n$ [mm] | $B_s$ [mm] | $F_x$ [kN] | $M_x$ [Nm] | $M_y$ [Nm] | SCF [-] |
|---|---|---|---|---|---|---|---|---|---|---|
| 250 | 196,004 | 213 | 100.7 | 100.5 | 100.6 | 100.4 | 2,480 | -1,500 | 1,500 | 2.15 |

Note:

### Fracture Surface

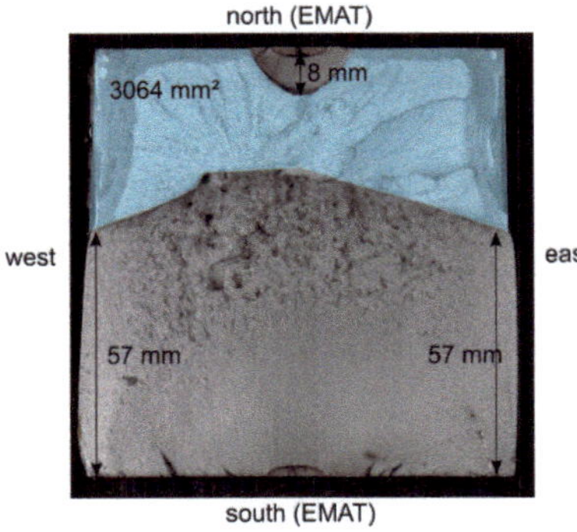


### Positions of Strain Gauges

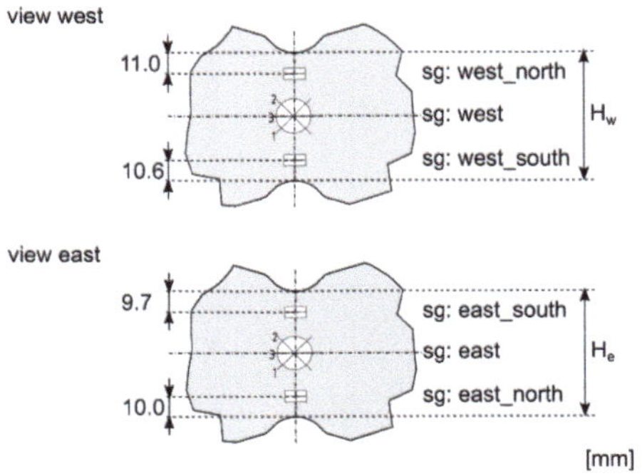


### Strain Ranges

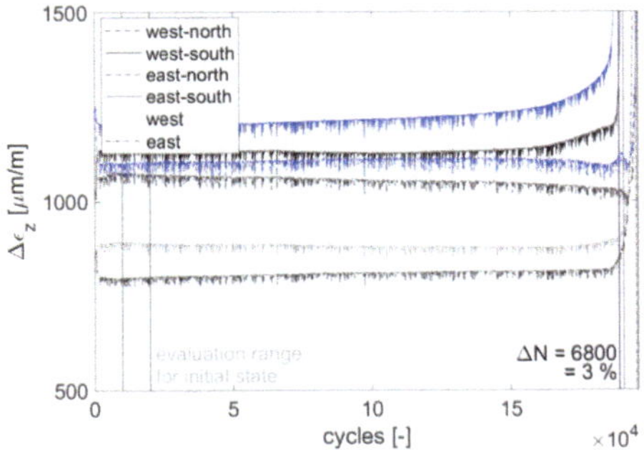


### Deviation from Initial Strain Ranges

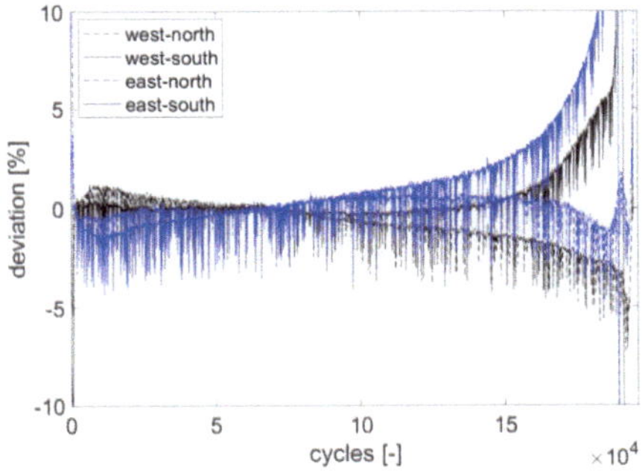


### Initial Strain Ranges

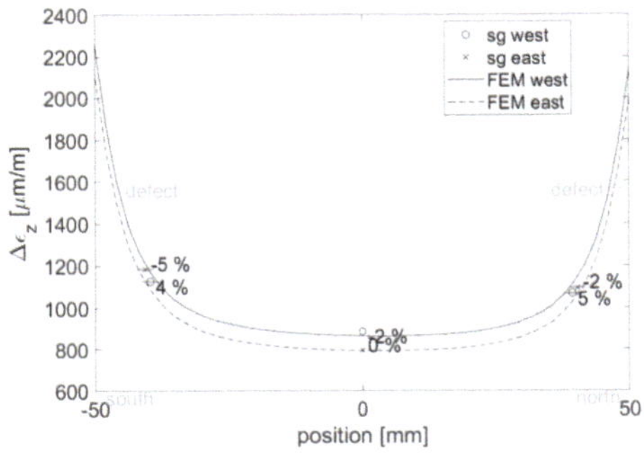


### Initial 3D Stress Field

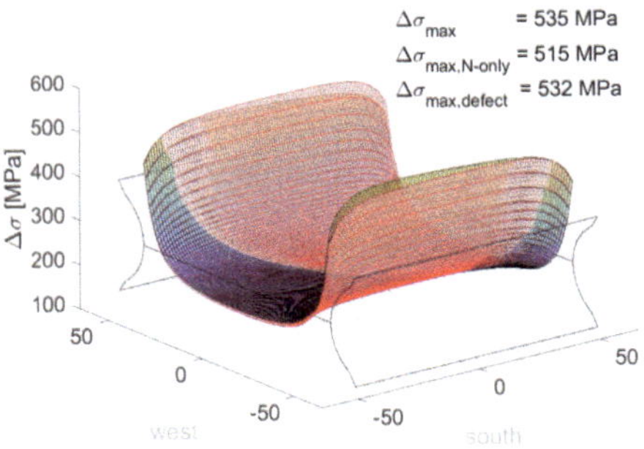

## GGA3

| $\Delta\sigma$ [MPa] | cycles to failure [-] | $\sigma_{max}/\sigma_y$ [%] | $H_w$ [mm] | $H_e$ [mm] | $B_n$ [mm] | $B_s$ [mm] | $F_x$ [kN] | $M_x$ [Nm] | $M_y$ [Nm] | SCF [-] |
|---|---|---|---|---|---|---|---|---|---|---|
| 165 | 3,879,843 | 205 | 100.4 | 99.9 | 99.5 | 99.8 | 1.640 | 0 | -3.000 | 2.25 |

Note: -

### Fracture Surface

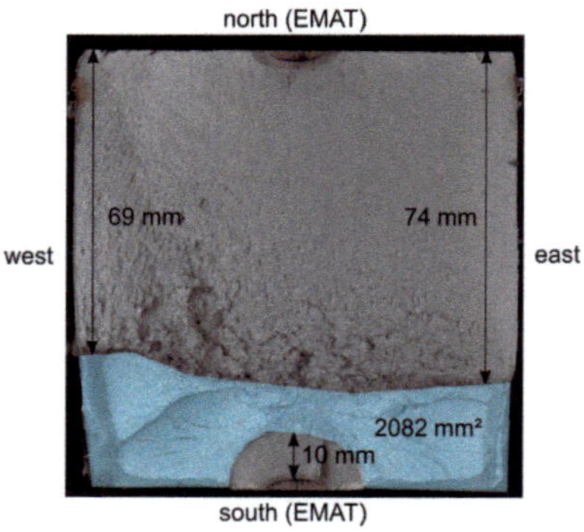


### Positions of Strain Gauges

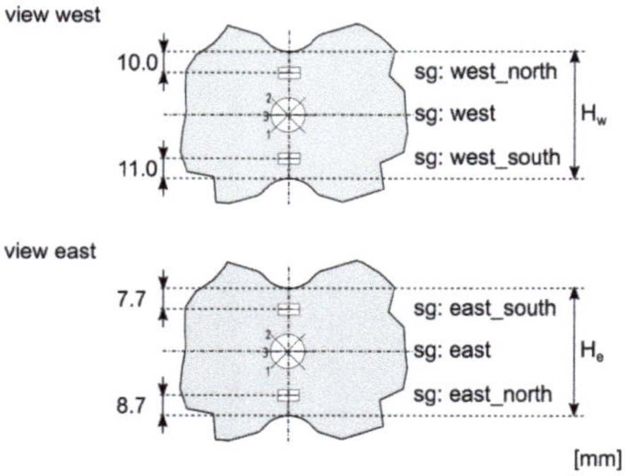


### Strain Ranges

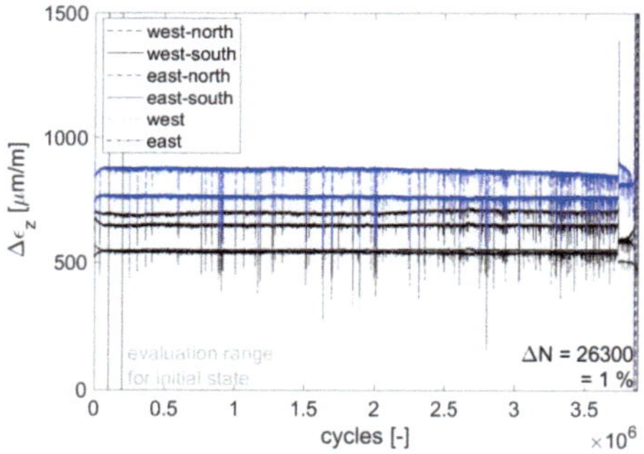


### Deviation from Initial Strain Ranges

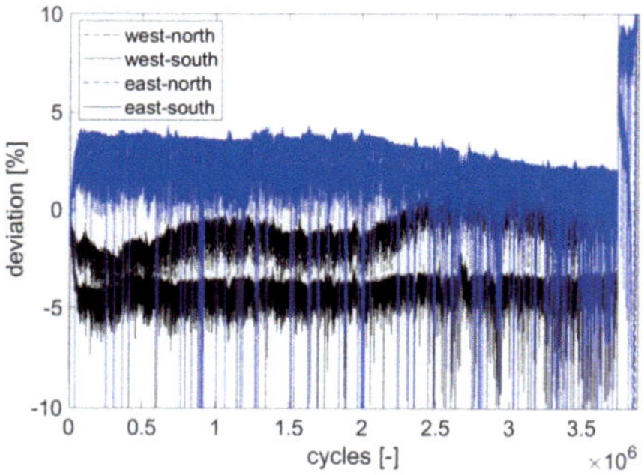


### Initial Strain Ranges

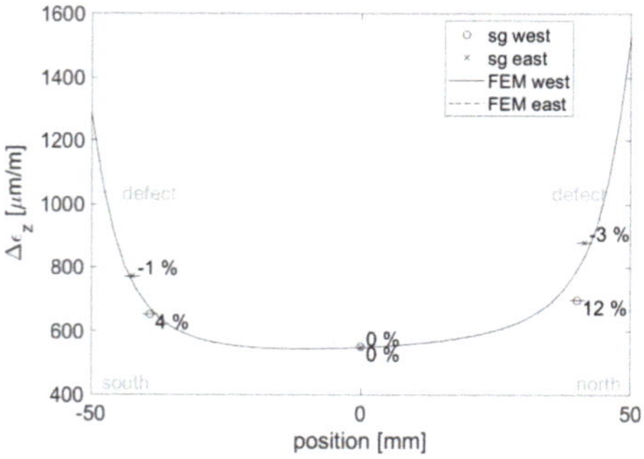


### Initial 3D Stress Field

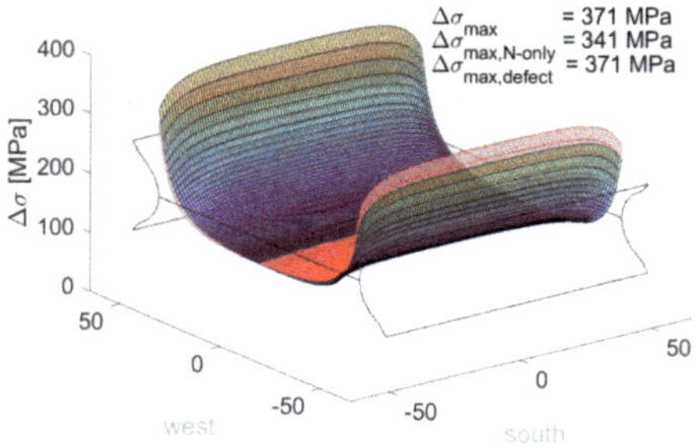

## GGB1

| $\Delta\sigma$ [MPa] | cycles to failure [-] | $\sigma_{max}/\sigma_y$ [%] | $H_w$ [mm] | $H_e$ [mm] | $B_n$ [mm] | $B_s$ [mm] | $F_x$ [kN] | $M_x$ [Nm] | $M_y$ [Nm] | SCF [-] |
|---|---|---|---|---|---|---|---|---|---|---|
| 147 | 410,145 | - | 100.4 | 100.5 | 102.3 | 102.2 | 1,463 | 8,900 | -1,500 | 2.64 |

Note: Early termination of the experiment. Failure was induced under monotonic loading at $F_{max}$ = 2,320 kN

### Fracture Surface

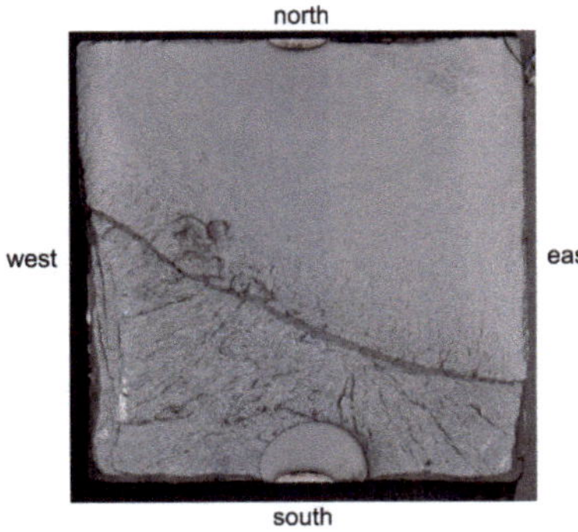


### Positions of Strain Gauges

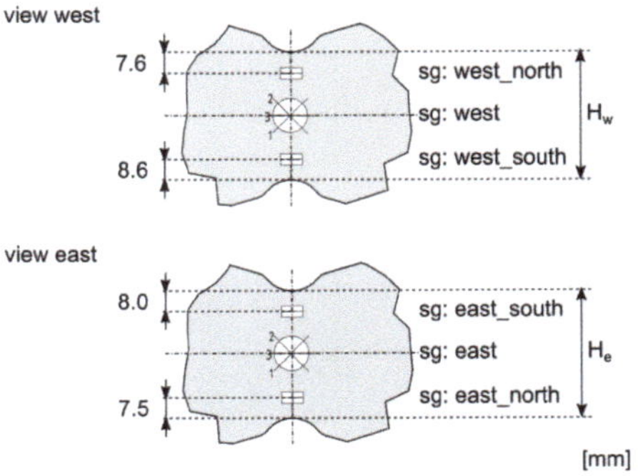


### Strain Ranges

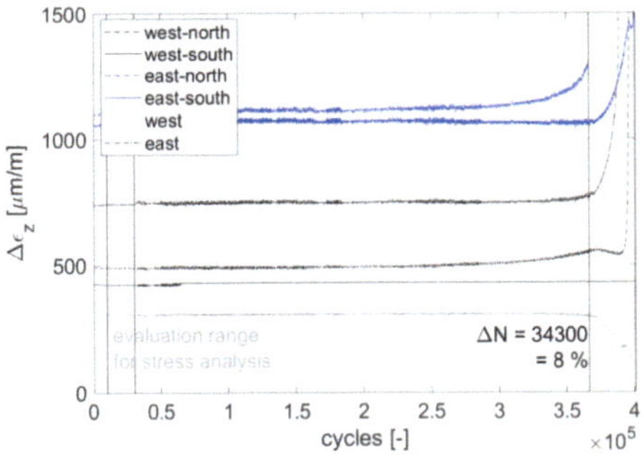


### Deviation from Initial Strain Ranges

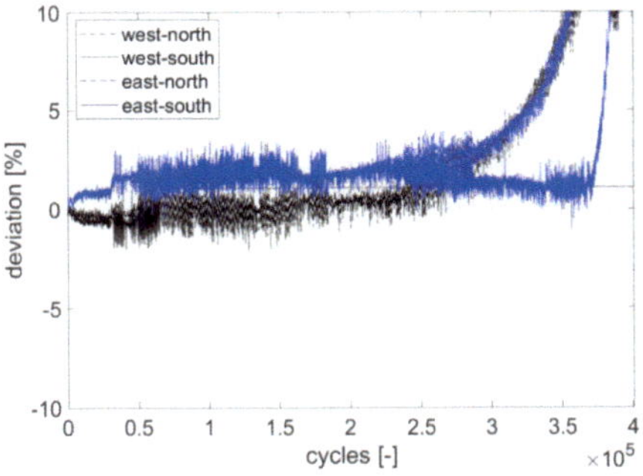


### Initial Strain Ranges

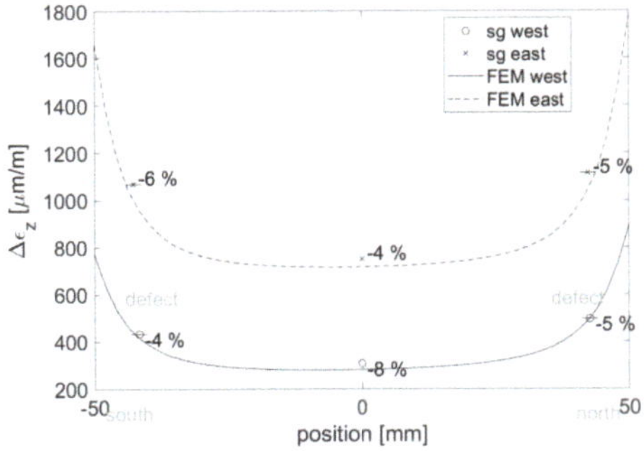


### Initial 3D Stress Field

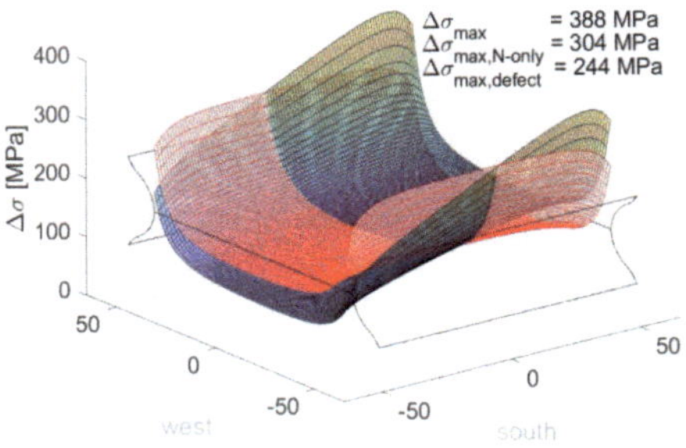

## GGB2

| $\Delta\sigma$ [MPa] | cycles to failure [-] | $\sigma_{max}/\sigma_y$ [%] | $H_w$ [mm] | $H_e$ [mm] | $B_n$ [mm] | $B_s$ [mm] | $F_x$ [kN] | $M_x$ [Nm] | $M_y$ [Nm] | SCF [-] |
|---|---|---|---|---|---|---|---|---|---|---|
| 100 | 794,838 | 230 | 100.6 | 100.7 | 99.6 | 99.8 | 997 | 10,090 | -2,000 | 3.00 |

Note: The first crack was not caused by the artificial defect but by a groove in the highly stressed specimen edge. No agreement between FEM and SG

### Fracture Surface

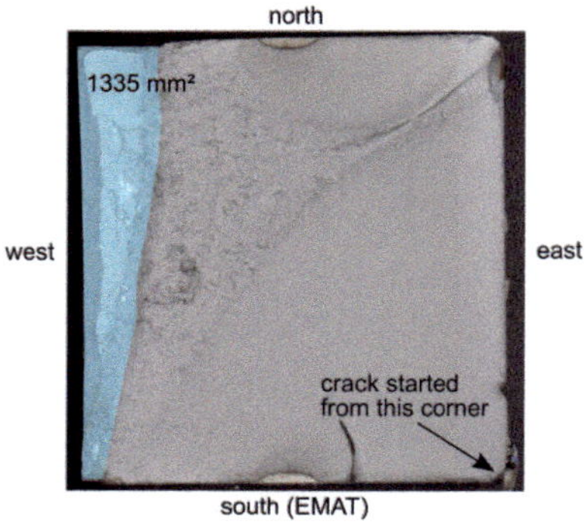


### Positions of Strain Gauges

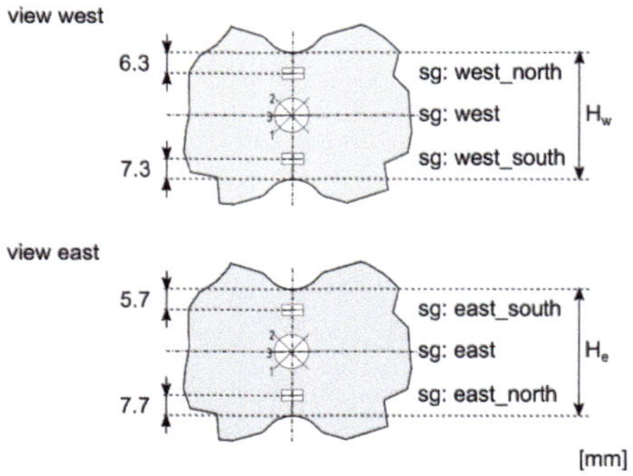


### Strain Ranges

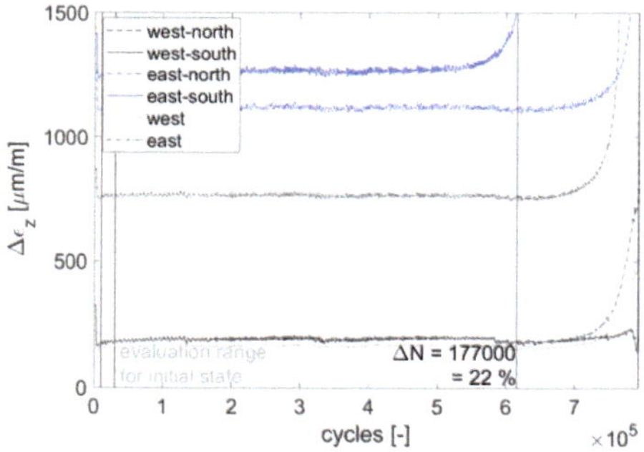


### Deviation from Initial Strain Ranges

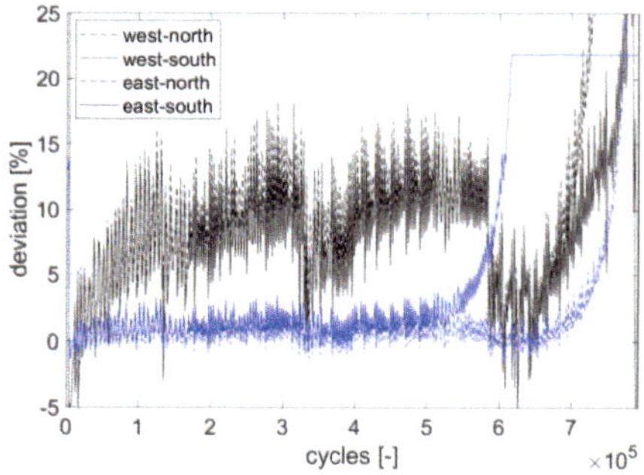


### Initial Strain Ranges

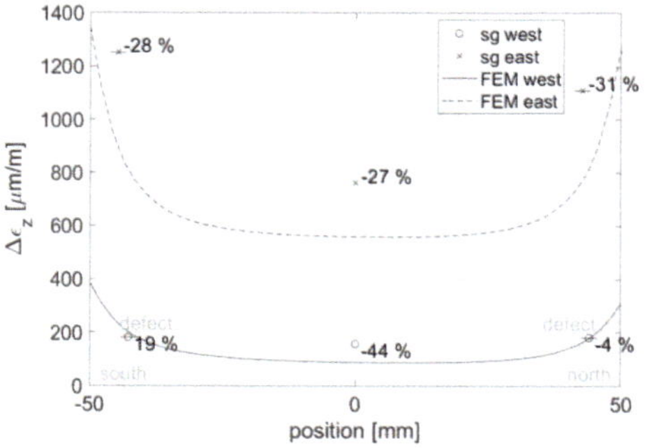


### Initial 3D Stress Field

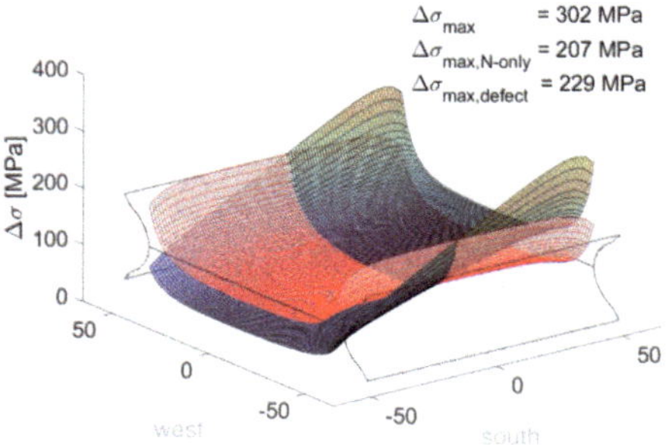

## GGB3

| $\Delta\sigma$ [MPa] | cycles to failure [-] | $\sigma_{max}/\sigma_y$ [%] | $H_w$ [mm] | $H_e$ [mm] | $B_n$ [mm] | $B_s$ [mm] | $F_x$ [kN] | $M_x$ [Nm] | $M_y$ [Nm] | SCF [-] |
|---|---|---|---|---|---|---|---|---|---|---|
| 220 | 116,352 | 129 | 101.0 | 100.8 | 100.3 | 100.2 | 2,192 | 11,250 | -5,000 | 2.69 |

Note: -

Fracture Surface

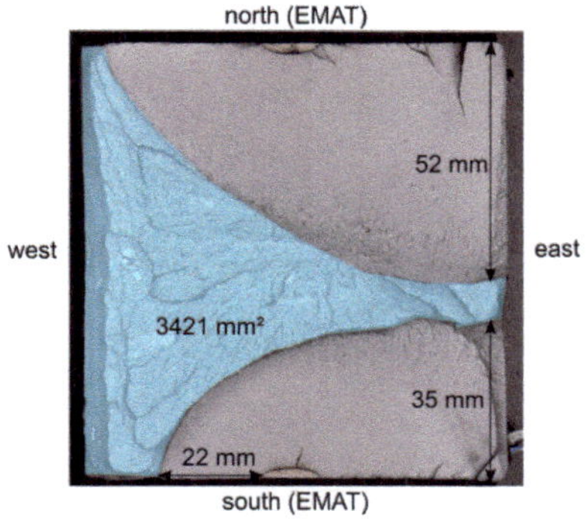


Positions of Strain Gauges

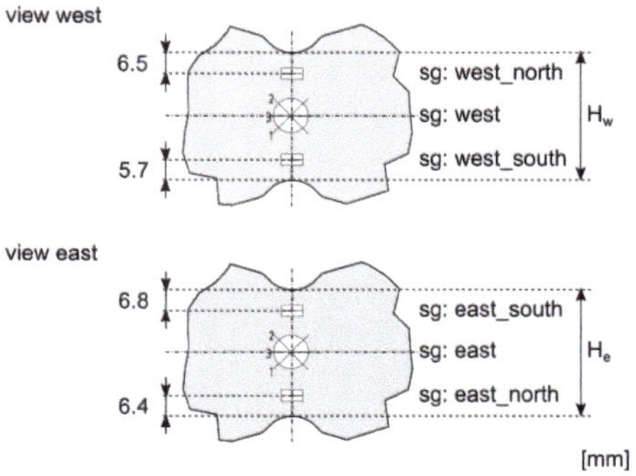


Strain Ranges

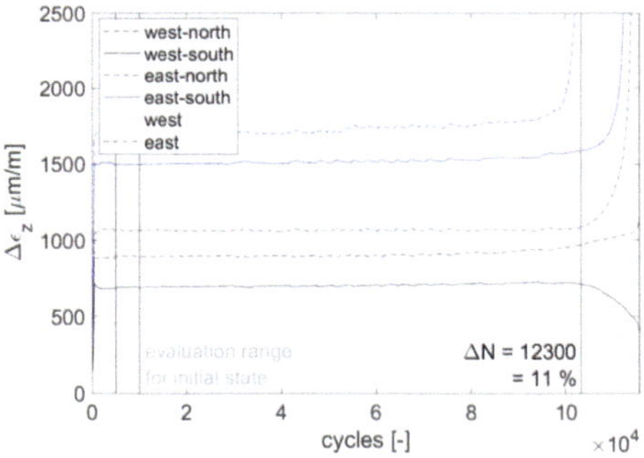


Deviation from Initial Strain Ranges

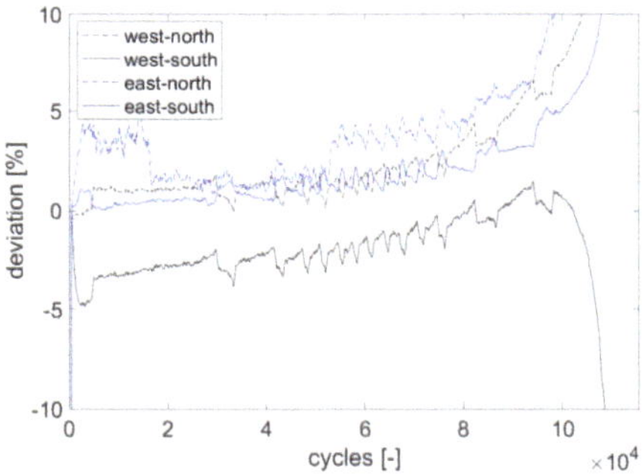


Initial Strain Ranges

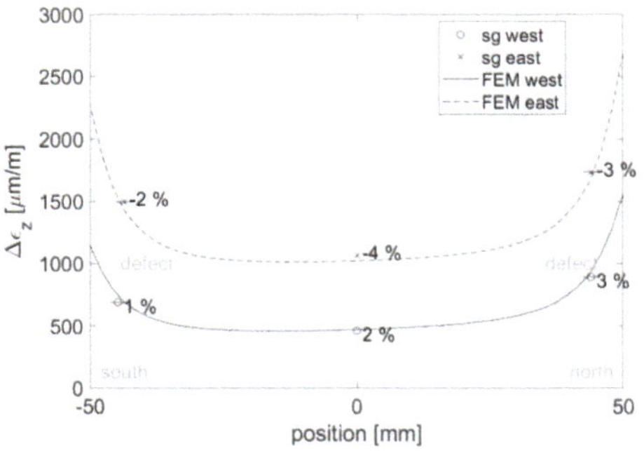


Initial 3D Stress Field

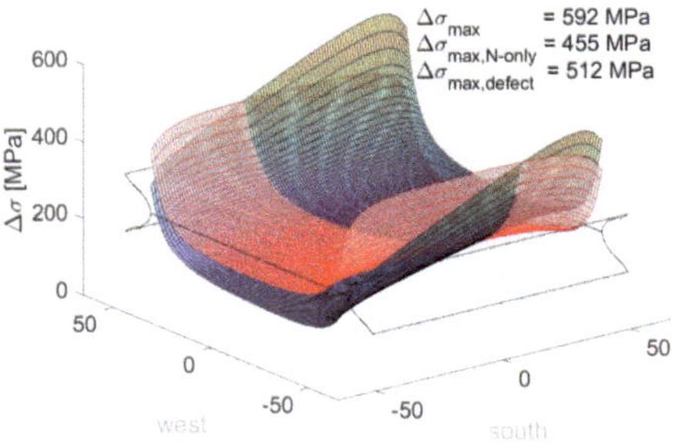

# E Supplementary Data on Internal Defects

In addition, the intermediate results of the basic case and the internal defects according to Section 5.2.3 are listed. The contents and statements correspond to those of the surface defects. The presentation is without further comments.

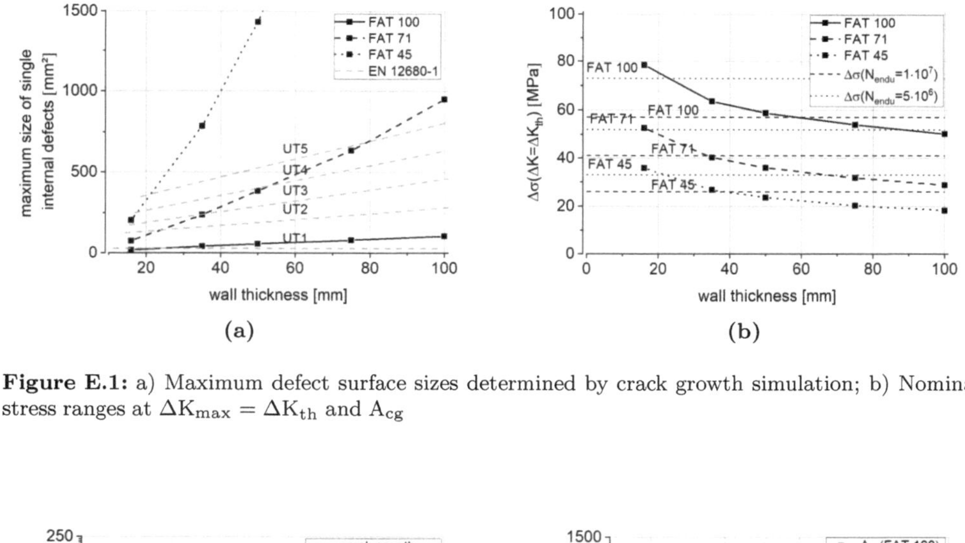


(a) (b)

**Figure E.1:** a) Maximum defect surface sizes determined by crack growth simulation; b) Nominal stress ranges at $\Delta K_{max} = \Delta K_{th}$ and $A_{cg}$

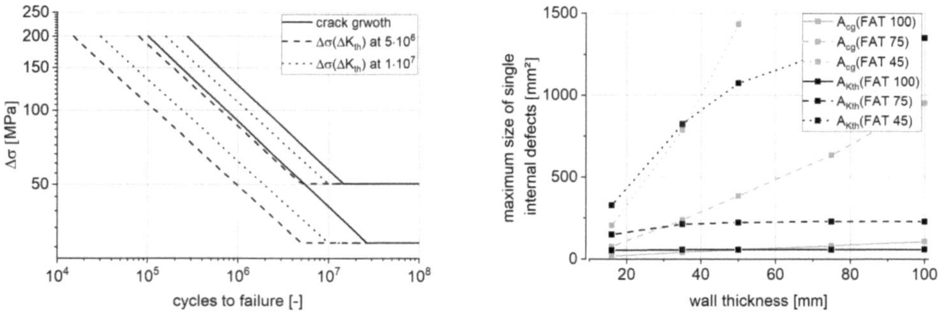


(a) (b)

**Figure E.2:** a) SN curves with $\Delta\sigma(\Delta K_{th})$ at different reference points compared to the results of crack growth simulations; b) Comparison of $A_{cg}$ and $A_{Kth}$

# F Determination of Residual Stresses at the Component Test

To determine residual stresses in the specimen of the component tests from Section 3.3, the hole-drilling strain-gauge method according to ASTM E837:2013 [134] was used. This investigation intended to explain the differences between the specimens from foundries A and B. With this goal, only four measurements two from each foundry – of the large-type specimens were carried out after the fatigue tests were completed.

The drilling procedure for thick work pieces with uniform stresses was applied using a strain gauge rosette type 062 RE (type A according to ASTEM E837:2013) [134] with a mean diameter of 5.13 mm. The holes were drilled with carbide cutters 1.6 mm in diameter to a maximum depth of 1 mm in 10 equally spaced increments. The drilling was carried out using a milling guide on the surface perpendicular to the geometric notches, as shown in Figure F.1. Table F.1 gives an overview of the tested specimen and the final measured diameters of the holes ($D_0$).

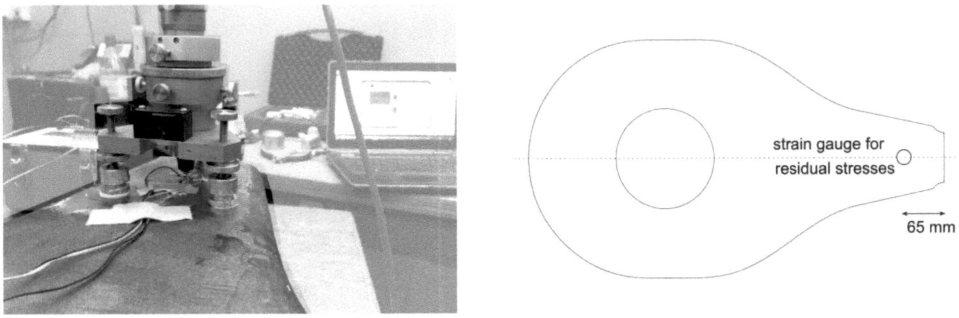


(a) (b)

**Figure F.1:** a) Test setup and milling guide for residual stress measurement using the hole-drilling strain-gauge method; b) Position of the strain measurement in the large specimen

**Table F.1:** Summary of the specimen considered in the residual stress analysis

| Specimen [-] | GGA1 | GGA3 | GGB1 | GGB3 |
|---|---|---|---|---|
| $D_0$ [mm] | 1.76 | 1.62 | 1.80 | 1.81 |

The stress evaluation was done using the software H-Drill [92]. Figure F.2a to F.5a shows the measured strain data over the depth from the surface along with the theoretical distribution according to ASTM E837:2013 [134]. A linear stress distribution led to the best fit between the theory and the measured data according to the evaluation criteria of ASTM E837:2013 [134]. Only for specimen GGA1 could no satisfactory correlation between the theoretical distributions and measured strains be reached. Even approaches of higher orders brought no significant improvements. As this investigation only aimed to confirm the assumption of different residual stress states and insufficient stress relief annealing for the specimen from foundry A, this was deemed sufficiently accurate. It is noted, that for practical applications, these compressive residual stresses cause higher fatigue resistances, but to evaluate the basic principles of cast steel components without residual stresses as normally intended, these stresses are of limited use. In the case of specimens GGA1 and GGA3, the strains showed their minimums in the center of the hole depth and increased slightly upon reaching deeper layers. For specimens GGB1 and GGB3, the strains decreased continuously from the surface to the final depth of the holes.

The resulting principal stress distributions are shown in Figure F.2b to F.5b. At the surface, not yet explainable tensile stresses occurred that vanished within less than 1 mm of depth from the surface. In the case of foundry A, these stresses transformed into significant compressive stresses almost as high as the material's yield strength.

In both, the strain curves as well as the calculated residual stresses below the components' surface show major differences between both foundries and support the assumption of additional compressive residual stresses in the foundry A specimen. Comparing these findings with the artificially induced defects at a depth of 2.7 mm, these compressive residual stresses are likely to be the reason for the deviations between the specimen of both foundries.

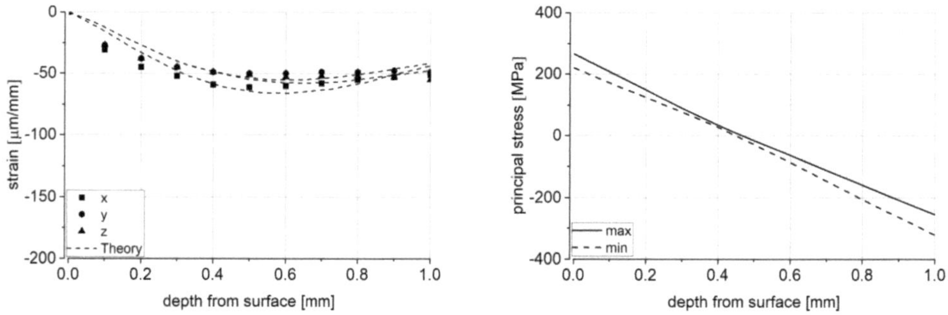


(a) (b)

**Figure F.2:** Specimen GGA1: a) Measured strains and b) Residual principal stresses

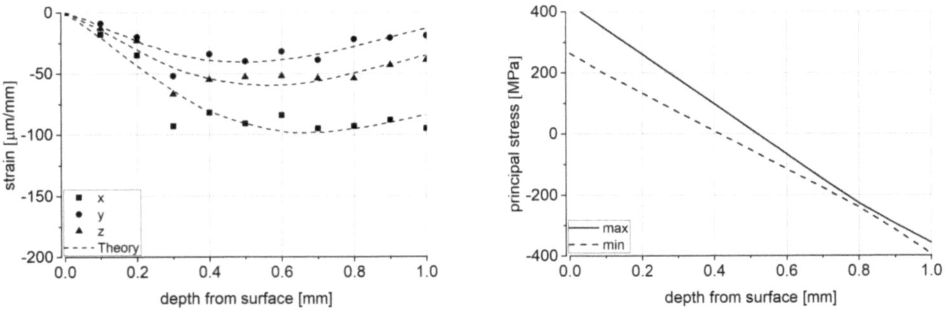


(a) (b)

**Figure F.3:** Specimen GGA3: a) Measured strains and b) Residual principal stresses

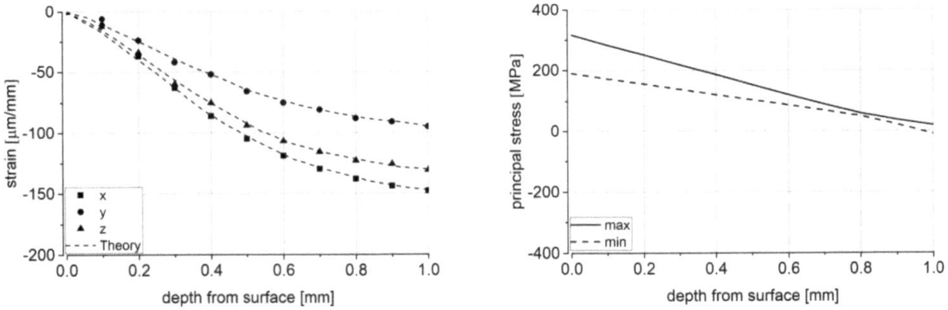


(a) (b)

**Figure F.4:** Specimen GGB1: a) Measured strains and b) Residual principal stresses

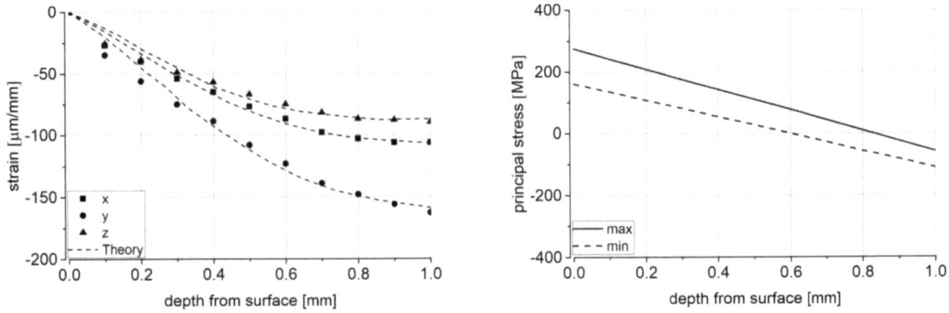


(a) (b)

**Figure F.5:** Specimen GGB3: a) Measured strains and b) Residual principal stresses

**BERICHTE ZUM STAHL- UND LEICHTBAU (ISSN 2198-7912)**
Versuchsanstalt für Stahl, Holz und Steine, Stahl- und Leichtbau
Karlsruher Institut für Technologie (KIT)

---

Eine Übersicht der Berichte der Versuchsanstalt für Stahl, Holz und Steine ab dem Jahr 1963 finden Sie unter folgender URL: http://stahl.vaka.kit.edu/berichte.php

Band 1 OLIVER FLEISCHER
Axial beanspruchte K-Knoten aus dünnwandigen Rechteckhohlprofilen.
ISBN 978-3-7315-0190-9

Band 2 THOMAS REINKE
Tragverhalten von biegebeanspruchten Stahlmasten mit polygonalen Querschnitten.
ISBN 978-3-7315-0398-9

Band 3 ROBIN MARC PLUM
Fatigue crack detection on structural steel members by using ultrasound excited thermography. Erkennung von Ermüdungsrissen in Stahlbauteilen durch ultraschallangeregte Thermografie.
ISBN 978-3-7315-0417-7

Band 4 TIM ZINKE
Nachhaltigkeit von Infrastrukturbauwerken – Ganzheitliche Bewertung von Autobahnbrücken unter besonderer Berücksichtigung externer Effekte.
ISBN 978-3-7315-0509-9

Band 5 MAX JONAS SPANNAUS
Bemessung von Erzeugnissen aus Stahlguss unter vorwiegend ruhender Beanspruchung.
ISBN 978-3-7315-0560-0

Band 6 MATTHIAS FRIEDRICH ALBIEZ
Zur statischen Tragfähigkeit geklebter Kreishohlprofilverbindungen im Stahlbau.
ISBN 978-3-7315-0561-7

Band 7 ANDREAS LIPP
Kreishohlprofil-X-Knoten aus nichtrostenden Stählen unter Axialbeanspruchung.
ISBN 978-3-7315-0569-3

Band 8 PAUL DARIO TOASA CAIZA
Consideration of runouts by the evaluation of fatigue experiments.
ISBN 978-3-7315-0900-4

Band 9 JENNIFER C. HRABOWSKI
Ermüdungsverhalten von Schweißverbindungen aus höchstfestem Stahl im Kurzzeitfestigkeitsbereich.
ISBN 978-3-7315-0931-8

---

Die Bände sind unter www.ksp.kit.edu als PDF frei verfügbar oder als Druckausgabe bestellbar.

Band 10 **STEFAN RACK**
Formadaptive Tensegrity-Strukturen – Ein Beitrag zur numerischen Simulation.
ISBN 978-3-7315-0574-7

Band 11 **ACHIM STRUVE**
Analysis of a Rotatable Wind Turbine Tower by means of Aero-Servo-Elastic Load Simulations.
ISBN 978-3-7315-1045-1

Band 12 **PHILIPP WEIDNER**
Zum Ermüdungsverhalten einseitig geschweißter Stumpfstöße von Kreishohlprofilen.
ISBN 978-3-7315-1017-8

Band 13 **STEFANOS GKATZOGIANNIS**
Finite Element Simulation of Residual Stresses from Welding and High Frequency Hammer Peening.
ISBN 978-3-7315-1066-6

Band 14 **MARTIN DIRK BUES**
Ein Beitrag zur Auslegung tragender Klebverbindungen im Fassadenbau.
ISBN 978-3-7315-1122-9

Band 15 **SVEN NAGEL**
Design of Cast Steel Components under Cyclic Loading.
ISBN 978-3-7315-1126-7